T0183318

Coimbra Mathematical Texts

Volume 1

Series Editors

Ana Paula Santana, Department of Mathematics, University of Coimbra, Coimbra, Portugal

Júlio S. Neves, Department of Mathematics, University of Coimbra, Coimbra, Portugal

Marcelo Viana, IMPA, Rio de Janeiro, Brazil

Maria Paula Serra de Oliveira, Department of Mathematics, University of Coimbra, Coimbra, Portugal

Rui Loja Fernandes, Department of Mathematics, University of Illinois, Urbana, IL, USA

More information about this series at http://www.springer.com/series/16723

Maria Manuel Clementino ·
Alberto Facchini · Marino Gran

Editors

New Perspectives in Algebra, Topology and Categories

Summer School, Louvain-la-Neuve, Belgium,
September 12–15, 2018 and
September 11–14, 2019

 Springer

Editors
Maria Manuel Clementino
Departamento de Matemática
Universidade de Coimbra
Coimbra, Portugal

Alberto Facchini
Dipartimento di Matematica
Università di Padova
Padova, Italy

Marino Gran
Institut de Recherche en Mathématique
et Physique
Université catholique de Louvain
Louvain-la-Neuve, Belgium

Coimbra Mathematical Texts
ISBN 978-3-030-84321-2   ISBN 978-3-030-84319-9   (eBook)
https://doi.org/10.1007/978-3-030-84319-9

Mathematics Subject Classification: 08A30, 16-01, 16B50, 16S90, 18A20, 18E13, 18E08, 18B10, 06D22, 18F70, 18A32, 20M14, 17-01, 54H11, 22-01

© The Editor(s) (if applicable) and The Author(s), under exclusive license
to Springer Nature Switzerland AG 2021
This work is subject to copyright. All rights are solely and exclusively licensed by the Publisher, whether the whole or part of the material is concerned, specifically the rights of translation, reprinting, reuse of illustrations, recitation, broadcasting, reproduction on microfilms or in any other physical way, and transmission or information storage and retrieval, electronic adaptation, computer software, or by similar or dissimilar methodology now known or hereafter developed.
The use of general descriptive names, registered names, trademarks, service marks, etc. in this publication does not imply, even in the absence of a specific statement, that such names are exempt from the relevant protective laws and regulations and therefore free for general use.
The publisher, the authors and the editors are safe to assume that the advice and information in this book are believed to be true and accurate at the date of publication. Neither the publisher nor the authors or the editors give a warranty, expressed or implied, with respect to the material contained herein or for any errors or omissions that may have been made. The publisher remains neutral with regard to jurisdictional claims in published maps and institutional affiliations.

This Springer imprint is published by the registered company Springer Nature Switzerland AG
The registered company address is: Gewerbestrasse 11, 6330 Cham, Switzerland

Editor's Preface

This book is addressed to a new generation of students and researchers in algebra, topology and category theory who are willing to learn the fundamental notions and principles of some of the active research fields in these areas.

The content of each of the seven chapters of this book fully covers the corresponding specialized course that was taught at the international "Summer Schools in Algebra and Topology". The schools were hosted by the Institut de Recherche en Mathématique et Physique (IRMP) of the Université catholique de Louvain (UCLouvain) in September 2018 and 2019.

These events were organized in the framework of a project Fonds d'Appui à l'Internationalisation of the Université catholique de Louvain together with the universities of Coimbra, Padova and Poitiers, aiming at strengthening the research collaborations in mathematics among institutions of the Coimbra Group. The speakers of the summer schools gave some introductory courses on various topics of interest related to their own research fields, addressed to an audience including both master's and PhD students and more experienced mathematicians who were interested in these subjects. The participants in the summer schools came from many different countries in Europe, North America and Asia.

These summer schools greatly facilitated scientific discussions leading to the development of new connections among the various concepts and theories presented during the lectures.

The main subjects in algebra and topology the reader will be introduced to in this volume are:

- the theory of comodules and contramodules;
- the theory of topological semi-abelian algebras;
- the theories of commutative monoids and of non-commutative rings;
- regular categories and the calculus of relations;
- the categorical approach to commutator theory;
- locale theory and the point-free perspective on topology;
- the theory of non-associative algebras.

We are very grateful to the many colleagues who have contributed in making this project possible and successful. In particular, we would like to express our gratitude to

- Ms. Dana Samson, *Prorectrice à l'international de l'UCLouvain*;
- Ms. Ana Fernandez-Gacio, Coordinator of the international relations of the *Science and Technology Sector* of the UCLouvain;
- Ms. Anne Querinjean, director of the *Musée universitaire de Louvain*;
- Ms. Carine Baras, Secrétaire de recherche;
- Ms. Martine Furnemont, Secrétaire of the IRMP;
- Mr. Michel Devillers, Vice-rector of the Science and Technology Sector;
- Mr. Tim van der Linden, Professor at the UCLouvain.
- The editors of the "Coimbra Mathematical Texts", Ana Paula Santana, Júlio S. Neves, Marcelo Viana, Maria Paula Serra de Oliveira and Rui Loja Fernandes.

The project leading to the publication of this book has been generously supported by the following institutions:

- UCLouvain (via the *Fonds d'Appui à l'Internationalisation*)
- The Institut de Recherche en Mathématique et Physique;
- The Centre for Mathematics of the University of Coimbra;
- the *Coimbra Group;*
- the *Ecole doctorale en Mathématique du FNRS;*
- the *Erasmus+* programme of the European Union.

Sincerely,

Maria Manuel Clementino
Alberto Facchini
Marino Gran

Preface to the Book Series Coimbra Mathematical Texts

This new Springer series "Coimbra Mathematical Texts" is a direct successor of a collection entitled "Textos de Matemática", published by the University of Coimbra, which started in 1993.

The aim of that collection was the publication of advanced mathematics texts resulting from events held at the University of Coimbra—Mathematics Department, and included conference proceedings as well as monographs related to graduate-level short courses. The first volume in the collection was "Classical Invariants", by J. A. Green (1993), and the last one, no. 48, was "13th Young Researchers Workshop on Geometry, Mechanics and Control: three mini-courses" (2019). The complete list of volumes, covering a wide range of mathematical topics, may be found at

https://www.uc.pt/fctuc/dmat/seccoes/publicacoes/textosDeMatematica

All volumes went through a rigorous refereeing process.

The series now gains a new life with the launching of "Coimbra Mathematical Texts", a partnership between Springer Nature and the University of Coimbra. The new collection maintains the same spirit but aims at publishing advanced texts from more diverse origins. We therefore invite the wider mathematical community to submit quality contributions to the series, both monographs and proceedings, in all branches of mathematics.

Inquiries regarding the submission process should be sent to cmt@mat.uc.pt.

Ana Paula Santana
Júlio S. Neves
Marcelo Viana
Maria Paula Serra de Oliveira
Rui Loja Fernandes
The CMT Series Editors

Contents

1 Ring Epimorphisms, Gabriel Topologies and Contramodules 1
Silvana Bazzoni

   1   Preliminaries 3
   2   Ring Epimorphisms 4
   3   Gabriel Topologies, Torsion Pairs and the Ring of Quotients . . . 8
   4   Comodules and Contramodules 13
   5   First Matlis Category Equivalence 20
   References 25

2 An Invitation to Topological Semi-abelian Algebras 27
Maria Manuel Clementino

   1   Semi-abelian Algebras 28
   2   Topological Semi-abelian Algebras 35
   3   The Categorical Behaviour of Topological Semi-abelian
        Algebras 42
   4   Split Extensions: Semi-direct Products 46
   5   Split Extensions: Classifiers 53
   6   Some Open Problems 61
   References 65

3 Commutative Monoids, Noncommutative Rings and Modules 67
Alberto Facchini

   1   Commutative Monoids 68
   2   Preordered Groups, Positive Cones 76
   3   Some Set Theory 78
   4   The Monoid $V(\mathscr{C})$, Discrete Valuations, Krull Monoids 82
   5   Modules 84
   6   Representations/Modules/Actions of Other
        Algebraic Structures 89
   7   Free Modules 94

8   IBN Rings   98
9   Simple Modules, Semisimple Modules   99
10   Projective Modules   100
11   Superfluous Submodules and Radical of a Module   102
12   The Jacobson Radical of a Ring   103
13   Injective Modules   103
14   Projective Covers   105
15   Injective Envelopes   106
16   The Monoid $V(R)$ 107
References   110

4   An Introduction to Regular Categories   113
Marino Gran
1   Regular Categories   114
2   Relations in Regular Categories   125
3   Calculus of Relations and Mal'tsev Categories   129
4   Goursat Categories   136
References   144

5   Categorical Commutator Theory   147
Sandra Mantovani and Andrea Montoli
1   Commutators of Groups   148
2   The Case of Ω-groups   150
3   The Categorical Higgins Commutator   152
4   The Huq Commutator   161
5   Abelian Objects   165
References   171

6   Notes on Point-Free Topology   173
Jorge Picado and Aleš Pultr
1   Prologue   174
2   Background   177
3   Frames and Spaces   183
4   Categorical Remarks   191
5   Loc as a Concrete Category. Localic Maps and Sublocales   197
6   Images and Preimages. Localic Maps as Continuous Ones.
        Open Maps   206
7   Examples   212
References   221

7   Non-associative Algebras   225
Tim Van der Linden
1   Non-associative Algebras   226
2   Examples   227
3   The Zero Algebra; Kernels and Cokernels   231

4 Kernels and Ideals, Cokernels and Quotients 233
5 Short Exact Sequences and Protomodularity 236
6 Polynomials and Free Non-associative Algebras 238
7 Varieties of Non-associative Algebras 241
8 Regularity, Exact Sequences 244
9 Semi-abelian Categories 248
10 Birkhoff Subcategories 250
11 Homogeneous Identities 253
12 Some Recent Results 254
13 Bibliography 257
References 257

Chapter 1
Ring Epimorphisms, Gabriel Topologies and Contramodules

Silvana Bazzoni

Abstract During the 1960s considerable work was done in order to understand the meaning of "epimorphism". The notion plays an important role in categories of rings where the abstract category-theoretic meaning is now of common use.

The notion of ring epimorphism has relations with torsion theory and localisation theory. In particular, perfect right Gabriel topologies (in Stenström's terminology) correspond bijectively to left flat ring epimorphisms.

In these notes we will consider two classes of modules defined in terms of a ring epimorphism: the comodules and the contramodules as introduced by Leonid Positselski. Adding mild conditions on the ring epimorphism we will extend classical results proved by Matlis for commutative rings by showing an equivalence between suitable subcategories of the two classes of comodules and contramodules.

Keywords Associative rings and modules · Commutative rings · Ring epimorphisms · Torsion modules · Divisible modules · Comodules · Contramodules · Harrison-Matlis category equivalences · Derived categories.

Math. Subj. Classification   16E30 · 16E35 · 16D90 · 13D05 · 13D30 · 18E10

Introduction

During the 1960s considerable effort was done in order to understand the meaning of epimorphisms in various concrete categories. The notion plays an important role in categories of rings. Localisations of commutative rings with respect to multiplicative subsets are important examples of ring epimorphisms which are moreover, flat ring epimorphisms. A generalisation to noncommutative rings is accomplished by localisations with respect to Gabriel topologies, and flat ring epimorphisms correspond bijectively to a particular class of Gabriel topologies.

S. Bazzoni (✉)
Dipartimento di Matematica "Tullio Levi -Civita", Università di Padova, Via Trieste 63,
35121 Padova, Italy
e-mail: bazzoni@math.unipd.it

© The Author(s), under exclusive license to Springer Nature Switzerland AG 2021
M. M. Clementino et al. (eds.), *New Perspectives in Algebra, Topology and Categories*, Coimbra Mathematical Texts 1,
https://doi.org/10.1007/978-3-030-84319-9_1

In Sect. 2 we will present a characterisation of ring epimorphisms by means of five equivalent conditions. Afterward in Sect. 3 we will introduce Gabriel topologies and their bijective correspondence with hereditary torsion pairs. Furthermore we will define rings and modules of quotients with respect to a Gabriel topology and outline some of their properties.

The purpose of our investigation is to generalise classical equivalences between subcategories of modules over rings to the case of subcategories of modules arising from a ring epimorphism between associative rings.

An important example of such equivalences is provided by the famous Brenner and Butler's Theorem: A finitely generated tilting module T over an artin algebra Λ gives rise to a torsion pair $(\mathcal{T}, \mathcal{F})$, where $\mathcal{T}$ is the class of modules generated by T. If D denotes the standard duality and Γ is the endomorphism ring of T, then $D(T)$ is a cotilting Γ-module with an associated torsion pair $(\mathcal{X}, \mathcal{Y})$ where $\mathcal{Y}$ is the class of modules cogenerated by $D(T)$. The Brenner and Butler's Theorem states that the functor $\mathrm{Hom}_\Lambda(T, -)$ induces an equivalence between the categories $\mathcal{T}$ and $\mathcal{Y}$ with inverse the functor $- \otimes_\Gamma T$, and the functor $\mathrm{Ext}^1_\Lambda(T, -)$ induces an equivalence between $\mathcal{F}$ and $\mathcal{X}$ with inverse the functor $\mathrm{Tor}^\Gamma_1(-, T)$.

For infinitely generated modules a first example of equivalences was provided by Harrison [6] in the category of abelian groups. One equivalence is provided by the tensor product functor $\mathbb{Q}/\mathbb{Z} \otimes_\mathbb{Z} -$, with the functor $\mathrm{Hom}_\mathbb{Z}(\mathbb{Q}/\mathbb{Z}, -)$ as inverse. In Matlis' memoir [8, Sect. 3], the setting was generalised to the case of a commutative domain R and its quotient field Q establishing two kinds of equivalences between certain full additive subcategories of the category of R-modules. The first equivalence is provided by the functor of tensor product with the R-module Q/R with inverse the functor $\mathrm{Hom}_R(Q/R, -)$. The second equivalence is given by the pair of functors $\mathrm{Tor}^R_1(Q/R, -)$ and $\mathrm{Ext}^1_R(Q/R, -)$, which are mutually inverse if restricted to suitable subcategories. Moreover, in the book [9] Matlis extended the first one of his two category equivalences to the setting of an arbitrary commutative ring R and its total ring of quotients Q.

Two further generalisations of the Matlis category equivalences appeared in the two recent papers [4, 12]. In the paper [12], Matlis category equivalences were constructed for a localisation $R[S^{-1}]$ of a commutative ring R with respect to a multiplicative subset $S \subset R$. Injectivity of the map $R \longrightarrow R[S^{-1}]$ was not assumed, but it was assumed that the projective dimension of the R-module $R[S^{-1}]$ was at most one. In the paper [4], Matlis category equivalence was constructed for certain injective epimorphisms of noncommutative rings $R \longrightarrow Q$, where Q is the localisation of R with respect to a one-sided Ore subset of regular elements.

In the paper [3] the first Matlis additive category equivalence is constructed for any ring epimorphism $f: R \longrightarrow U$ such that $\mathrm{Tor}^R_1(U, U) = 0$, and the second Matlis category equivalence is constructed under the assumption $\mathrm{Tor}^R_1(U, U) = 0 = \mathrm{Tor}^R_2(U, U)$. Let us emphasize that *neither* injectivity of f, *nor* any condition on the projective or flat dimension of the R-module U is required for these results. Commutativity of the rings R and U is not assumed, either.

In these notes we present only some of the results proved in [3]. More precisely, in Sect. 5 we give details for the construction of the first Matlis additive category equivalence (Theorem 5.6).

Our key tools are the notions of comodules and contramodules. In Sect. 4 we introduce and discuss the subcategories of comodules and contramodules associated to a ring epimorphism. In Subsect. 4.1 we justify the terminology starting with classical definitions of coalgebras and comodules over coalgebras and show a natural way to introduce the notion of contramodules over coalgebras.

1 Preliminaries

We will assume familiarity with basic notions on category theory like functors, natural transformations, equivalences of categories (see e.g. [15, Ch.1] or [7, Ch.1]).

We will mostly consider categories of modules over associative rings R with unit and we denote by R-Mod, or Mod-R the categories of left, respectively right R-modules.

For every right, or left R-modules M and N, $\mathrm{Hom}_R(M, N)$ denotes the abelian group of all R-linear maps from M to N.

For every right R-module M and a left R-module N, $M \otimes_R N$ denotes the tensor product between M and N.

We will assume that the properties of the functor $\mathrm{Hom}_R(-, -)$ and of the tensor product functor $- \otimes_R -$ are well known.

A left (right) R-module P is *projective* if $\mathrm{Hom}_R(P, -)$ is an exact functor and a module M has projective dimension (p. dim) at most one if it is an epimorphic image of a projective module with kernel a projective module.

A left (right) R-module E is *injective* if $\mathrm{Hom}_R(-, E)$ is an exact functor.

A right (left) R-module F is said to be *flat* if the functor $F \otimes_R - (- \otimes_R F)$ is exact and a module M has flat dimension (f. dim) at most one if it is an epimorphic image of a flat module with kernel a flat module.

Moreover we will use the adjunction between the tensor product functor and the Hom functor. More specifically, if R and S are rings, $_S E_R$ is an S-R-bimodule, the pair of functors $(E \otimes_R -, \mathrm{Hom}_S(E, -))$ is an adjoint pair, that is:

$$E \otimes_R - : R\text{-Mod} \longrightarrow S\text{-Mod}; \quad \mathrm{Hom}_S(E, -) : S\text{-Mod} \longrightarrow R\text{-Mod},$$

and for every left R-module M and left S-module N there is an isomorphism of abelian groups, called the adjunction isomorphism

$$\mathrm{Hom}_S(E \otimes_R M, N) \xrightarrow{\quad \phi(M,N) \quad} \mathrm{Hom}_R(M, \mathrm{Hom}_S(E, N)) ,$$

natural in M and N.

We recall the definition of the unit and the counit of this adjoint pair. The unit is the natural transformation

$$\eta \colon \mathrm{id}_{R\text{-Mod}} \to \mathrm{Hom}_S(E, -) \circ E \otimes_R - ,$$

where for every $M \in R$-Mod the morphism η_M is given by:

$$\mathrm{Hom}_S(E \otimes_R M, E \otimes_R M) \xrightarrow{\phi(M, E \otimes_R M)} \mathrm{Hom}_R(M, \mathrm{Hom}_S(E, E \otimes_R M)) ,$$

$$1_{E \otimes_R M} \xrightarrow{\phi(M, E \otimes_R M)} \eta_M .$$

The counit is the natural transformation

$$\xi \colon E \otimes_R - \circ \mathrm{Hom}_S(E, -) \to \mathrm{id}_{S\text{-Mod}}$$

where for every $N \in S$-Mod the morphism ξ_N is given by

$$\mathrm{Hom}_S(E \otimes_R \mathrm{Hom}_S(E, N), N) \xrightarrow{\phi(\mathrm{Hom}_S(E, N), N)} \mathrm{Hom}_R(\mathrm{Hom}_S(E, N), \mathrm{Hom}_S(E, N))$$

$$\xi_N \xrightarrow{\phi(\mathrm{Hom}_S(E, N), N)} 1_{\mathrm{Hom}_S(E, N)} .$$

The morphism η_M and ξ_N will be used in Sect. 5.

For more details on all these notions see e.g. [15, Ch. 1] or [1, Ch. 1 and 5] or [14, Ch. 2 and 3].

We will make use of some tools in homological algebra, namely the derived functors. In particular, we will deal with the left derived functors Tor_i^R of the tensor product functor, and the right derived functors Ext_R^i of the Hom_R functor.

For their construction and their properties see e.g. [16, Ch. 2 and 3].

In particular, for a right (left) R-module M, we have p. dim $M \leq 1$ if and only if $\mathrm{Ext}^2(M, -) = 0$ and f. dim $M \leq 1$ if and only if $\mathrm{Tor}_2^R(M, -) = 0$ ($\mathrm{Tor}_2^R(-, M) = 0$).

2   Ring Epimorphisms

Definition 2.1 Let $\mathscr{C}$ be a category and $f \colon A \longrightarrow B$ be a morphism between two objects of $\mathscr{C}$.

f is an *epimorphism* if for every object $C \in \mathscr{C}$ and morphisms $g, h \colon B \longrightarrow C$, $g \circ f = h \circ f$ implies $g = h$.

A category $\mathscr{C}$ is concrete if there is a faithful functor F from $\mathscr{C}$ to the category of sets. The functor F makes it possible to think of the objects of the category as sets, possibly with additional structure, and of its morphisms as structure-preserving maps. Examples of concrete categories include trivially the category of sets, the category of topological spaces, the category of groups, the category of rings and the category of modules over a ring R. Every morphism in a concrete category whose underlying map is surjective is an epimorphism. In many concrete categories of interest the converse is also true.

Example 2.2 For instance if $f: X \longrightarrow Y$ is an epimorphism in the category of sets, consider $g: Y \longrightarrow \{0, 1\}$ the characteristic function of $f(X)$ (i.e. $g(f(x)) = 1$, for every $x \in X$ and $g(y) = 0$ for every $y \in Y \setminus f(X)$) and let $h: Y \longrightarrow \{0, 1\}$ be the constant function such that $h(y) = 1$ for every $y \in Y$. Then $g \circ f = h \circ f$, hence $g = h$ and $f(X) = Y$.

There are examples of concrete categories for which epimorphisms are not necessarily surjective maps as we are going to show.

Denote by Rng the category of associative unital rings. A ring homomorphism $f: R \longrightarrow U$ between two rings R, U is a ring epimorphism if it is an epimorphism in the category Rng. That is, for every ring V and every ring homomorphisms $v, w: U \longrightarrow V, v \circ f = w \circ f$ implies $v = w$.

Example 2.3 (1) If I is a two sided ideal of a ring R, then the natural quotient morphism $q: R \longrightarrow R/I$ is a surjective map, hence a ring epimorphism.

(2) If R is a commutative ring, S a multiplicative subset of R, consider the ring of fractions $R_S = R[S^{-1}]$.

Recall that $R_S = \{\left[\frac{r}{s}\right] \mid r \in R, s \in S\}$ where $\left[\frac{r}{s}\right]$ is the equivalence class of the fraction $\frac{r}{s}$ under the equivalence relation defined by $\frac{r}{s} \sim \frac{r'}{s'}$ if and only if there is $t \in S$ such that $t(rs' - sr') = 0$. The ring of fractions R_S becomes a ring with the obvious ring operations (see [2, Ch 3]).

The natural localisation map $\psi: R \longrightarrow R_S$, $\psi(r) = \left[\frac{r}{1}\right]$, is a ring epimorphism. Indeed, if U is a ring and $g, h: R_S \longrightarrow U$ are two ring homomorphisms such that $g \circ \psi = h \circ \psi$, then for every element $r \in R$ and every $s \in S$, we have $g(\left[\frac{r}{s}\right]) = g(\left[\frac{r}{1}\right])g(\left[\frac{s}{1}\right]^{-1})$, where $g(\left[\frac{s}{1}\right]^{-1})$ is the inverse in U of $g(\left[\frac{s}{1}\right]) = h(\left[\frac{s}{1}\right])$, hence $g(\left[\frac{s}{1}\right]^{-1}) = h(\left[\frac{s}{1}\right]^{-1})$. Thus $g(\left[\frac{r}{s}\right]) = h(\left[\frac{r}{1}\right])h(\left[\frac{s}{1}\right]^{-1}) = h(\left[\frac{r}{s}\right])$. That is $g = h$.

(3) The above example shows that there are many ring epimorphisms which are not surjective. In fact, if S is a multiplicative subset of a commutative ring R such that the elements of S are not all invertible in R, then the localisation map $R \longrightarrow R_S$ is a non-surjective ring epimorphism. This applies in particular to the inclusion $\mathbb{Z} \hookrightarrow \mathbb{Q}$.

What do epimorphisms of rings look like? There is a list of equivalent conditions for a ring homomorphism to be an epimorphism which allow to have a better understanding of the notion.

We first note the following. Let $f: R \longrightarrow U$ be a ring homomorphism and let M be a left U-module. Then M is also a left R-module via the scalar multiplication $rx = f(r)x$ for every $x \in M$ and every $r \in R$. Similarly, a right U-module inherits the structure of right R-module via f.

In this way we can define a functor

$$f_* : U\text{-Mod} \longrightarrow R\text{-Mod}; \quad {}_U M \mapsto f_*({}_U M)$$

where $f_*({}_U M)$ is M viewed as a left R-module via f. The functor f_* is called the restriction functor. Similarly, for right U-modules and right R-modules.

In particular, the ring U is a left and right R-module and even an R-R-bimodule.

Consider the tensor product $U \otimes_R U$ which becomes an R-R bimodule and the morphisms

$$i_1 : U \longrightarrow U \otimes_R U, \quad u \mapsto u \otimes 1,$$

$$i_2 : U \longrightarrow U \otimes_R U, \quad u \mapsto 1 \otimes u,$$

$$p : U \otimes_R U \longrightarrow U, \quad u \otimes v \mapsto uv,$$

for every $u, v \in U$.

Proposition 2.4 *Let $f : R \longrightarrow U$ be a ring homomorphism. The following conditions are equivalent:*

1. *f is a ring epimorphism.*
2. *For every U-U-bimodule M,*

$$\{x \in M \mid xr = rx, \forall r \in R\} = \{x \in M \mid xu = ux, \forall u \in U\}.$$

3. *$i_1 = i_2$.*
4. *The restriction functor f_* is fully faithful.*
5. *$p : U \otimes_R U \longrightarrow U$ is an isomorphism as U-U-bimodules.*

Proof (1) $\Rightarrow$ (2) Let M be an U-U-bimodule. Consider the trivial extension of U by M, i.e. the ring

$$U \propto M = \left\{ \begin{pmatrix} u & x \\ 0 & u \end{pmatrix} \mid u \in U, x \in M \right\},$$

with matrix operations. Fix $x \in M$ such that $xr = rx$, for every $r \in R$ and define two ring homomorphisms $g, h : U \longrightarrow U \propto M$ by

$$g(u) = \begin{pmatrix} u & 0 \\ 0 & u \end{pmatrix}, \quad h(u) = \begin{pmatrix} u & xu - ux \\ 0 & u \end{pmatrix}, \text{ for every } u \in U.$$

Then $gf(r) = h(f(r))$ for every $r \in R$.

By (1) $g = h$, that is $xu = ux$ for every $u \in U$.

Conversely, fixing $x \in M$, if $xu = ux$ for every $u \in U$, then $xr = rx$ for every $r \in R$ by the way in which the R-module structure of M is defined via f.

(2) $\Rightarrow$ (3) Let $1 \otimes_R 1 = x \in U \otimes_R U$. Then, for every $r \in R$, $xr = 1 \otimes_R r = r \otimes_R 1 = rx$. Applying (2) to the U-U-bimodule $U \otimes_R U$ we conclude that $xu = ux$ for every $u \in U$, hence $(1 \otimes_R 1)u = 1 \otimes_R u = u(1 \otimes_R 1) = u \otimes_R 1$.

(3) $\Rightarrow$ (4) f_* fully faithful means that for every left U-modules M and N there is an isomorphism of abelian groups

$$\mathrm{Hom}_U(M, N) \xrightarrow{\phi} \mathrm{Hom}_R(f_*(M), f_*(M)).$$

The morphism ϕ is easily seen to be injective and ϕ is surjective if every R-linear morphism between left U-modules is also U-linear. Let M, N be left U-modules and let $\alpha: M \longrightarrow N$ be an R-linear morphism. Fix $x \in M$, $u, v \in U$ and define $\beta: U \otimes_R U \longrightarrow N$ by $\beta(u \otimes_R v) = u\alpha(vx)$. It is easy to check that β is well defined since α is R-linear. By condition (3), $\beta(1 \otimes_R u) = \beta(u \otimes_R 1)$ which yields $\alpha(ux) = u\alpha(x)$, hence α is U-linear.

(4) $\Rightarrow$ (1) Let V be a ring and $g, h: U \longrightarrow V$ be ring homomorphisms such that $g \circ f = h \circ f$. Then V can be viewed as a left U-module via h, that is for every $v \in V$ and $u \in U$ one has $uv = h(u)v$; but V can also be viewed as a left R-module via $g \circ f$, that is $rv = g(f(r))v$, for every $v \in V$ and every $r \in R$. Then g is R-linear; indeed $g(ru) = g(f(r)u) = g(f(r))g(u) = rg(u)$. By condition (4) g is also U-linear, hence $g(u) = ug(1) = uh(1)$ and by the left U-module structure on V via h we have $uh(1) = h(u)h(1) = h(u)$. We conclude that $g = h$.

(4) $\Rightarrow$ (5) The morphism i_2 is R-linear. Indeed, $i_2(ru) = 1 \otimes_R ru = r \otimes_R u = r(1 \otimes_R u)$. By assumption f_* is fully faithful, thus i_2 is also U-linear, that is $i_2(uv) = 1 \otimes_R uv = ui_2(v) = u(1 \otimes_R v) = u \otimes_R v$. We conclude that i_2 is the inverse of p.

(5) $\Rightarrow$ (3) By definition $p(u \otimes_R 1) = u = p(1 \otimes_R u)$. Thus $u \otimes_R 1 = 1 \otimes_R u$. $\square$

Remark 2.5 Clearly, the equivalent conditions in Proposition 2.4 can be stated and proved for right R-modules and right U-modules.

Definition 2.6 A ring epimorphism $f: R \longrightarrow U$ between associative rings R, U is said to be a *homological ring epimorphism* if $\mathrm{Tor}_i^R(U, U) = 0$ for every $i \geq 1$ and it is called a left (right) *flat ring epimorphism* if U is flat as a left (right) R-module.

Example 2.7 Let R be a commutative ring and $R_S = R[S^{-1}]$ be the localisation of R at a multiplicative subset S of R. Then the localisation map $\psi: R \longrightarrow R_S$ is a flat ring epimorphism (see [2, Proposition 3.3]).

In particular, if $\mathfrak{p}$ is a prime ideal of R and $R_\mathfrak{p} = R[(R \setminus \mathfrak{p})^{-1}]$ the localisation map $R \longrightarrow R_\mathfrak{p}$ is a flat ring epimorphism.

Thus flat ring epimorphisms can be viewed as generalisations of localisations of commutative rings at multiplicative sets. As mentioned in the Introduction, Gabriel topologies allow to generalise to the non-commutative setting the notion of localisation and flat ring epimorphisms correspond to localisations with respect to a particular type of Gabriel topologies as we will explain next.

3 Gabriel Topologies, Torsion Pairs and the Ring of Quotients

A *topological ring* is a ring with a topology for which the ring operations are continuous functions. A topological ring is *right linearly topological* if it has a basis of neighbourhoods of zero consisting of right ideals.

A set $\mathcal{F}$ of right ideals of a ring R is the collection of open right ideals of the linearly topological ring R if and only if it satisfies the following conditions:

- (T1) If $I \in \mathcal{F}$ and $I \subseteq J$, then $J \in \mathcal{F}$.
- (T2) If $I, J \in \mathcal{F}$, then $I \cap J \in \mathcal{F}$.
- (T3) If $I \in \mathcal{F}$ and $r \in R$ then $I : r = \{s \in R \mid rs \in I\}$ belongs to $\mathcal{F}$.

The first two conditions just say that $\mathcal{F}$ is a filter of right ideals of R and if R is commutative $I : r$ contains I, thus condition (T3) follows by (T1).

Definition 3.1 A *(right) Gabriel topology* on R, denoted by $\mathcal{G}$, is a filter of open right ideals of a linearly topological ring R (thus satisfying (T1), (T2), (T3)) such that the following additional condition holds.

- (T4) If I is a right ideal of R and there exists $J \in \mathcal{G}$ such that $I : r \in \mathcal{G}$ for every $r \in J$, then $I \in \mathcal{G}$.

Example 3.2 (1) If R is a commutative ring and S is a multiplicative subset of R, then $\mathcal{G} = \{J \leq R \mid S \cap J \neq \emptyset\}$ is a Gabriel topology.

Indeed, (T1) is obvious and (T2) follows since for $s, t \in S$, $st \in S$. As for (T4), if $J \in \mathcal{G}$ and I is an ideal of R such that $I : r \in \mathcal{G}$ for every $r \in J$, let $s \in J \cap S$. Then there exist an element $t \in S$ such that $t \in I : s$, so $st \in I$ and thus $I \in \mathcal{G}$.

(2) If R is a commutative ring and I is a finitely generated ideal of R, then $\mathcal{G} = \{J \leq R \mid J \supseteq I^n, \exists n \in \mathbb{N}\}$ is a Gabriel topology.

Indeed, (T1), (T2) and (T3) are obvious. Let $J \in \mathcal{G}$ and let $L \leq R$ be such that $L : r \in \mathcal{G}$ for every $r \in J$. There is $n_0 \in \mathbb{N}$ such that $J \geq I^{n_0}$. Let $(a_1, a_2, , \ldots, a_k)$ be a set of generators of I^{n_0}. For every $i = 1, 2, \ldots, k$ there is $n_i \in \mathbb{N}$ such that $L : a_i \geq I^{n_i}$. Then there is $m \in \mathbb{N}$ such that $L \geq I^m$ (take e.g. $m = n_0 n$ where n is the supremum of the n_i's).

Definition 3.3 A right R-module over a topological ring R is called *discrete* if the scalar multiplication $M \times R \to M$ is continuous with respect to the discrete topology on M and the topology on R, that is M is a topological R-module in the discrete topology. If R is a right linearly topological ring and $\mathcal{F}$ is the filter of open right ideals, a discrete right R-module M is called $\mathcal{F}$-*discrete*. This amounts to have that for every $x \in M$ the annihilator ideal of x, $\operatorname{Ann}_R x = \{r \in R \mid xr = 0\}$, belongs to $\mathcal{F}$.

Recall that a class $\mathscr{C}$ of R-modules is closed under extensions if for every short exact sequence $0 \to A \to B \to C \to 0$ with $A, C \in \mathscr{C}$, also B is in $\mathscr{C}$.

Proposition 3.4 *Let $\mathcal{F}$ be a set of right ideals of R satisfying* (T1), (T2) *and* (T3). *Then $\mathcal{F}$ satisfies* (T4) *if and only if the class of $\mathcal{F}$-discrete modules is closed under extensions.*

Proof Assume that $\mathcal{F}$ satisfies (T4).

Let $0 \to A \to B \to C \to 0$ be a short exact sequence of R-modules with A and C $\mathcal{F}$-discrete modules. W.l.o.g. we may assume that $A \leq B$ and $C = B/A$. Let $x \in B$ and let $I = \operatorname{Ann}_R x$. If $x \in A$, then $I \in \mathcal{F}$. If $x \notin A$ consider the element $x + A \in C$. The annihilator $\operatorname{Ann}_R(x + A) = J$ is in $\mathcal{F}$ and $xJ \subseteq A$. Thus, for every $r \in J$, $xr \in A$ and the annihilator $\operatorname{Ann}_R xr = J_r$ of the element xr is in $\mathcal{F}$. We have $I : r \supseteq J_r$, so $I \in \mathcal{F}$ by (T4), hence B is $\mathcal{F}$-discrete.

Conversely, assume that the class of $\mathcal{F}$-discrete modules is closed under extensions and let $I \leq R$, $J \in \mathcal{F}$ be such that $I : r \in \mathcal{F}$ for every $r \in J$. We must show that $I \in \mathcal{F}$. Consider the short exact sequence

$$0 \to J/(I \cap J) \cong (I + J)/I \to R/I \to R/(I + J) \to 0.$$

We show that $J/(I \cap J)$ and $R/(I + J)$ are $\mathcal{F}$-discrete modules.

$I + J \in \mathcal{F}$, since $(I + J) \geq J$ and $J \in \mathcal{F}$. The annihilator of an element $a + (I + J) \in R/(I + J)$ is $(I + J) : a$ which is in $\mathcal{F}$ by (T3), hence $R/(I + J)$ is $\mathcal{F}$-discrete. If $r \in J$, then $\operatorname{Ann}_R(r + (I \cap J)) = I : r$ which is assumed to be in $\mathcal{F}$. By assumption R/I is $\mathcal{F}$-discrete, hence $I = \operatorname{Ann}_R(1 + I)$ is in $\mathcal{F}$. $\qquad\square$

For a right Gabriel topology $\mathcal{G}$, denote by $\mathcal{T}_\mathcal{G}$ the class of $\mathcal{G}$-discrete modules.

Lemma 3.5 *Let $\mathcal{G}$ be a right Gabriel topology. The class $\mathcal{T}_\mathcal{G}$ of $\mathcal{G}$-discrete modules is closed under submodules, direct sums, epimorphic images, and extensions.*

Proof The closure under submodules follows immediately by the definition. If $f : M \longrightarrow N$ is an R-linear map, then $\operatorname{Ann}_R x \leq \operatorname{Ann}_R f(x)$ for every $x \in M$, thus the closure under epimorphic images follows by (T1). The annihilator of an element in a direct sum $\oplus_i M_i$ of modules M_i contains the finite intersection of the annihilators of its finitely many non-zero components, hence the closure under direct sums follows by (T1) and (T2). The closure under extensions follows by Proposition 3.4. $\qquad\square$

The above lemma actually says that $\mathcal{T}_\mathcal{G}$ is a hereditary torsion class as we are going to explain next. For a reference and more details on the notion of torsion pairs in module categories see [15, Ch. VI].

Definition 3.6 A *torsion pair* $(\mathcal{T}, \mathcal{F})$ in Mod-R is a pair of classes of modules which are mutually orthogonal with respect to the Hom-functor and maximal with respect to this property. That is,

$$\mathcal{T} = \{T \in \text{Mod-}R \mid \operatorname{Hom}_R(T, F) = 0 \text{ for every } F \in \mathcal{F}\},$$

$$\mathcal{F} = \{F \in \text{Mod-}R \mid \operatorname{Hom}_R(T, F) = 0 \text{ for every } T \in \mathcal{T}\}.$$

The class $\mathcal{T}$ is called a *torsion class* and $\mathcal{F}$ a *torsion-free class*.

A torsion pair $(\mathcal{T}, \mathcal{F})$ is called *hereditary* if $\mathcal{T}$ is also closed under submodules (which is equivalent to $\mathcal{F}$ being closed under injective envelopes).

We show that a torsion class is characterised by its closure properties.

Proposition 3.7 *A class $\mathcal{T}$ of right R-modules is a torsion class if and only if it is closed under direct sums, epimorphic images and extensions.*

Proof The necessary condition follows by the properties of the Hom-functor and the definition of a torsion class. For the sufficiency, assume that a class $\mathcal{T}$ has the stated closure properties. Let

$$\mathcal{F} = \mathcal{T}^{\perp_0} = \{F \in \text{Mod-}R \mid \text{Hom}_R(T, F) = 0, \forall T \in \mathcal{T}\} \text{ and}$$

$$\mathcal{T}' = {}^{\perp_0}\mathcal{F} = \{X \in \text{Mod-}R \mid \text{Hom}_R(X, F) = 0, \forall F \in \mathcal{F}\}$$

we show that $\mathcal{T}' = \mathcal{T}$.

For every $X \in \text{Mod-}R$ let $\mathcal{H}(X) = \{Z \leq X \mid Z \in \mathcal{T}\}$ be the class of the sub-modules of X belonging to $\mathcal{T}$. Consider the submodule $t(X)$ of X defined as $t(X) = \sum\limits_{Z \in \mathcal{H}(X)} Z$. Then $t(X) \in \mathcal{T}$ since it is the image in X of the natural map from the direct sum $\bigoplus\limits_{Z \in \mathcal{H}(X)} Z$ to X. Clearly $t(X)$ is the maximal submodule of X contained in $\mathcal{T}$.

We show now that for every $X \in \mathcal{T}'$, the module $X/t(X)$ belongs to $\mathcal{F}$. Indeed, if $T \in \mathcal{T}$ and $f : T \to X/t(X)$ is a nonzero morphism, let $0 \neq Y/t(X)$ be the nonzero image of f. Then $Y/t(X) \in \mathcal{T}$, since it is an epimorphic image of $T \in \mathcal{T}$ and from the short exact sequence

$$0 \to t(X) \to Y \to Y/t(X) \to 0$$

and the closure under extensions of $\mathcal{T}$ we conclude that $Y \in \mathcal{T}$, that is $Y = t(X)$ contradicting the maximality of $t(X)$. Thus $f = 0$ and $X/t(X) \in \mathcal{F}$. The definition of $\mathcal{T}'$ yields that $X = t(X)$, that is $X \in \mathcal{T}$. $\qquad\qquad \square$

Remark 3.8 Let $(\mathcal{T}, \mathcal{F})$ be a torsion pair. For every R-module M there is a short exact sequence

$$0 \to T \to M \to M/T \to 0,$$

with $T \in \mathcal{T}$ and $M/T \in \mathcal{F}$. T is the *torsion submodule* of M, that is

$$T = t(M) = \sum\{Z \leq M \mid Z \in \mathcal{T}\}.$$

Example 3.9 (1) Let R be a commutative ring and S a multiplicative subset of R. Let $\mathcal{T}$ be the class of the R-modules X such that for every $x \in X$ there is an element $s \in S$ satisfying $xs = 0$. Note that $\mathcal{T}$ coincides with the class of R-modules X such that $X \otimes_R R_S = 0$. Let $\mathcal{F}$ be the class of R-modules Y such that for every $0 \neq y \in Y$, $ys \neq 0$ for every $s \in S$. Then $(\mathcal{T}, \mathcal{F})$ is a hereditary torsion pair.

(2) If S is the set of regular elements r of R, that is the nonzero divisors (i.e. for every $a \in R$, $ra = 0$ or $ar = 0$ implies $a = 0$) the localisation R_S is denoted by Q and called the total quotient ring of R. In case R is a commutative domain, then Q is the quotient field of R. An R-module is simply called a torsion module if it belongs to the torsion class $\mathcal{T}$ in the hereditary torsion pair described in example (1) above.

Theorem 3.10 *Let R be a ring. There is a bijective correspondence:*

$$\left\{ \begin{matrix} \text{right Gabriel topologies} \\ \text{on } R \end{matrix} \right\} \underset{\Psi}{\overset{\Phi}{\rightleftarrows}} \left\{ \begin{matrix} \text{hereditary torsion} \\ \text{pairs in Mod-}R \end{matrix} \right\}.$$

1. If $\mathcal{G}$ is a right Gabriel topology $\Phi(\mathcal{G}) = (\mathcal{T}_{\mathcal{G}}, \mathcal{F}_{\mathcal{G}})$ where $\mathcal{T}_{\mathcal{G}}$ is the class of the $\mathcal{G}$-discrete modules and

$$\mathcal{F}_{\mathcal{G}} = \{Y_R \in \text{Mod-}R \mid \text{Hom}_R(R/J, Y) = 0, \forall J \in \mathcal{G}\}.$$

2. If $(\mathcal{T}, \mathcal{F})$ is a hereditary torsion pair, $\Psi((\mathcal{T}, \mathcal{F})) = \{J_R \leq R \mid R/J \in \mathcal{T}\}$.

Proof (1) For every Gabriel topology $\mathcal{G}$, the class of $\mathcal{G}$-discrete modules is a hereditary torsion class by Lemma 3.5 and Proposition 3.7. A module $T \in \mathcal{T}_{\mathcal{G}}$ is an epimorphic image of a direct sum of copies of modules R/J, for some $J \in \mathcal{G}$. Hence the description of the torsion-free class follows.

(2) Let $(\mathcal{T}, \mathcal{F})$ be a hereditary torsion pair and $\mathcal{G} = \{J_R \leq R \mid R/J \in \mathcal{T}\}$. The closure of $\mathcal{T}$ under epimorphic images implies that $\mathcal{G}$ satisfies condition (T1). If I, J are in $\mathcal{G}$, then $R/I \oplus R/J \in \mathcal{T}$ and $\text{Ann}_R(1 + I, 1 + J) = I \cap J$; hence (T2) is satisfied by $\mathcal{G}$. As for (T3), let $J \in \mathcal{G}$ and $r \in R$. Then $J : r = \{s \in R \mid rs \in J\} = \text{Ann}_R(r + J)$. Since $\mathcal{T}$ is hereditary, the cyclic module $(r + J)R$ belongs to $\mathcal{T}$, thus $J : r \in \mathcal{G}$. At this point (T4) follows by Proposition 3.4 since $\mathcal{T}$ is closed under extensions.

If $\mathcal{G}$ is a Gabriel topology, it is clear by construction that $\Psi \circ \Phi(\mathcal{G}) = \mathcal{G}$, since $J \in \mathcal{G}$ if and only if $R/J \in \mathcal{T}_{\mathcal{G}}$.

If $(\mathcal{T}, \mathcal{F})$ is a hereditary torsion pair and $\mathcal{G} = \Psi((\mathcal{T}, \mathcal{F}))$, then a module $N \in \mathcal{T}_{\mathcal{G}}$ is an epimorphic image of a direct sum of cyclic modules of the form R/J for some $J \in \mathcal{G}$, hence $N \in \mathcal{T}$. Conversely, if $M \in \mathcal{T}$, then every cyclic submodule $xR \cong R/J$ of M is in $\mathcal{T}$, since $\mathcal{T}$ is hereditary, thus $J \in \mathcal{G}$ and consequently $M \in \mathcal{T}_{\mathcal{G}}$. We conclude that $\Phi \circ \Psi((\mathcal{T}, \mathcal{F})) = (\mathcal{T}, \mathcal{F})$. □

Remark 3.11 Note that the hereditary torsion pair defined in Examples 3.9 (1) corresponds under the bijection of Theorem 3.10 to the Gabriel topology defined in Examples 3.2 (1).

If G is a Gabriel topology with corresponding torsion pair $(\mathcal{T}_G, \mathcal{F}_G)$, a G-discrete module is also called G-torsion and a module in $\mathcal{F}_G$ is called G-torsion-free.

A Gabriel topology allows to generalise localisations of commutative rings to the case of non-commutative rings and as already mentioned, we will see that flat ring epimorphisms are localizations of particular types of Gabriel topologies.

In this section we state some notions and results on rings and modules of quotients with respect to a Gabriel topology. For their proofs we refer to [15, Chapter IX].

On a Gabriel topology G consider the partial order given by inclusion and for an arbitrary R-module N consider the direct system

$$\{\mathrm{Hom}_R(J, N); f_{IJ}\}_{J \in G, I \leq J}$$

where for every $I \leq J$ the morphism

$$f_{IJ} : \mathrm{Hom}_R(J, N) \longrightarrow \mathrm{Hom}_R(I, N)$$

is the restriction map.

Given a module M, the *module of quotients* with respect to a Gabriel topology G is defined by:

$$M_G := \varinjlim_{J \in G} \mathrm{Hom}_R(J, M/t_G(M))$$

where $t_G(M)$ is the torsion submodule of M in the torsion pair $(\mathcal{T}_G, \mathcal{F}_G)$ corresponding to G under Theorem 3.10.

Furthermore, there is a natural homomorphism

$$\psi_M : M \cong \mathrm{Hom}_R(R, M) \longrightarrow M_G$$

For each R-module M, both the kernel and cokernel of the map ψ_M are G-torsion R-modules.

If $M = R$, then

$$R_G := \varinjlim_{J \in G} \mathrm{Hom}_R(J, R/t_G(R))$$

is a ring and is called the *ring of quotients* of R with respect to the Gabriel topology G and the morphism $\psi_R : R \longrightarrow R_G$ is a ring homomorphism. Moreover, for each R-module M the module M_G is both an R-module and an R_G-module.

A right R-module is *G-closed* if the natural homomorphisms

$$M \cong \mathrm{Hom}_R(R, M) \longrightarrow \mathrm{Hom}_R(J, M)$$

are all isomorphisms for each $J \in G$.

This amounts to saying that $\mathrm{Hom}_R(R/J, M) = 0$ for every $J \in G$ (i.e. M is *G-torsion-free*) and $\mathrm{Ext}^1_R(R/J, M) = 0$ for every $J \in G$ (i.e. M is *G-injective*). Moreover, if M is G-closed then M is isomorphic to its module of quotients M_G

via ψ_M. Conversely, every R-module of the form $M_\mathcal{G}$ is $\mathcal{G}$-closed. The $\mathcal{G}$-closed modules form a full subcategory of both Mod-R and Mod-$R_\mathcal{G}$. In fact, every R-linear morphism of $\mathcal{G}$-closed modules is also $R_\mathcal{G}$-linear.

Remark 3.12 In general the natural ring homomorphism $\psi_R \colon R \longrightarrow R_\mathcal{G}$ is not a ring epimorphism, but in some important cases ψ_R is even a flat ring epimorphism.

The following two results characterise when a ring homomorphism is a flat ring epimorphism and describe the associated Gabriel topology.

Theorem 3.13 [15, Theorem XI.2.1] *Suppose* $f \colon R \longrightarrow U$ *is a ring homomorphism. Then the following are equivalent.*

(i) *f is an epimorphism of rings which makes U into a flat left R-module.*
(ii) *The family $\mathcal{G}$ of right ideals J such that $JU = U$ is a Gabriel topology, and the natural ring homomorphism $\psi \colon R \longrightarrow R_\mathcal{G}$ is equivalent to $f \colon R \longrightarrow U$. That is, there is a ring isomorphism $\sigma \colon U \longrightarrow R_\mathcal{G}$ such that $\sigma \circ f \colon R \longrightarrow R_\mathcal{G}$ is the natural homomorphism $\psi_R \colon R \longrightarrow R_\mathcal{G}$.*

Proposition 3.14 [15, Proposition XI.3.4] *Let $\mathcal{G}$ be a right Gabriel topology. Then the following conditions are equivalent.*

1. *$\psi_R \colon R \longrightarrow R_\mathcal{G}$ is a flat ring epimorphism and $\mathcal{G} = \{J \le R \mid JR_\mathcal{G} = R_\mathcal{G}\}$.*
2. *$R_\mathcal{G}$ is $\mathcal{G}$-divisible, i.e. $JR_\mathcal{G} = R_\mathcal{G}$ for every $J \in \mathcal{G}$.*
3. *For every right R-module M, $\mathrm{Ker}(M \to M \otimes_R R_\mathcal{G})$ is the $\mathcal{G}$-torsion submodule of M.*

Definition 3.15 A right Gabriel topology satisfying the equivalent conditions of Proposition 3.14 is called a *perfect Gabriel topology*.

In particular, the right Gabriel topology $\mathcal{G}$ associated to a flat ring epimorphism $R \xrightarrow{u} U$ is finitely generated and the $\mathcal{G}$-torsion submodule $t_\mathcal{G}(M)$ of a right R-module M is the kernel of the natural homomorphism $M \to M \otimes_R U$. Additionally, $K = U/u(R)$ is $\mathcal{G}$-torsion, hence $\mathrm{Hom}_R(K, U) = 0$.

4 Comodules and Contramodules

We first introduce the definitions of "comodules and contramodules" via ring epimorphisms and in the next subsection we will explain how the terminology is borrowed from the coalgebras setting.

From now on $f \colon R \longrightarrow U$ will always denote a *ring epimorphism* of associative rings.

Recall from Sect. 2 that the functor of restriction of scalars

$$f_* \colon U\text{-Mod} \to R\text{-Mod}$$

is fully faithful. Similar assertions hold for the categories of right modules.

We will say that a certain R-module "is a U-module" if it belongs to the image of the functor f_*.

We will use the notation U/R for the cokernel of the map $f : R \longrightarrow U$, so U/R is an R-R-bimodule.

Definition 4.1   1. A left R-module M is called a *left U-comodule* if

$$U \otimes_R M = 0 = \operatorname{Tor}_1^R(U, M).$$

Similarly, a right R-module N is said to be a right U-comodule if

$$N \otimes_R U = 0 = \operatorname{Tor}_1^R(N, U).$$

2. A left (right) R-module C is called a *left (right) U-contramodule* if

$$\operatorname{Hom}_R(U, C) = 0 = \operatorname{Ext}_R^1(U, C).$$

Proposition 4.2 [5, dual of Proposition 1.1] *Let $\mathscr{C} \subset R$-Mod be the class of all left U-comodules. Then:*

1. *$\mathscr{C}$ is closed under direct sums, cokernels of morphisms, and extensions in R-Mod.*
2. *If f. $\dim_R U \leq 1$ as a right R-module, then $\mathscr{C}$ is closed also under kernels of morphisms.*

Proof (1) The closure under direct sums and extensions of left U-comodules follows by the properties of the tensor product functor and Tor functor.

Let $g : N \to L$ be a morphism between left U-comodules and consider the associated short exact sequences

(a) $0 \to \operatorname{Ker} g \to N \to \operatorname{Im} g \to 0$; (b) $0 \to \operatorname{Im} g \to L \to \operatorname{Coker} g \to 0$

Applying the right exact functor $U \otimes_R -$ to sequence (b) we obtain

$$0 = U \otimes_R L \to U \otimes_R \operatorname{Coker} g \to 0,$$

so $U \otimes_R \operatorname{Coker} g = 0$. From sequence (a) we get

$$0 = U \otimes_R N \to U \otimes_R \operatorname{Im} g \to 0,$$

hence $U \otimes_R \operatorname{Im} g = 0$. The long exact sequence associated to (b) yields

$$0 = \operatorname{Tor}_1^R(U, L) \to \operatorname{Tor}_1^R(U, \operatorname{Coker} g) \to U \otimes_R \operatorname{Im} g = 0.$$

Thus also $\operatorname{Tor}_1^R(U, \operatorname{Coker} g) = 0$ and $\operatorname{Coker} g$ is a left U-comodule.

(2) The assumption f. $\dim_R U \leq 1$ is equivalent to $\mathrm{Tor}_2^R(U, -) = 0$. Let $g: N \to L$ be a morphism between left U-comodules and consider the exact sequences as in part (1). From sequence (b) we get

$$0 = \mathrm{Tor}_2^R(U, \mathrm{Coker}\, g) \to \mathrm{Tor}_1^R(U, \mathrm{Im}\, g) \to \mathrm{Tor}_1^R(U, L) = 0.$$

So $\mathrm{Tor}_1^R(U, \mathrm{Im}\, g) = 0$ and sequence (a) gives

$$0 = \mathrm{Tor}_1^R(U, \mathrm{Im}\, g) \to U \otimes_R \mathrm{Ker}\, g \to U \otimes_R N = 0$$

and

$$\mathrm{Tor}_2^R(U, \mathrm{Im}\, g) = 0 \to \mathrm{Tor}_1^R(U, \mathrm{Ker}\, g) \to \mathrm{Tor}_1^R(U, N) = 0.$$

We conclude that $\mathrm{Ker}\, g$ is a left U-comodule. $\qquad\square$

The dual situation is expressed by the following.

Proposition 4.3 [5, Proposition 1.1] *Let $\mathscr{C} \subset R$-Mod be the class of all left U-contramodules. Then:*

1. *$\mathscr{C}$ is closed under products, kernels of morphisms, and extensions in R-Mod.*
2. *If* p. $\dim_R U \leq 1$ *as a left R-module, then $\mathscr{C}$ is closed also under cokernels of morphisms.*

Proof (1) The closure under direct products and extensions follows by the closure properties of the functors Hom_R and Ext_R^1.

Let $g: C \to D$ be a morphism between left U-contramodules. The proof that $\mathrm{Ker}\, g \in \mathscr{C}$ is analogous to the proof of Proposition 4.2 (1) applying the functors Hom_R and Ext_R^1 to the short exact sequences

$$0 \to \mathrm{Ker}\, g \to C \to \mathrm{Im}\, g \to 0; \qquad 0 \to \mathrm{Im}\, g \to D \to \mathrm{Coker}\, g \to 0$$

(2) p. $\dim_R U \leq 1$ is equivalent to $\mathrm{Ext}_R^2(U, -) = 0$. Thus the proof follows similarly to the proof of Proposition 4.2 (2). $\qquad\square$

4.1 Coalgebras, Comodules, Contramodules

We follow the presentation developed in [13, Section 1.1].

Let k be a field. Recall that a k-algebra A is a k-vector space with k-linear maps

$$A \otimes_k A \xrightarrow{m} A, \qquad k \xrightarrow{e} A,$$

m is the multiplication map and e the unit satisfying *associativity*, i.e.:

$$m \circ (m \otimes 1_A) = m \circ (1_A \otimes m) : \quad A \otimes_k A \otimes_k A \xrightarrow[m \otimes 1_A]{1_A \otimes m} A \otimes_k A \xrightarrow{m} A$$

and *unitality*, i.e.::

$$m \circ (e \otimes 1_A) = 1_A = m \circ (1_A \otimes e) : \quad A \xrightarrow[e \otimes 1_A]{1_A \otimes e} A \otimes_k A \xrightarrow{m} A$$

A left A-module is a k-vector space with a k-linear map (scalar multiplication)

$$A \otimes_k M \xrightarrow{\lambda} M$$

satisfying associativity, i.e.:

$$\lambda \circ (m \otimes_k 1_M) = \lambda \circ (1_A \otimes_k \lambda), \quad A \otimes_k A \otimes_k M \xrightarrow[m \otimes 1_M]{1_A \otimes \lambda} A \otimes_k M \xrightarrow{\lambda} M$$

and unitality, i.e.:

$$\lambda \circ (e \otimes 1_M) = 1_M, \quad k \otimes_k M \cong M \xrightarrow{e \otimes 1_M} A \otimes_k M \xrightarrow{\lambda} M$$

Dualising the above diagrams we get the notions of coalgebras and comodules.

Definition 4.4 A *coalgebra* $\mathscr{C}$ is a k-vector space with k linear maps

$$\mathscr{C} \xrightarrow{\mu} \mathscr{C} \otimes_k \mathscr{C}, \quad \mathscr{C} \xrightarrow{\varepsilon} k,$$

μ the *comultiplication* and ε the *counit* satisfying *coassociativity*, i.e.:

$$(1_{\mathscr{C}} \otimes \mu) \circ \mu = (\mu \otimes 1_{\mathscr{C}}) \circ \mu : \quad \mathscr{C} \xrightarrow{\mu} \mathscr{C} \otimes_k \mathscr{C} \xrightarrow[\mu \otimes 1_{\mathscr{C}}]{1_{\mathscr{C}} \otimes \mu} \mathscr{C} \otimes_k \mathscr{C} \otimes_k \mathscr{C}$$

and *counitality*, i.e.:

$$(\varepsilon \otimes 1_{\mathscr{C}}) \circ \mu = 1_{\mathscr{C}} = (1_{\mathscr{C}} \otimes \varepsilon) \circ \mu : \quad \mathscr{C} \xrightarrow{\mu} \mathscr{C} \otimes_k \mathscr{C} \xrightarrow[\varepsilon \otimes 1_{\mathscr{C}}]{1_{\mathscr{C}} \otimes \varepsilon} \mathscr{C}$$

Definition 4.5 A left $\mathscr{C}$-*comodule* over a coalgebra $\mathscr{C}$ is a k-vector space N with a k-linear map (coaction map)

$$N \xrightarrow{\ \upsilon\ } \mathscr{C} \otimes_k N$$

satisfying *coassociativity*:

$$(1_{\mathscr{C}} \otimes \upsilon) \circ \upsilon = (\mu \otimes 1_N) \circ \upsilon : \quad N \xrightarrow{\ \upsilon\ } \mathscr{C} \otimes_k N \begin{array}{c} \xrightarrow{1_{\mathscr{C}} \otimes \upsilon} \\[-2pt] \xrightarrow[\mu \otimes 1_N]{} \end{array} \mathscr{C} \otimes_k \mathscr{C} \otimes_k N$$

and *counitality*:

$$(\varepsilon \otimes 1_N) \circ \upsilon = 1_N : \quad N \xrightarrow{\ \upsilon\ } \mathscr{C} \otimes_k N \xrightarrow{\varepsilon \otimes 1_N} N \cong k \otimes_k N.$$

A right $\mathscr{C}$-*comodule* is defined as a k-vector space with a k-linear map

$$N \xrightarrow{\ \upsilon\ } N \otimes_k \mathscr{C}$$

satisfying the corresponding coassociativity and counitality conditions.

Note that having a left A-module M with a scalar multiplication λ is the same as having a k-linear map

$$M \xrightarrow{\ p\ } \mathrm{Hom}_k(A, M) \qquad x \mapsto \dot{x} : a \longrightarrow ax = \lambda(a \otimes_R x)$$

which satisfies the associativity:

$\mathrm{Hom}_k(m, M) \circ p = \mathrm{Hom}(A, p) \circ p$ (via the adjunction isomorphism):

$$M \xrightarrow{\ \ p\ \ } \mathrm{Hom}_k(A, M) \xrightarrow{\ \mathrm{Hom}_k(m,M)\ } \mathrm{Hom}_k(A \otimes_k A, M)$$
$$\searrow_{\mathrm{Hom}(A,p)} \qquad \downarrow^{\cong}$$
$$\mathrm{Hom}_k(A, \mathrm{Hom}_k(A, M))$$

and unitality:

$\mathrm{Hom}(e, M) \circ p = 1_M$:

$$M \xrightarrow{\ p\ } \mathrm{Hom}_k(A, M) \xrightarrow{\mathrm{Hom}(e,M)} M = \mathrm{Hom}_k(k, M).$$

The notion of a left $\mathscr{C}$-contramodule over a coalgebra $\mathscr{C}$ is obtained by dualizing the above description of a left A-module over a k-algebra A.

Definition 4.6 A left $\mathscr{C}$-*contramodule* over a coalgebra $\mathscr{C}$ is a k-vector space B with a k-linear map (contraaction map)

$$\mathrm{Hom}_k(\mathscr{C}, B) \xrightarrow{\pi_B} B$$

satisfying the *contraassociativity* which means:
$\pi_B \circ \mathrm{Hom}_k(\mu, B) = \pi_B \circ \mathrm{Hom}_k(\mathscr{C}, \pi_B)$ (via the adjunction isomorphism):

$$
\begin{array}{ccc}
\mathrm{Hom}_k(\mathscr{C} \otimes_k \mathscr{C}, B) & \xrightarrow{\mathrm{Hom}_k(\mu, B)} & \mathrm{Hom}_k(\mathscr{C}, B) & \xrightarrow{\pi_B} & B \\
\cong \Big\downarrow & \nearrow{}^{\mathrm{Hom}(\mathscr{C}, \pi_B)} & & & \\
\mathrm{Hom}_k(\mathscr{C}, \mathrm{Hom}_k(\mathscr{C}, B)) & & & &
\end{array}
$$

and *contraunitality* meaning:
$\pi_B \circ \mathrm{Hom}_k(\varepsilon, B) = 1_B$:

$$\mathrm{Hom}_k(k, B) \cong B \xrightarrow{\mathrm{Hom}_k(\varepsilon, B)} \mathrm{Hom}_k(\mathscr{C}, B) \xrightarrow{\pi_B} B.$$

An easy way to construct a left $\mathscr{C}$-contramodule is via a right $\mathscr{C}$-comodule.
Let M be a right $\mathscr{C}$-comodule with $M \xrightarrow{\nu_M} M \otimes_k \mathscr{C}$ the right coaction map.

Let V be a k-vector space and let $B = \mathrm{Hom}_k(M, V)$. Then B is a left $\mathscr{C}$-contramodule with left contraaction map π_B defined by the diagram:

$$
\begin{array}{ccc}
\mathrm{Hom}_k(\mathscr{C}, \mathrm{Hom}_k(M, V)) & \dashrightarrow^{\pi_B} & \mathrm{Hom}_k(M, V) = B \ . \\
\cong \Big\downarrow & & = \Big\downarrow \\
\mathrm{Hom}_k(\mathscr{C} \otimes_k M, V) & \xrightarrow[\mathrm{Hom}(\nu_M, V)]{} & \mathrm{Hom}_k(M, V) = B
\end{array}
$$

Remark 4.7 ([13, Sections 1.3–1.4]) The k-duality functor identifies the opposite of the category of vector spaces with the category of linearly compact vector spaces. Thus, up to inverting the arrows, every coalgebra $\mathscr{C}$ can be thought as a linearly compact topological algebra $\mathscr{C}^*$, called the dual topological algebra. Then the category of left $\mathscr{C}$-comodules is the full subcategory of discrete left $\mathscr{C}^*$-modules.

We illustrate now a particular example of a coalgebra $\mathscr{C}$, its associated dual topological algebra and describe the categories of $\mathscr{C}$-comodules and $\mathscr{C}$-contramodules.

Example of a coalgebra 4.8 [13, Section 1.3] Let k be a field. Let $\mathscr{C}$ be a k-vector space with countable basis denoted by the symbols $1^*, x^*, (x^2)^* \ldots (x^n)^* \ldots$ with comultiplication and counit given by

$$\mathscr{C} \xrightarrow{\mu} \mathscr{C} \otimes_k \mathscr{C}; \quad (x^n)^* \longmapsto \sum_{i+j=n} (x^i)^* \otimes (x^j)^*$$

$$\mathscr{C} \xrightarrow{\varepsilon} k; \quad 1^* \longmapsto 1, \quad (x^n)^* \longmapsto 0, \ \forall n \geq 1.$$

The dual topological algebra $\mathscr{C}^*$ is isomorphic to the ring of formal power series $k[[x]]$.

By Remark 4.7 a $\mathscr{C}$-comodule is a torsion $k[[x]]$-module.

Indeed a $\mathscr{C}$-comodule M is a k-vector space with a locally nilpotent operator i.e. a k-linear map $x : M \to M$ such that for every $z \in M$ there exists $m \in \mathbb{N}$ satisfying $x^m(z) = 0$ so that M becomes a $\mathscr{C}$-comodule via

$$\nu_M : M \longrightarrow \mathscr{C} \otimes M; \qquad z \longmapsto \sum_{n \geq 0} (x^n)^* \otimes x^n(z)$$

A $\mathscr{C}$-contramodule B is the datum of a k-vector space with a k-linear map $\mathrm{Hom}_k(\mathscr{C}, B) \xrightarrow{\pi_B} B$ satisfying the contraassociativity and the contraunitality which in our case means that for every sequence $b_0, b_1, \ldots, b_n \ldots$ of elements of B, there is an element $b \in B$ written formally as $\sum_{n \geq 0} x^n b_n$ satisfying the axiom of linearity:

$$\sum_{n \geq 0} x^n (\alpha b_n + \beta c_n) = \alpha \sum_{n \geq 0} x^n b_n + \beta \sum_{n \geq 0} x^n c_n; \quad \forall \alpha, \beta \in k, \quad b_n, c_n \in B,$$

the axiom of unitality:

$$\sum_{n \geq 0} x^n b_n = b_0, \quad \text{if } b_1 = b_2 = \cdots = 0$$

and the axiom of contraassociativity:

$$\sum_{i \geq 0} x^i \sum_{j \geq 0} x^j b_{ij} = \sum_{n \geq 0} x^n \sum_{i+j=n} b_{ij}, \quad \forall b_{ij} \in B, i, j \in \mathbb{N}.$$

Thus, a $\mathscr{C}$-contramodule B is determined by a single linear operator $x : B \to B$ such that $x(b) = 1 \cdot 0 + x \cdot b + x^2 \cdot 0 + x^3 \cdot 0 \ldots$ (see [13, Section 1.6] or [11, Section 3]).

Now we justify the definitions of U-comodules and U-contramodules given above by exhibiting an example of a ring epimorphism $f : R \to U$ such that the U-comodules and U-contramodules correspond exactly to the $\mathscr{C}$-comodules and $\mathscr{C}$-contramodules for the coalgebra described in Example 4.8.

Example 4.9 [13, Section 1.3] Let $R = k[x]$ be the ring of polynomials in one variable over a field k, let $U = k[x, x^{-1}]$ be the ring of Laurent polynomials, and let $f : R \longrightarrow U$ be the natural inclusion. So U is obtained from R by inverting the single element x.

Let $\mathscr{C}$ be the coalgebra constructed in Example 4.8.

Since U is a flat R-module, and the $\mathscr{C}$-comodules are the torsion $k[[x]]$-modules, one sees that the full subcategory of U-comodules in R-Mod is equivalent to the category of $\mathscr{C}$-comodules.

An application of [11, Theorem 3.3] and the description of $\mathscr{C}$-contramodules illustrated in Example 4.8 yields that the full subcategory of U-contramodules in R-Mod is equivalent to the category of $\mathscr{C}$-contramodules.

5 First Matlis Category Equivalence

In this section we present some results obtained by using the notion of ring epimorphism as well as the notions of comodules and contramodules.

We show that these notions are useful tools allowing to achieve relevant results like for instance, a generalisation of classical equivalences between subcategories of the module category over commutative rings.

Indeed, by Theorem 5.6 we extend the first Matlis equivalence to a much more general setting and under much weaker assumptions ([3]).

We borrow the terminology going back to Harrison [6] and Matlis [10].

Definition 5.1 Let $f : R \longrightarrow U$ be a ring epimorphism.

1. A left R-module A is U-torsion-free if it is an R-submodule of a left U-module, or equivalently, if the morphism $A \xrightarrow{1_A \otimes f} A \otimes_R U$ is injective.
2. A left R-module B is U-divisible if it is a quotient module of a left U-module, or equivalently, if the map $\operatorname{Hom}_R(U, B) \xrightarrow{\operatorname{Hom}(f, B)} B$ is surjective.

Remark 5.2 It is easy to check that the class of all U-torsion-free left R-modules is closed under subobjects, direct sums, and products in R-Mod. Any left R-module A has a unique maximal U-torsion-free quotient module, which is the image of the morphism $A \xrightarrow{1_A \otimes f} A \otimes_R U$.

The class of all U-divisible left R-modules is closed under quotients, direct sums, and products. Any left R-module B has a unique maximal U-divisible submodule, which is the image of the morphism $\operatorname{Hom}_R(U, B) \xrightarrow{\operatorname{Hom}(f, B)} B$.

Definition 5.3 1. A left R-module A is said to be U-torsion if its maximal U-torsion-free quotient is zero, or equivalently, if $A \otimes_R U = 0$.
2. A left R-module B is said to be U-reduced if its maximal U-divisible submodule is zero, or equivalently, if $\operatorname{Hom}_R(U, B) = 0$.

We first state a useful homological result which has interest in its own and which will be used later on.

Lemma 5.4 *1. For any associative rings R and S, left R-module L, S-R-bimodule E, and left S-module M such that $\operatorname{Tor}_1^R(E, L) = 0$, there is a natural injective map of abelian groups*

$$\operatorname{Ext}_R^1(L, \operatorname{Hom}_S(E, M)) \longhookrightarrow \operatorname{Ext}_S^1(E \otimes_R L, M).$$

2. *Dually, for any associative rings R and S, right R-module B, R-S-bimodule E, and left S-module C such that $\mathrm{Tor}_1^R(B, E) = 0$, there is a natural surjective map of abelian groups*

$$\mathrm{Tor}_1^S(B \otimes_R E, C) \twoheadrightarrow \mathrm{Tor}_1^R(B, E \otimes_S C).$$

Proof (1) Let (*a*) $0 \longrightarrow H \longrightarrow P \longrightarrow L \longrightarrow 0$ be a short exact sequence in R-Mod with P a projective left R-module. Apply the functor $E \otimes_R -$ to sequence (*a*) obtaining

(*b*) $0 = \mathrm{Tor}_1^R(E, L) \longrightarrow E \otimes_R H \longrightarrow E \otimes_R P \longrightarrow E \otimes_R L \longrightarrow 0.$

Apply the functor $\mathrm{Hom}_R(-, \mathrm{Hom}_S(E, M))$ to sequence (*a*) and the functor $\mathrm{Hom}_S(-, M)$ to sequence (*b*) obtaining a diagram

$$\begin{array}{ccccccc}
\mathrm{Hom}_R(P, \mathrm{Hom}_S(E, M)) & \twoheadrightarrow & \mathrm{Hom}_R(H, \mathrm{Hom}_S(E, M)) & \twoheadrightarrow & \mathrm{Ext}_R^1(L, \mathrm{Hom}_S(E, M)) & \longrightarrow & 0 \\
\cong \downarrow & & \cong \downarrow & & \alpha \downarrow & & \\
\mathrm{Hom}_R(E \otimes_R P, M) & \longrightarrow & \mathrm{Hom}_R(E \otimes H, M) & \longrightarrow & \mathrm{Ext}_R^1(E \otimes_R L, M) & \longrightarrow & \mathrm{Ext}_R^1(E \otimes_R P, M)
\end{array}$$

where the left and central vertical arrows are the natural isomorphisms for the adjoint pair $(E \otimes_R -, \mathrm{Hom}_E(E, -))$ and the morphism α exists since $\mathrm{Ext}_R^1(L, \mathrm{Hom}_S(E, M))$ is a cokernel.

By diagram chasing the commutativity of the diagram yields that α is injective. The details are left to the reader.

For the dual statement (2) start with a projective presentation

(*c*) $0 \longrightarrow H \longrightarrow P \longrightarrow B \longrightarrow 0$

of B in Mod-R with P a projective right R-module. Apply the functor $- \otimes_R E$ to sequence (*c*) obtaining

(*d*) $\mathrm{Tor}_1^R(B, E) = 0 \longrightarrow H \otimes_R E \longrightarrow P \otimes_R E \longrightarrow B \otimes_R E \longrightarrow 0.$

Apply the functor $- \otimes_S C$ to sequence (*d*) and the functor $- \otimes_R (E \otimes_S C)$ to sequence (*c*) obtaining a diagram

$$\begin{array}{ccccccc}
\mathrm{Tor}_1^R(B \otimes_R E, C) & \twoheadrightarrow & (H \otimes_R E) \otimes_S C & \twoheadrightarrow & (P \otimes_R E) \otimes_S C & \twoheadrightarrow & (B \otimes_R E) \otimes_S C \\
\beta \downarrow & & \cong \downarrow & & \cong \downarrow & & \cong \downarrow \\
0 \twoheadrightarrow \mathrm{Tor}_1^R(B, E \otimes_S C) & \twoheadrightarrow & H \otimes_R (E \otimes_S C) & \twoheadrightarrow & P \otimes_R (E \otimes_S C) & \twoheadrightarrow & B \otimes_R (E \otimes_S C)
\end{array}$$

where the morphism β exists since $\mathrm{Tor}_1^R(B, E \otimes_S C)$ is a kernel.

By diagram chasing the commutativity of the diagram yields that β is surjective.   □

Remark 5.5 In [9] Matlis considers the flat injective ring epimorphism $R \to Q$ where Q is the total quotient ring of a commutative ring R, that is the localisation of R at the multiplicative set of all the regular elements of R (see Examples 3.9). In [9] a module C satisfying $\operatorname{Hom}_R(Q, C) = 0 = \operatorname{Ext}_R^1(Q, C)$ is called *cotorsion* and a module D such that $\operatorname{Hom}_R(Q, D) \to D$ is surjective is called *h-divisible*.

Then in [9, Corollary 2.4] the first Matlis equivalence states that the functors $Q/R \otimes_R -$ and $\operatorname{Hom}_E(Q/R, -)$ induce the equivalence:

$$\left\{ \begin{array}{c} \text{torsion-free cotorsion} \\ \text{R-modules} \end{array} \right\} \underset{\operatorname{Hom}_R(Q/R, -)}{\overset{(Q/R)\otimes_R -}{\rightleftarrows}} \left\{ \begin{array}{c} \text{h-divisible torsion} \\ \text{R-modules} \end{array} \right\},$$

where the notion of torsion is the classical one. That is an R-module M is torsion if for every element $x \in M$ there is a regular element $r \in R$ such that $rx = 0$.

The following theorem relaxes as much as possible the assumptions in [9, Corollary 2.4] to provide what appears to be the best possible generalisation for the first of the two classical Matlis category equivalences (going back to Harrison's [6, Proposition 2.1]).

Theorem 5.6 *Let $f : R \longrightarrow U$ be a ring epimorphism, $U/R = \operatorname{Coker} f$. Assume $\operatorname{Tor}_1^R(U, U) = 0$.*
Then the functors $(U/R) \otimes_R -$ and $\operatorname{Hom}_R(U/R, -)$ induce mutually inverse equivalences

$$\left\{ \begin{array}{c} \text{left U-torsion-free} \\ \text{U-contramodules} \end{array} \right\} \underset{\operatorname{Hom}_R(U/R, -)}{\overset{(U/R)\otimes_R -}{\rightleftarrows}} \left\{ \begin{array}{c} \text{left U-divisible} \\ \text{U-comodules} \end{array} \right\}.$$

Before proving the theorem we state a lemma showing that the functors $(U/R) \otimes_R -$ and $\operatorname{Hom}_R(U/R, -)$ take values in the pertinent classes.

Lemma 5.7 *If $\operatorname{Tor}_1^R(U, U) = 0$, then*

1. *For any left R-module M, the left R-module $\operatorname{Hom}_R(U/R, M)$ is a U-torsion-free U-contramodule;*
2. *For any left R-module C, the left R-module $(U/R) \otimes_R C$ is a U-divisible U-comodule.*

Proof (1) From the surjection $U \longrightarrow U/R \to 0$ one sees that the left R-module $\operatorname{Hom}_R(U/R, M)$ is U-torsion-free as an R-submodule of the left U-module $\operatorname{Hom}_R(U, M)$.
Furthermore, since $U \otimes_R U = U$, we have $(U/R) \otimes_R U = 0$, and therefore $\operatorname{Hom}_R(U, \operatorname{Hom}_R(U/R, M)) \cong \operatorname{Hom}_R((U/R) \otimes_R U, M) = 0$.

To show that $\mathrm{Ext}^1_R(U, \mathrm{Hom}_R(U/R, M)) = 0$, observe that our assumptions $U \otimes_R U = U$ and $\mathrm{Tor}^R_1(U, U) = 0$ imply $\mathrm{Tor}^R_1(U/R, U) = 0$, because the map $(R/\mathrm{Ker}(f)) \otimes_R U \longrightarrow U \otimes_R U$ is an isomorphism.

We apply Lemma 5.4 (1) letting $L =_R U$ and E the R-R-bimodule U/R to get that

$$\mathrm{Ext}^1_R(U, \mathrm{Hom}_R(U/R, M)) \hookrightarrow \mathrm{Ext}^1_R((U/R) \otimes_R U, \; M) = 0,$$

hence $\mathrm{Hom}_R(U/R, M)$ is a left U-contramodule.

The proof of part (2) is dual-analogous. The left R-module $(U/R) \otimes_R C$ is U-divisible as a quotient R-module of the left U-module $U \otimes_R C$. Since $U \otimes_R (U/R) = 0$, we have $U \otimes_R (U/R) \otimes_R C = 0$.

Apply Lemma 5.4 (2) letting $B = U_R$ and E the R-R-bimodule U/R to get

$$0 = \mathrm{Tor}^R_1(U \otimes_R (U/R), \; C) \longrightarrow \mathrm{Tor}^R_1(U, \; (U/R) \otimes_R C) \to 0.$$

Hence $(U/R) \otimes_R C$ is a left U-comodule.   □

Proof of Theorem 5.6 By Lemma 5.7, the functor $M \longmapsto \mathrm{Hom}_R(U/R, M)$ takes U-divisible left U-comodules to U-torsion-free left U-contramodules and the functor $(U/R) \otimes_R -$ takes U-torsion-free left U-contramodules to U-divisible left U-comodules (in fact, they take arbitrary left R-modules to left R-modules from these two classes). It remains to show that the restrictions of these functors to these two full subcategories in R-Mod are mutually inverse equivalences between them.

First we consider the case of a U-divisible left U-comodule M and show that the counit morphism

$$\xi_M : (U/R) \otimes_R \mathrm{Hom}_R(U/R, M) \longrightarrow M$$

is an isomorphism.

Since M is U-divisible, we have a natural short exact sequence of left R-modules

(1) $0 \longrightarrow \mathrm{Hom}_R(U/R, M) \longrightarrow \mathrm{Hom}_R(U, M) \longrightarrow M \longrightarrow 0.$

Since the left R-module $\mathrm{Hom}_R(U/R, M)$ is U-torsion-free, applying the functor $- \otimes_R \mathrm{Hom}_R(U/R, M)$ to the sequence $R \to U \to U/R \to 0$ we also have a natural short exact sequence of left R-modules

(2) $0 \to \mathrm{Hom}_R(U/R, M) \to U \otimes_R \mathrm{Hom}_R(U/R, M) \to (U/R) \otimes_R \mathrm{Hom}_R(U/R, M) \to 0.$

Since M is a U-comodule, applying the functor $U \otimes_R -$ to the short exact sequence (1) produces an isomorphism

$$U \otimes_R \mathrm{Hom}_R(U/R, M) \cong U \otimes_R \mathrm{Hom}_R(U, M) \cong \mathrm{Hom}_R(U, M).$$

Now the commutative diagram

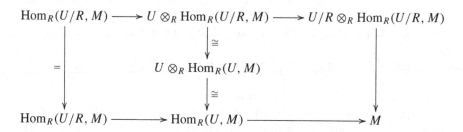

shows that we have a morphism from the short exact sequence (2) to the short exact sequence (1) that is the identity on the leftmost terms, an isomorphism on the middle terms, and the counit morphism ξ_M on the rightmost terms. Therefore, the counit morphism ξ_M is an isomorphism.

Next we consider a U-torsion-free left U-contramodule C and show that the unit morphism

$$\eta_C : C \longrightarrow \operatorname{Hom}_R(U/R, \ (U/R) \otimes_R C)$$

is an isomorphism.

Since C is U-torsion-free, we have a natural short exact sequence of left R-modules

$$(3) \quad 0 \longrightarrow C \longrightarrow U \otimes_R C \longrightarrow (U/R) \otimes_R C \longrightarrow 0.$$

Since the left R-module $(U/R) \otimes_R C$ is U-divisible, applying the functor $\operatorname{Hom}_R(-, (U/R) \otimes_R C)$ to the sequence $R \to U \to U/R \to 0$ we also have a natural short exact sequence of left R-modules

$$(4) \quad 0 \to \operatorname{Hom}_R(U/R, \ (U/R) \otimes_R C) \to \operatorname{Hom}_R(U, \ (U/R) \otimes_R C) \to (U/R) \otimes_R C \to 0.$$

Since C is a U-contramodule, applying the functor $\operatorname{Hom}_R(U, -)$ to the short exact sequence (3) produces an isomorphism

$$U \otimes_R C = \operatorname{Hom}_R(U, \ U \otimes_R C) \cong \operatorname{Hom}_R(U, \ (U/R) \otimes_R C).$$

Now the commutative diagram

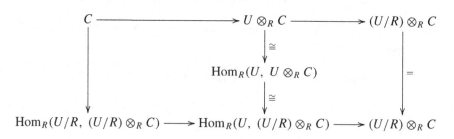

shows that we have a morphism from the short exact sequence (3) to the short exact sequence (4) that is the identity on the rightmost terms, an isomorphism on the middle terms, and the unit morphism η_C on the leftmost terms. Therefore, the unit morphism η_C is an isomorphism.   □

Further developments As noticed in the Introduction, the second Matlis category equivalence can be constructed in case $f : R \to U$ is a ring epimorphism such that $\operatorname{Tor}_1^R(U, U) = 0 = \operatorname{Tor}_2^R(U, U)$ (see [3, Theorem 2.3]).

Further results in the setting of derived categories are obtained in [3] in case f is a homological ring epimorphism. Indeed, assuming that U has projective dimension at most 1 as a left R-module and flat dimension at most one as a right R-module, it is shown that there is what may be called the *triangulated Matlis equivalence* in [12], that is an equivalence between the (bounded or unbounded) derived category of complexes of R-modules with U-comodule cohomology modules and the similar derived category of complexes of R-modules with U-contramodule cohomology modules.

Finally, under certain additional assumptions (which hold for instance when f is injective) the exact embedding functors of the full subcategories of U-comodules and U-contramodules into the category R-Mod induce fully faithful functors between the corresponding derived categories and also an equivalence between the derived categories of the categories of U-comodules and U-contramodules.

References

1. Anderson, F.W., Fuller, K.R.: Rings and categories of modules. In: Graduate Texts in Mathematics. Springer-Verlag, New York (1974)
2. Atiyah, M.F., Macdonald, I.G.: Introduction to Commutative Algebra. Addison-Wesley Publishing Co., Reading, Mass.-London-Don Mills (1969)
3. Bazzoni, S., Positselski, L.: Matlis category equivalences for a ring epimorphism. J. Pure Appl. Algebra **224**(10), 106398, 25 (2020). https://doi.org/10.1016/j.jpaa.2020.106398
4. Facchini, A., Nazemian, Z.: Equivalence of some homological conditions for ring epimorphisms. J. Pure Appl. Algebra **223**(4), 1440–1455 (2019). https://doi.org/10.1016/j.jpaa.2018.06.013
5. Geigle, W., Lenzing, H.: Perpendicular categories with applications to representations and sheaves. J. Algebra **144**(2), 273–343 (1991). https://doi.org/10.1016/0021-8693(91)90107-J
6. Harrison, D.K.: Infinite abelian groups and homological methods. Ann. of Math. 2(69), 366–391 (1959). https://doi.org/10.2307/1970188
7. Mac Lane, S.: Categories for the working mathematician. In: Graduate Texts in Mathematics, vol. 5, 2nd edn. Springer-Verlag, New York (1998)
8. Matlis, E.: Cotorsion Modules. American Mathematical Society, Memoirs Series, Providence (1964)
9. Matlis, E.: 1-dimensional Cohen-Macaulay Rings. Lecture Notes in Mathematics, vol. 327. Springer-Verlag, Berlin-New York (1973)
10. Matsumura, H.: Commutative Ring Theory, Cambridge Studies in Advanced Mathematics, vol. 8, 2nd edn. Cambridge University Press, Cambridge (1989). Translated from the Japanese by M. Reid
11. Positselski, L.: Contraadjusted modules, contramodules, and reduced cotorsion modules. Mosc. Math. J. **17**(3), 385–455 (2017)

12. Positselski, L.: Triangulated Matlis equivalence. J. Algebra Appl. **17**(4), 1850067, 44 (2018). https://doi.org/10.1142/S0219498818500676
13. Positselski, L.: Contramodules (2019). Electronic preprint arXiv:1503.00991
14. Rotman, J.J.: An introduction to Homological Algebra, 2nd edn. Universitext. Springer, New York (2009). https://doi.org/10.1007/b98977
15. Stenström, B.: Rings of quotients Die Grundlehren der Mathematischen Wissenschaften, In: An Introduction to Methods of Ring Theory, vol. 217. Springer-Verlag, New York-Heidelberg (1975)
16. Weibel, C.A.: An Introduction to Homological Algebra, Cambridge Studies in Advanced Mathematics, vol. 38. Cambridge University Press, Cambridge (1994). https://doi.org/10.1017/CBO9781139644136

Chapter 2
An Invitation to
Topological Semi-abelian Algebras

Maria Manuel Clementino

Abstract In this text we present the fundamental results that show that both classical topological properties of topological groups and categorical properties of the category of topological groups and continuous homomorphisms can be extended to the more general setting of topological semi-abelian algebras.

Keywords Semi-abelian theories · Topological algebras · Semidirect products · Split extension classifiers

Math. Subj. Classification 18E13 · 54H11 · 22-01 · 18A32

Introduction

The introduction of protomodular categories by Bourn in [11] and, subsequently, of semi-abelian categories by G. Janelidze, Márki, and Tholen in [26] are fundamental milestones in the study of categories that 'behave like' the category of groups. Citing G. Janelidze, Márki and Tholen in [26], *the notion of semi-abelian category... is designed to capture typical algebraic properties valid for groups, rings and algebras, say, just as abelian categories allow for a generalized treatment of abelian-group and module theory*. A few years later, together with Borceux, in [4] we showed that, in addition, the algebraic properties captured by the notion of semi-abelian category are crucial in the study of topological algebras, capturing both typical topological properties valid for topological groups and typical categorical properties of the category of topological groups. These results have been used since then in various contexts, and showed to be particularly interesting in the study of split extensions. Here we invite a non-specialist reader to go through these results and, moreover, to think about some interesting open problems. This material was presented in September 2018, in Louvain-la-Neuve, in a course entitled *Topological Algebras*, for a diverse audience including MSc and PhD students, but also senior researchers.

M. M. Clementino (✉)
CMUC, Department of Mathematics, University of Coimbra, 3001-501 Coimbra, Portugal
e-mail: mmc@mat.uc.pt

© The Author(s), under exclusive license to Springer Nature Switzerland AG 2021   27
M. M. Clementino et al. (eds.), *New Perspectives in Algebra, Topology and Categories*, Coimbra Mathematical Texts 1,
https://doi.org/10.1007/978-3-030-84319-9_2

We assume that the reader has a basic knowledge of category theory. Still, we chose to postpone the use of category theory to Sect. 3 (with the unavoidable exception of Theorem 1.6), and so one may essentially follow the first two sections without any knowledge of categories. In Sect. 1 we start by presenting the characterization of the varieties (in the sense of universal algebra) which are semi-abelian categories. We proceed by presenting some auxiliary results on their algebras, and conclude with establishing basic properties of the varieties of topological algebras. In Sect. 2 we focus on topological properties of topological semi-abelian algebras, while in Sect. 3 we focus on categorical properties of the category $\mathsf{Top}^{\mathbb{T}}$ of topological semi-abelian algebras and continuous homomorphisms. (The categorical background needed in this section can be found in [29].) Sect. 4 studies semidirect products in $\mathsf{Top}^{\mathbb{T}}$. First it is shown that, as for topological groups, every split extension is isomorphic to a split extension given by a semidirect product. (Here we avoid to use the categorical notion of semidirect product, which would oblige us to introduce monadicity.) Next we introduce the notion of split extension classifier [9, 8], and show that the category of topological groups has split extension classifiers [7, 15].

We end this chapter by presenting some open problems. We point out that a complete description of the topology on coproducts of topological algebras is a long-standing problem, and it also constitutes the principal obstacle for solving some of the other open problems formulated.

In this text we chose to treat semi-abelian categories very briefly. A more detailed study of these categories would make this text longer than intended, and, mostly, the literature on semi-abelian categories is abundant and diverse. For a reader interested in the subject we refer to [26, 3, 12], or the short introductory text [18].

1 Semi-abelian Algebras

We start by studying the varieties (in the sense of universal algebra) which, as categories, are semi-abelian, and which we call simply *semi-abelian varieties*.

1.1 Semi-abelian Theories

The algebraic theories whose varieties are semi-abelian were characterized by Bourn and G. Janelidze in [14] as those containing a unique constant 0 and, for some natural number n, having

(SA1) n binary operations $\alpha_1, \ldots, \alpha_n$ satisfying $\alpha_i(x, x) = 0$,
(SA2) an $(n+1)$-ary operation θ satisfying

$$\theta(\alpha_1(x, y), \ldots, \alpha_n(x, y), y) = x.$$

Such theories will be called *semi-abelian theories*.

Remarks 1.1 1. We point out that these theories were studied by Ursini in 1972 under the name *BIT speciali* in [32].

2. The theory of groups is semi-abelian: in the characterization above, let $n = 1$, $\alpha(x, y) = x - y, \theta(x, y) = x + y$. (Throughout we will use the additive notation although our groups need not be commutative.)

3. A semi-abelian algebraic theory may have different data identifying it as semi-abelian; that is, there may be different $(n, (\alpha_i), \theta)$ satisfying conditions (SA) above. We will see an example of this situation in 4.2.2.

4. For simplicity, we will sometimes use the abbreviations $\underline{\alpha}$ for $\alpha_1, \ldots, \alpha_n$, $\underline{x}$ for $x_1, \ldots, x_n$, and $\underline{0}$ for $0, \ldots, 0$ (n times); for instance, we may write condition (SA2) as

$$\theta(\underline{\alpha}(x, y), y) = x.$$

5. In a semi-abelian theory $\mathbb{T}$ the formula

$$p(x, y, z) = \theta(\alpha_1(x, y), \ldots, \alpha_n(x, y), z) = \theta(\underline{\alpha}(x, y), z)$$

defines a *Mal'tsev operation*, that is, a ternary operation p such that $p(x, z, z) = x$ and $p(x, x, z) = z$; indeed,

$$p(x, z, z) = \theta(\alpha_1(x, z), \ldots, \alpha_n(x, z), z) = x,$$

and

$$p(x, x, z) = \theta(\underline{\alpha}(x, x), z) = \theta(\underline{0}, z) = \theta(\underline{\alpha}(z, z), z) = z.$$

In case $\mathbb{T}$ is the theory of groups, the Mal'tsev operation p is given by

$$p(x, y, z) = x - y + z.$$

For each algebraic theory $\mathbb{T}$, we will denote by $\mathsf{Set}^{\mathbb{T}}$ the category of $\mathbb{T}$-algebras and $\mathbb{T}$-homomorphisms, i.e. the variety of $\mathbb{T}$-algebras. The category $\mathsf{Set}^{\mathbb{T}}$ is complete and cocomplete, with limits built like in Set, with the corresponding operations. Colimits are more difficult to build; throughout we will describe them when necessary. In Sect. 3 we will study the categorical properties of $\mathsf{Set}^{\mathbb{T}}$ in more detail.

1.2   Semi-abelian Algebras: Examples

There are plenty of examples of semi-abelian varieties. You can find them in, for instance, [3, 4, 7, 19]. Here we list very briefly some of the examples that will be used throughout. We point out that more examples will appear in Sect. 4.

1.2.1   Varieties with a Group Operation Every algebraic theory containing a unique constant and a group operation is semi-abelian: as in the remark above, take $n = 1$, $\alpha(x, y) = x - y$, and $\theta(x, y) = x + y$. In particular, abelian groups, Ω-groups [24], modules on a ring, rings or algebras without a unit, Lie algebras, Jordan algebras are examples of semi-abelian theories.

1.2.2   Right Ω-Loops The theory of right Ω-loops has a constant 0 and two binary operations, $+$ and $-$, satisfying the following conditions:

($\Omega 1$) $x + 0 = x$,
($\Omega 2$) $0 + x = x$,
($\Omega 3$) $(x - y) + y = x$,
($\Omega 4$) $(x + y) - y = x$.

Then, for $n = 1$, $\alpha = -$ and $\theta = +$, ($\Omega 2$) and ($\Omega 4$) give us $\alpha(x, x) = 0$, while ($\Omega 3$) states that $\theta(\alpha(x, y), y) = x$.

1.2.3   Heyting Semilattices A Heyting semilattice is a $\wedge$-semilattice with top element 1 and a binary operation $\Rightarrow$ satisfying the property

$$a \wedge b \leq c \text{ if and only if } a \leq (b \Rightarrow c).$$

As shown in [27], Heyting semilattices form a semi-abelian variety: take $n = 2$,

$$\alpha_1(x, y) = x \Rightarrow y, \quad \alpha_2(x, y) = ((x \Rightarrow y) \Rightarrow y) \Rightarrow x, \quad \theta(x, y, z) = (x \Rightarrow z) \wedge y.$$

1.3   Semi-abelian Algebras: Some Properties

First we give a very useful characterization of normal subalgebra in $\mathsf{Set}^{\mathbb{T}}$. For groups we know that a subgroup X of A is normal if and only if it is closed under conjugation. This property can be stated using the operation $\tau(a, x) = a + x - a$, saying that $\tau(a, x) \in X$ whenever $x \in X$. This can be generalized for algebraic theories with a Mal'tsev operation, where, of course, by *normal subalgebra* we mean that X is the kernel of some morphism $f : A \to B$ of $\mathbb{T}$-algebras. We point out that in this result *we do not assume that the theory $\mathbb{T}$ is semi-abelian.*

Theorem 1.2 *Let $\mathbb{T}$ be an algebraic theory with a unique constant 0 and a Mal'tsev operation. For a subalgebra X of A the following assertions are equivalent:*

(i). X is a normal subalgebra;
(ii). for every $(k + l)$-operation τ of the theory such that,

$$\textit{if, for all } a_1, \ldots, a_k \in A, \ \tau(a_1, \ldots, a_k, 0, \ldots, 0) = 0,$$
$$\textit{then, for all } x_1, \ldots, x_l \in X, \ \ \tau(a_1, \ldots, a_k, x_1, \ldots, x_l) \in X. \tag{1}$$

Proof Assume that X is the kernel of f. Then condition (1) is clearly necessary, since

$$f(\tau(a_1, \ldots, a_k, x_1, \ldots, x_l)) = \tau(f(a_1), \ldots, f(x_l)) = f(\tau(a_1, \ldots, a_k, 0, \ldots, 0)).$$

To prove the converse, we construct a morphism $f : A \to B$ whose kernel is X. First we recall that every reflexive relation in $\mathbf{Set}^{\mathbb{T}}$ is a congruence, due to the existence of a Mal'tsev operation in $\mathbb{T}$ (see [16]). Therefore the subalgebra $R \subseteq A \times A$ generated by the pairs (a, a), for $a \in A$, and $(x, 0)$, for $x \in X$, is a reflexive relation, hence a congruence. Let $f : A \to B$ be the quotient of A by R. We will check now that the kernel of f is X. By construction, X is contained in the kernel of f. Conversely, if $a \in A$ is such that $f(a) = 0$, then $(a, 0) \in R$, and so it is obtained as

$$(a, 0) = \gamma((a_1, a_1), \ldots, (a_k, a_k), (x_1, 0), \ldots, (x_l, 0))$$
$$= (\gamma(a_1, \ldots, a_k, x_1, \ldots, x_l), \gamma(a_1, \ldots, a_k, 0, \ldots, 0)),$$

for some operation γ, $a_1, \ldots, a_k \in A$, and $x_1, \ldots, x_l \in X$. Using the Mal'tsev operation p, we define then the operation τ by

$$\tau(a_1, \ldots, a_k, y_1, \ldots, y_l) = p(\gamma(a_1, \ldots, a_k, y_1, \ldots, y_l), \gamma(a_1, \ldots, a_k, 0, \ldots, 0), 0).$$

It is easily checked that τ satisfies the hypothesis of (1), and therefore

$$x = \gamma(a_1, \ldots, a_k, x_1, \ldots, x_l) = p(\gamma(a_1, \ldots, a_k, x_1, \ldots, x_l), 0, 0)$$
$$= p(\gamma(a_1, \ldots, a_k, x_1, \ldots, x_l), \gamma(a_1, \ldots, a_k, 0, \ldots, 0), 0)$$
$$= \tau(a_1, \ldots, a_k, x_1, \ldots, x_l)$$

belongs to X by condition (1).   □

From now on, *for a semi-abelian theory* $\mathbb{T}$, *we will use the notation* 0, $(\alpha_i)_{i=1,\ldots,n}$, θ *for fixed operations in* $\mathbb{T}$ *satisfying conditions* (SA).

Lemma 1.3 *Let* $\mathbb{T}$ *be a semi-abelian theory, and* A *a* $\mathbb{T}$-*algebra.*

1. *The family* $\alpha_i(-, a) : A \to A$ *is jointly monomorphic, that is, for* $x, y \in A$, $x = y$ *provided that* $\alpha_i(x, a) = \alpha_i(y, a)$ *for every* $i = 1, \ldots, n$.
2. *For* $x, y \in A$, $x = y$ *provided that* $\alpha_i(x, y) = 0$ *for every* $i = 1, \ldots, n$.

Proof 1. follows from the equality $\theta(\alpha_1(x, a), \ldots, \alpha_n(x, a), a) = x$, while 2. follows directly from 1.   □

We point out that, in general, the family $(\alpha_i(a, -))_i$ does not need to be jointly monomorphic (as, for instance, in the theory of right Ω-loops).

Finally we describe the quotient maps, i.e. the cokernels, in $\mathrm{Set}^{\mathbb{T}}$ in terms of their kernels.

Lemma 1.4 *Let $\mathbb{T}$ be a semi-abelian theory and X a normal subalgebra of A, corresponding to the quotient map $q\colon A \to A/X$. With $[a] = \{y \in A\,;\, q(y) = q(a)\}$, for any $a, b \in A$,*

$$[a] = [b] \in A/X \iff \forall i\ \alpha_i(a, b) \in X \iff \forall i\ \alpha_i(b, a) \in X.$$

Proof For $a, b \in A$,

$$[a] = [b] \iff (\forall i)\ 0 = \alpha_i([a], [b]) = [\alpha_i(a, b)] \iff (\forall i)\ \alpha_i(a, b) \in X.$$

□

Proposition 1.5 *Let $\mathbb{T}$ be a semi-abelian theory, let X be a normal subalgebra of A, and $q\colon A \to A/X$ its cokernel. For any $S \subseteq A$,*

$$\begin{aligned}
q^{-1}(q(S)) &= \{a \in A\,;\, \exists s \in S\ \forall i\ \alpha_i(a, s) \in X\} \\
&= \{a \in A\,;\, \exists s \in S\ \forall i\ \alpha_i(s, a) \in X\} \\
&= \{a \in A\,;\, \exists x_1, \ldots, x_n \in X\ \theta(x_1, \ldots, x_n, a) \in S\} \\
&= \theta(X^n \times S);
\end{aligned}$$

in particular, for any $a \in A$,

$$[a] = \theta(X^n, a).$$

Proof We only have to check the description of $q^{-1}(q(S))$. The first two equalities follow from the previous lemma. To show that

$$q^{-1}(q(S)) = \{a \in A\,;\, \exists x_1, \ldots, x_n \in X\ \theta(x_1, \ldots, x_n, a) \in S\},$$

let $\theta(x_1, \ldots, x_n, a) \in S$; then

$$[a] = [\theta(\underline{0}, a)] = \theta([\underline{0}], [a]) = \theta([x_1], \ldots, [x_n], [a]) = [\theta(x_1, \ldots, x_n, a)] \in q(S),$$

and therefore $a \in q^{-1}(q(S))$. Conversely, if $a \in q^{-1}(q(S))$, then $[a] = [s]$ for some $s \in S$. By the previous lemma, $\alpha_i(s, a) \in X$ for each i. Therefore, with $x_i = \alpha_i(s, a) \in X$, we get $\theta(x_1, \ldots, x_n, a) = s \in S$ as claimed.

It remains to show that $q^{-1}(q(S)) = \theta(X^n \times S)$. If $[a] = [s]$, with $s \in S$, then $\alpha_i(a, s) \in X$ and so $a = \theta(\alpha_1(a, s), \ldots, \alpha_n(a, s), s) \in \theta(X^n \times S)$. Conversely, if $a = \theta(x_1, \ldots, x_n, s)$ with $x_1, \ldots, x_n \in X$ and $s \in S$, then $q(a) = \theta(0, \ldots, 0, q(s)) = q(s)$.

□

1.4 Topological Algebras

Throughout this section we do not assume that the algebraic theory $\mathbb{T}$ is semi-abelian. The results presented here are valid for any algebraic theory.

A *topological group* A is a group equipped with a topology making both the addition and the inversion continuous maps; that is, the maps

$$i: A \longrightarrow A \quad \text{and} \quad \theta: A \times A \longrightarrow A$$
$$a \longmapsto -a \qquad\qquad (a, b) \longmapsto a + b$$

are continuous (where $A \times A$ has the product topology). The *right* maps between topological groups – their *morphisms* – are those maps which are both group homomorphisms and continuous.

For an algebraic theory $\mathbb{T}$, topological $\mathbb{T}$-algebras are defined similarly. A *topological $\mathbb{T}$-algebra* (or simply a *topological algebra*) A is a $\mathbb{T}$-algebra equipped with a topology making the operations of the theory $\mathbb{T}$ continuous. A morphism between $\mathbb{T}$-algebras is a continuous $\mathbb{T}$-homomorphism. Topological algebras and their morphisms form the category $\mathsf{Top}^{\mathbb{T}}$.

A brief analysis of the behaviour of the forgetful functor $U: \mathsf{Top}^{\mathbb{T}} \to \mathsf{Set}^{\mathbb{T}}$, which assigns to each topological algebra its underlying algebra and to each morphism its underlying homomorphism, allows us to understand how to lift special constructions from algebras to topological algebras. In fact, this functor is *topological*, that is, U *has initial lifts*: for each algebra A and each family $(f_i: A \to A_i)_{i \in I}$ of homomorphisms with topological algebras (A_i, OA_i), there is a topology OA on A making $(f_i: (A, OA) \to (A_i, OA_i))_I$ a family of morphisms in $\mathsf{Top}^{\mathbb{T}}$ with the following universal property: for any family $(g_i: (Y, OY) \to (A_i, OA_i))_I$ in $\mathsf{Top}^{\mathbb{T}}$ and homomorphism $h: Y \to A$ with $f_i \cdot h = g_i$ for every $i \in I$, $h: (Y, OY) \to (A, OA)$ is continuous. The family $(f_i: (A, OA) \to (A_i, OA_i))_I$ is said to be the *initial lift* of $(f_i: A \to (A_i, OA_i))_I$. (Here I may be a proper class; for details see [1; Chapter 21].)

Theorem 1.6 *The functor $U: \mathsf{Top}^{\mathbb{T}} \to \mathsf{Set}^{\mathbb{T}}$ is topological.*

Proof Given a family $(f_i: A \to (A_i, OA_i))_I$, the initial topology OA on A with respect to $(f_i)_I$, that is, the topology generated by $\{f_i^{-1}(U_i); \ U_i \in OA_i, \ i \in I\}$, makes A a topological algebra: for any operation τ of $\mathbb{T}$, the following diagram commutes

$$
\begin{array}{ccc}
A^n & \xrightarrow{\ f_i^n\ } & A_i^n \\
\tau \downarrow & & \downarrow \tau \\
A & \xrightarrow[\ f_i\ ]{} & A_i
\end{array}
$$

and so the continuity of $f_i \cdot \tau = \tau \cdot f_i^n$, together with the initiality of OA, implies that $\tau: A^n \to A$ is continuous. Moreover, the definition of initial topology guarantees that this is an initial lift as claimed. $\qquad\qquad\qquad\qquad\qquad\qquad\qquad\qquad\qquad\qquad\qquad\qquad\qquad\square$

Remark 1.7 This Theorem gives immediately, via U, several important properties of $\mathsf{Top}^{\mathbb{T}}$. Namely:

1. The functor U has also *final lifts* (i.e. the dual construction).
2. U has both left and right adjoints, L and R, built via the final and initial liftings of the families $(f: (A, OA) \to X)$ and $(g: X \to (B, OB))$ of all such homomorphisms f, g, respectively. That is, LX is X endowed with the indiscrete topology, while RX is X endowed with the discrete topology. Note that $U \cdot L = U \cdot R = \mathrm{Id}_{\mathsf{Set}^{\mathbb{T}}}$.
3. From 2. it follows that U preserves both limits and colimits. Therefore a product of topological algebras is their product as algebras equipped with the initial topology with respect to the projections, i.e. the product topology. Likewise, a kernel in $\mathsf{Top}^{\mathbb{T}}$ is the kernel in $\mathsf{Set}^{\mathbb{T}}$ endowed with the subspace topology. Dually, a coproduct of topological algebras $(A_i, OA_i)_I$ is their coproduct $\coprod A_i$ as algebras equipped with the final topology for the inclusions $((A_j, OA_j) \to \coprod A_i)_{j \in I}$, that is, the finest topology making these maps continuous. (We will make additional comments on this topology in 6.1.) Analogously, the cokernel of a morphism in $\mathsf{Top}^{\mathbb{T}}$ is the corresponding cokernel of algebras with the final topology.
4. Consequently, if A is a topological semi-abelian algebra and X is a normal subalgebra of the underlying algebra of A, with cokernel A/X,

$$X \xrightarrow{\ k\ } A \xrightarrow{\ q\ } A/X,$$

   thanks to Theorem 1.6 we know that both X and A/X can be endowed with topologies so that k and q are continuous, and, moreover, $k = \ker q$ and $q = \mathrm{coker} k$ in $\mathsf{Top}^{\mathbb{T}}$.
5. For any algebraic theory $\mathbb{T}$, in $\mathsf{Set}^{\mathbb{T}}$ every morphism $f: A \to B$ can be factorized, essentially in a unique way, as a surjective homomorphism e followed by an injective one m:

$$A \xrightarrow{\ e\ } f(A) \xrightarrow{\ m\ } B.$$

   In $\mathsf{Set}^{\mathbb{T}}$ surjective morphisms coincide with the regular epimorphisms, that is, those morphisms which are coequalizers of some pair of morphisms. Indeed, if $f: A \to B$ is surjective, then it can be written as $f: A \to A/R$, where R is the equivalence relation $\{(x, x') \in A \times A ;\ f(x) = f(x')\}$. Then R is a congruence (that is, a subalgebra of $A \times A$), and f is the coequalizer of $R \underset{\pi_2}{\overset{\pi_1}{\rightrightarrows}} A$.
   Conversely, if f is a coequalizer, in the factorization above m is necessarily an isomorphism, and so f is surjective.
   This (regular epi, mono)-factorization system lifts to $\mathsf{Top}^{\mathbb{T}}$: for each morphism $f: (A, OA) \to (B, OB)$ in $\mathsf{Top}^{\mathbb{T}}$, take the (regular epi, mono)-factorization of

the underlying homomorphism $f \colon A \to B$ in $\mathsf{Set}^{\mathbb{T}}$, $A \overset{e}{\longrightarrow} f(A) \overset{m}{\longrightarrow} B$; it is easily checked that, equipping $f(A)$ with the final topology $Of(A)$ with respect to e,

$$(A, OA) \overset{e}{\longrightarrow} (f(A), Of(A)) \overset{m}{\longrightarrow} (B, OB)$$

is the (regular epi, mono)-factorization of f in $\mathsf{Top}^{\mathbb{T}}$.

2 Topological Semi-abelian Algebras

Contrarily to the last section, here we only deal with topological $\mathbb{T}$-algebras, where $\mathbb{T}$ is a semi-abelian theory.

Topological groups are topologically very well behaved, mostly due to the following facts.

- They are *homogeneous spaces*, that is, for each pair of points x, y in the topological group A, there is an homeomorphism $h \colon A \to A$ with $h(x) = y$; indeed, defining h by $h(a) = a - x + y$ for every $a \in A$, as a composition of continuous maps

$$A \longrightarrow A^3 \overset{1 \times i \times 1}{\longrightarrow} A^3 \overset{\theta \times 1}{\longrightarrow} A^2 \overset{\theta}{\longrightarrow} A$$

$$a \longmapsto (a, x, y) \longmapsto (a, -x, y) \longmapsto (a - x, y) \longmapsto a - x + y$$

it is continuous, and it has a continuous inverse g, defined by $g(a) = a - y + x$.
- The Mal'tsev operation p defined by $p(x, y, z) = x - y + z$ is continuous and, moreover, it is preserved by any morphism, since they preserve both i and θ.

In this Section our goal is to study the topological properties of *topological semi-abelian algebras*, with topological groups as our main inspiration. From now on $\mathbb{T}$ is a semi-abelian theory, and A is an object in the category $\mathsf{Top}^{\mathbb{T}}$. We will say that A *is a topological semi-abelian algebra*.

Remarks 2.1 1. Most of these results can be extended to the more general setting of protomodular varieties (where there may be more than one constant). For details see [5].
2. Some of the results presented are valid in varieties with a Mal'tsev operation, as shown in [28].

2.1 How to Overcome Lack of Homogeneity

The underlying space of a topological semi-abelian algebra is in general no longer homogeneous, although the neighbourhoods of any point are still determined by the neighbourhoods of 0. Indeed, for any $a \in A$, the maps $\alpha_a : A \to A^n$ and $\theta_a : A^n \to A$, with $\alpha_a(x) = (\alpha_1(x, a), \dots, \alpha_n(x, a)) = \underline{\alpha}(x, a)$ and $\theta_a(\underline{x}) = \theta(\underline{x}, a)$ are continuous and make the following diagram

commute. Therefore α_a describes A as a subspace of A^n, mapping a into $\underline{0}$. Later it will be useful to consider also $\beta_a : A \to A^n$ defined by $\beta_a(x) = \underline{\alpha}(a, x)$. We note that, like α_a, also β_a is continuous and $\beta_a(a) = \underline{0}$; however, β_a does not need to be injective in general (e.g., in right Ω-loops).

Denoting by $O(a)$ the open subsets of A containing the point $a \in A$, one has:

Lemma 2.2 *The following sets are neighbourhood bases of $a \in A$:*

1. $\{\alpha_a^{-1}(U^n) ; \ U \in O(0)\}$;
2. $\{\theta_a(U^n) ; \ U \in O(0)\}$.

Proof 1. Every embedding preserves neighbourhood bases under inverse images. Therefore, since α_a is an embedding and $\{U^n ; \ U \in O(0)\}$ is a neighbourhood basis of $\underline{0}$, its inverse image is a neighbourhood basis of $\alpha_a^{-1}(\underline{0}) = a$.

2. For each $U \in O(0)$, $\alpha_a^{-1}(U^n) = \theta_a(\alpha_a(\alpha_a^{-1}(U^n))) \subseteq \theta_a(U^n)$, therefore $\theta_a(U^n)$ is a neighbourhood of a. Moreover, if $W \in O(a)$, then $\theta_a^{-1}(W) \in O(\underline{0})$, and so it contains U^n for some $U \in O(0)$. Therefore $W = \theta_a(\theta_a^{-1}(W)) \supseteq \theta_a(U^n)$ as claimed. $\square$

Lemma 2.3 *For every topological semi-abelian algebra A, $p : A^3 \to A$, defined by*

$$p(x, y, z) = \theta(\alpha_1(x, y), \dots, \alpha_n(x, y), z)$$

is a continuous Mal'tsev operation.

Proof As a composition of continuous maps, p is continuous. $\square$

2.2 The Closure on Subalgebras

For each semi-abelian topological algebra A, the topology on A assigns a closure $\overline{S}$ to each subset S of A, defining an endomap in the powerset of A, $\overline{(\ )} : \mathcal{P}A \to \mathcal{P}A$. Next we show that this map (co)restricts to subalgebras and to normal subalgebras.

Proposition 2.4 *Let X be a subalgebra of A. Then:*

1. $\overline{X}$ *is a subalgebra of A.*
2. *If X is a normal subalgebra, then $\overline{X}$ is a normal subalgebra.*
3. *If X is open, then it is closed.*

Proof 1. This is in fact valid for any topological algebra (not necessarily semi-abelian). Continuity of any operation $\tau : A^n \to A$ gives

$$\tau(\overline{X}^n) = \tau(\overline{X^n}) \subseteq \overline{\tau(X^n)} \subseteq \overline{X}.$$

2. To show that $\overline{X}$ is a normal subalgebra, we use Theorem 1.2. Let $\tau : A^{n+l} \to A$ be an operation such that $\tau(a_1, \ldots, a_k, 0, \ldots, 0) = 0$. Since τ is continuous and X is a normal subalgebra of A,

$$\tau(A^k \times \overline{X}^l) \subseteq \overline{\tau(A^k \times X^l)} \subseteq \overline{X}.$$

3. We show that $\overline{X} \subseteq X$. For every $a \in A$, $U = \beta_a^{-1}(X^n)$ is an open subset of A containing a, since $a = \beta_a^{-1}(0)$ and $X^n \in O(0)$. If $U \cap X \neq \emptyset$, that is, if there exists $y \in U \cap X$, then $a = \theta(\alpha_1(a, y), \ldots, \alpha_n(a, y), y) \in \theta(X^{n+1}) \subseteq X$. Therefore X is closed. $\qquad\square$

2.3 Quotient Maps

Thanks to Theorem 1.6, regular epimorphisms (which we will call *quotient maps* as it is usual for topological algebras) $f : A \to B$ in $\mathsf{Top}^\mathbb{T}$ are regular epimorphisms in $\mathsf{Set}^\mathbb{T}$ (that is, surjective homomorphisms) with B equipped with the topology

$$OB = \{U \subseteq B ; \ f^{-1}(U) \in OA\}.$$

Proposition 2.5 *Every quotient map $f : A \to B$ between topological semi-abelian algebras is open.*

Proof For U open in A, we want to show that $f^{-1}(f(U))$ is open. Let $x \in f^{-1}(f(U))$, so that $f(x) = f(u)$ for some $u \in U$. Since $p(x, x, u) = u \in U$ and U is open, there exists $V \in O(x)$ such that $p(V, x, u) \subseteq U$. For all $v \in V$, $f(v) = p(f(v), f(x), f(u)) = f(p(v, x, u)) \in f(U)$, hence $v \in f^{-1}(f(U))$. Therefore $f^{-1}(f(U))$ is open. $\qquad\square$

Corollary 2.6 *If $\mathbb{T}$ is a semi-abelian theory, then regular epimorphisms are pullback stable in $\mathsf{Top}^\mathbb{T}$.*

Proof Pullbacks in $\mathsf{Top}^\mathbb{T}$ are formed just like in Top and it is well known that in Top open surjections are stable under pullback. $\qquad\square$

2.4 Separation Properties

We recall that a topological space X is said to be *regular* if, for every closed subset S and point $x \in X \setminus S$, there exist disjoint open subsets U, V such that $S \subseteq U$ and $x \in V$. Regular spaces are also characterized as the topological spaces having a basis of closed neighbourhoods at every point. We will use this characterization throughout.

Proposition 2.7 *Every topological semi-abelian algebra A is a regular space.*

Proof We will show that the closed neighbourhoods of 0 form a neighbourhood basis of 0, and therefore, using the embedding α_a, we may conclude that the closed neighbourhoods form a neighbourhood basis of a.

Due to continuity of θ, for each $U \in O(0)$, there exists $V \in O(0)$ such that $\theta(V^{n+1}) \subseteq U$. We will show that $\overline{V} \subseteq U$. Let $a \in \overline{V}$. Then $\beta_a^{-1}(V^n) \in O(a)$, and so it intersects V. Given $x \in V \cap \beta_a^{-1}(V^n)$, $\alpha_1(a, x), \ldots, \alpha_n(a, x)$ belong to V, and so $a = \theta(\alpha_1(a, x), \ldots, \alpha_n(a, x), x)$ belongs to U as claimed. $\square$

Proposition 2.8 *For a topological semi-abelian algebra A, the following conditions are equivalent:*

(i). *A is a T0-space,*
(ii). *$\{0\}$ is closed,*
(iii). *A is a Hausdorff space.*

Proof Since A is regular, A is a Hausdorff space if, and only if, it is a T1-space. To show that A is T1, provided that it is a T0-space, is equivalent to showing that $\{0\}$ is closed, thanks to the embedding α_a. Let A be a T0-space, $a \in \overline{\{0\}}$, and let V be a neighbourhood of 0. By Lemma 2.2 we may take $V = \theta_0(U^n)$ for some $U \in O(0)$. Then $\beta_a^{-1}(U^n) \in O(a)$, and therefore, by our assumption, contains 0. Now we can write $a = \theta(\underline{\alpha}(a, 0), 0) \in \theta_0(U^n) = V$, and so $0 \in \overline{\{a\}}$, which implies $a = 0$. $\square$

Lemma 2.9 *For a topological semi-abelian algebra A, the following conditions are equivalent:*

(i). *A is discrete;*
(ii). *$\{0\}$ is open.*

Proof With $\{0\}$ open, also $\{\underline{0}\}$ is open in A^n. Therefore $\{a\} = \alpha_a^{-1}(\{\underline{0}\})$ is open in A, for every $a \in A$. $\square$

Proposition 2.10 *Let X be a normal subalgebra of A.*

1. *The following conditions are equivalent:*

    (i) *X is closed;*
    (ii) *A/X is Hausdorff.*

2. *The following conditions are equivalent:*

   (i) *X is open;*
   (ii) *A/X is discrete.*

Proof 1. Noting that $X = q^{-1}([0])$ and q is a quotient map, we get

$$X \text{ is closed } \Longleftrightarrow [0] \text{ is closed } \Longleftrightarrow A/X \text{ is Hausdorff.}$$

2. Analogously,

$$X \text{ is open } \Longleftrightarrow [0] \text{ is open } \Longleftrightarrow A/X \text{ is discrete.}$$

$\square$

2.5 (Local) Compactness

Next we study the behaviour of compact and locally compact topological semi-abelian algebras. In neither of these concepts we assume Hausdorffness: a space is *compact* if every open cover has a finite subcover, and it is *locally compact* if every point has a neighbourhood basis consisting of compact subsets.

2.5.1 Proper Maps First we recall that a space X is compact if, and only if, for any space Y, the projection $X \times Y \to Y$ is a closed map. Moreover, we recall that a continuous map $f : X \to Y$ is said to be *proper* (*a la Bourbaki* [10]) if it is closed and has compact fibres, or, equivalently, if it is closed and inverse images of compact subsets are compact. This is also equivalent to being *stably closed*, that is, if f' is the pullback of f along a continuous map, then f' is closed.

Proposition 2.11 *If X is a compact subalgebra of A, then the quotient $q : A \to A/X$ is a proper map.*

Proof Given a closed subset C of A, $q(C)$ is closed if, and only if, $q^{-1}(q(C))$ is closed. We consider the continuous maps

$$A \xleftarrow{\ p_A\ } X^n \times A \xrightarrow{\ \iota\ } A^{n+1} \xrightarrow{\ \theta\ } A,$$

where ι is the inclusion and p_A the projection, and note that, by Proposition 1.5,

$$q^{-1}(q(C)) = \{a \in A; \ \exists x_1, \ldots, x_n \in X \ \theta(x_1, \ldots, x_n, a) \in C\}$$
$$= p_A(\iota^{-1}(\theta^{-1}(C)));$$

since X^n is compact, p_A is closed, and therefore $q^{-1}(q(C))$ is closed.

Moreover, with X compact all the fibres of q are compact, since $q^{-1}(q(a)) = [a] = \theta_a(X^n)$ is the continuous image of the compact subset X^n.   □

Remark 2.12 It is worth to note that the classical proof of this result – in the special case of topological groups – is much more elaborate than this one.

Corollary 2.13 *Let X be a normal subalgebra of A. If X and A/X are compact, then A is compact as well.*

Proof The map q is proper and A/X is compact, hence $A = q^{-1}(A/X)$ is compact.   □

Corollary 2.14 *Compact Hausdorff semi-abelian algebras have the 2-out-of-3 property.*

Proof If X and A/X are compact and Hausdorff, then A is compact and X is closed in A. Therefore $\{0\}$ is closed in A because it is closed in X, and so A is Hausdorff.

If A and A/X are compact and Hausdorff, then X is a closed subalgebra of A, hence compact and Hausdorff.

If X and A are compact and Hausdorff, then A/X is compact because it is the image of a compact space under a continuous map, and it is Hausdorff because X, being compact, is closed in A.   □

Proposition 2.15 *For a topological semi-abelian algebra A the following conditions are equivalent:*

   (i). 0 has a compact neighbourhood;
   (ii). every point has a compact neighbourhood;
   (iii). A is locally compact.

Proof Obviously $(iii) \implies (ii) \implies (i)$

For $(i) \implies (ii)$ it is enough to observe that, if U is a compact neighbourhood of 0, then, for every $a \in A$, $\theta_a(U^n)$ is a compact neighbourhood of a.

$(ii) \implies (iii)$ is valid for any regular topological space. Indeed, let U be a compact neighbourhood of a. For V an arbitrary neighbourhood of a, let $U' \subseteq U$ and $V' \subseteq V$ be closed neighbourhoods of a. Then $U' \cap V'$ is a closed neighbourhood of a, it is closed in U, hence compact, and it is contained in V.   □

The result below is the algebraic version of the property that locally compact subspaces of Hausdorff spaces are intersections of closed and open subsets.

Proposition 2.16 *If A is a Hausdorff topological semi-abelian algebra, then every locally compact subalgebra X of A is closed.*

Proof Let $a \in \overline{X}$ and let Z be a compact neighbourhood of 0 in X. Then Z is closed in A and $Z \supseteq U \cap X$, with U open neighbourhood of 0 in A. The set $\beta_a^{-1}(U^n)$ is a neighbourhood of a, and therefore it meets X. Let $x \in X$ be such that $\alpha_i(a, x) \in U$ for every $i \in I$. Then, since $\overline{X}$ is also a subalgebra, $\alpha_i(a, x) \in U \cap \overline{X}$; but $U \cap \overline{X} \subseteq \overline{U \cap X}$ because U is open, $\overline{U \cap X} \subseteq Z$ because Z is closed, and $Z \subseteq X$ by our assumption. Therefore $a = \theta(\alpha_1(a, x), \dots, \alpha_n(a, x), x) \in X$ as claimed.   □

Proposition 2.17 *Locally compact topological semi-abelian algebras are stable under quotient maps.*

Proof This is valid for any locally compact space, since they are stable under open surjections. □

Proposition 2.18 *If X is a normal subalgebra of A, then A is locally compact provided that X is compact and A/X is locally compact.*

Proof By Proposition 2.11, $q : A \twoheadrightarrow A/X$ is proper. Hence the inverse image of a compact neighbourhood of $[0]$ in A/X will be a compact neighbourhood of 0 in A. □

2.6 Connectedness and Total Disconnectedness

Recall that a space X is *connected* if it cannot be written as a disjoint union of two non-empty subsets; or, equivalently, any subset of X which is both closed and open (shortly *clopen*) is either empty or X. A space is *totally disconnected* if its only connected subsets are the points (and $\emptyset$). Since the union of connected subsets with non-empty intersection is connected, for each $x \in X$ there is a largest connected subset containing x, $\Gamma(x)$. So, X is connected if, and only if, $\Gamma(x) = X$ for every $x \in X$, and totally disconnected if $\Gamma(x) = \{x\}$ for every $x \in X$.

Proposition 2.19 *If A is a topological semi-abelian algebra, then, for every $a \in A$, $\Gamma(a) = \theta_a(\Gamma(0)^n)$.*

Proof As a product of connected subsets, $\Gamma(0)^n$ is connected, and therefore $\theta_a(\Gamma(0)^n)$ is a connected subset containing a. To prove the converse inclusion, first note that $\alpha_i(\Gamma(a), a)$ is connected and contains 0. Thus, for $y \in \Gamma(a)$, $\alpha_i(y, a) \in \Gamma(0)$, and so $y = \theta(\alpha_1(y, a), \ldots, \alpha_n(y, a), a) \in \theta_a(\Gamma(0)^n)$. □

Corollary 2.20 *For a topological semi-abelian algebra A, the following conditions are equivalent:*

(i). $\Gamma(0) = \{0\}$;
(ii). A is totally disconnected. □

Proposition 2.21 $\Gamma(0)$ *is a closed normal subalgebra.*

Proof Any connected component is closed, hence $\Gamma(0)$ is closed. To show that $\Gamma(0)$ is a normal subalgebra, we check condition (1) of Theorem 1.2 via induction on l:

- When $l = 0$, (1) means $0 \in \Gamma(0)$.
- Assuming (1) for $l-1$, and considering the operation $\tau(a_1, \ldots, a_k, y_1, \ldots, y_{l-1}, 0)$, our assumption guarantees that, for all $a_1, \ldots, a_k \in A, x_1, \ldots, x_l \in \Gamma(0)$, $\tau(a_1, \ldots, a_k, x_1, \ldots, x_{l-1}, 0) \in \Gamma(0)$, and then $\tau(a_1, \ldots, a_k, x_1, \ldots, x_{l-1}, -)(\Gamma(0))$ is the image of a connected subset under a continuous map, hence it is connected, and therefore contained in $\Gamma(0)$.

□

Lemma 2.22 *If X is a connected normal subalgebra of the topological semi-abelian algebra A, with quotient $q: A \to A/X$, then:*

1. *$[a]$ is a connected subset of A, for every $a \in A$.*
2. *If $U \subseteq A$ is open and closed, then $q^{-1}(q(U)) = U$.*

Proof 1. By Proposition 1.5, for every $a \in A$, $[a] = \theta_a(X^n)$, hence connected.

2. For all $a \in A$, if $[a] \cap U \neq \emptyset$, then $[a]$ must be contained in U, since $[a]$ is connected and U is clopen.   □

Proposition 2.23 *Let X be a normal subalgebra of the topological semi-abelian algebra A. With X and A/X connected, A is connected as well.*

Proof If U is clopen in A, then $U = q^{-1}(q(U))$, and so $q(U)$ is also clopen. Hence $q(U)$ is either empty or A/X, and then $U = q^{-1}(q(U))$ is either empty or A.   □

Proposition 2.24 *Let X be a normal subalgebra of the topological semi-abelian algebra A. With X and A/X totally disconnected, A is totally disconnected as well.*

Proof The set $q(\Gamma(0))$ is connected and contains 0, hence $q(\Gamma(0)) = \{0\}$. Therefore $\Gamma(0) \subseteq X$. The only non-empty connected subsets of X are the singletons, hence $\Gamma(0) = \{0\}$.   □

3 The Categorical Behaviour of Topological Semi-abelian Algebras

3.1 Properties of the Category $\mathsf{Top}^{\mathbb{T}}$

3.1.1 $\mathsf{Top}^{\mathbb{T}}$ Is Pointed Since $\mathbb{T}$ has a unique constant, the singleton $\mathbb{T}$-algebra is a zero object of $\mathsf{Top}^{\mathbb{T}}$, that is, it is both an initial and a terminal object.

3.1.2 $\mathsf{Top}^{\mathbb{T}}$ Is Complete and Cocomplete Thanks to Theorem 1.6, we know that $\mathsf{Top}^{\mathbb{T}}$ is both complete and cocomplete, with limits and colimits built as in $\mathsf{Set}^{\mathbb{T}}$, and equipped with the initial and final topologies, respectively.

3.1.3 $\mathsf{Top}^{\mathbb{T}}$ Is Regular but Not Exact In Remark 1.7 we proved that $\mathsf{Top}^{\mathbb{T}}$ has the (regular epi, mono)-factorization system, and in Proposition 2.5 we showed that regular epimorphisms, being open surjections, are pullback stable. Since it is in particular finitely complete, we conclude that $\mathsf{Top}^{\mathbb{T}}$ is a regular category. (For more information on regular categories see [21].)

We point out that $\mathsf{Set}^{\mathbb{T}}$ is *Barr-exact* (for any theory $\mathbb{T}$), that is, it is regular and every internal equivalence relation $R \overset{r_1}{\underset{r_2}{\rightrightarrows}} A$ in $\mathsf{Set}^{\mathbb{T}}$ is a kernel pair (i.e. there is a $\mathbb{T}$-homomorphism $f : A \to B$ so that (r_1, r_2) is the pullback of f along f). This is no longer the case of $\mathsf{Top}^{\mathbb{T}}$: for every internal equivalence relation $R \overset{r_1}{\underset{r_2}{\rightrightarrows}} A$ in $\mathsf{Top}^{\mathbb{T}}$ there is a continuous morphism $f : A \to B$ so that

$$
\begin{array}{ccc}
R & \overset{r_1}{\longrightarrow} & A \\
{\scriptstyle r_2}\downarrow & & \downarrow{\scriptstyle f} \\
A & \underset{f}{\longrightarrow} & B
\end{array}
$$

is a pullback in $\mathsf{Set}^{\mathbb{T}}$, but R may not have the initial topology with respect to (r_1, r_2). For instance, let A be the additive group of real numbers with the Euclidean topology, and let R be the group $A \times A$ equipped with the topology $\mathcal{T}$ generated by the Euclidean topology and the sets

$$
O_s = \{(x, y) \in \mathbb{R}^2 \, ; \, x - y = s\},
$$

where $s \in \mathbb{R}$. Then the projections $R \overset{r_1}{\underset{r_2}{\rightrightarrows}} A$ are continuous, as well as the diagonal $A \to R$. Moreover, $\mathcal{T}$ makes R a topological group, since $i_R^{-1}(O_s) = O_{-s}$, and, for $s, s' \in \mathbb{R}$, $O_s + O_{s'} = O_{s+s'}$. Since $\mathcal{T}$ is not the initial topology for the projections r_1, r_2, $R \overset{r_1}{\underset{r_2}{\rightrightarrows}} A$ is not the kernel pair of its coequalizer.

We chose to refer only briefly to this exactness property, but we suggest our reader to consult [2, 3] for more information on this topic.

3.1.4 $\mathsf{Top}^{\mathbb{T}}$ Is Homological

A category is said to be *homological* [3] if it is pointed, regular, and protomodular, while it is said to be *semi-abelian* [26] if it is pointed, exact, and protomodular. Since here we are dealing only with pointed regular categories, we chose not to define protomodularity [11] in full generality, but instead to use the following characterizations of homological categories, which relate directly to the forthcoming topics of this text.

Proposition 3.1 *If $\mathscr{C}$ is a regular category, then the following assertions are equivalent:*

(i). $\mathscr{C}$ is protomodular;
(ii). In the following commutative diagram, with g a regular epimorphism,

$$
\begin{array}{ccc}
A \longrightarrow B \longrightarrow C \\
\downarrow \quad \boxed{1} \quad \downarrow g \quad \boxed{2} \quad \downarrow \\
A' \longrightarrow B' \longrightarrow C'
\end{array}
$$

if $\boxed{1}\,\boxed{2}$ and $\boxed{1}$ are both pullback diagrams, then $\boxed{2}$ is also a pullback.

The characterizations we want to focus on use the notion of short exact sequence. In a pointed category, a sequence of morphisms

$$
0 \longrightarrow X \xrightarrow{\ k\ } A \xrightarrow{\ f\ } B \longrightarrow 0 \tag{2}
$$

is said to be a *short exact sequence* if $k = \ker f$ and $f = \operatorname{coker} k$. A short exact sequence (2) where f is a split epimorphism, together with a section s of f

$$
0 \longrightarrow X \xrightarrow{\ k\ } A \underset{f}{\overset{s}{\rightleftarrows}} B \longrightarrow 0
$$

so that $f \cdot s = 1_B$, is called a *split short exact sequence* (also called *split extension*). A morphism of split short exact sequences is a triple (t, u, v) as in the *commutative* diagram

$$
\begin{array}{ccccccc}
0 & \longrightarrow & X & \xrightarrow{\ k\ } & A \underset{f}{\overset{s}{\rightleftarrows}} B & \longrightarrow & 0 \\
& & \downarrow t & & \downarrow u \qquad \downarrow v & & \\
0 & \longrightarrow & X' & \xrightarrow{\ k'\ } & A' \underset{f'}{\overset{s'}{\rightleftarrows}} B' & \longrightarrow & 0
\end{array} \tag{3}
$$

(that is, its three squares commute: $u \cdot k = k' \cdot t,\ f' \cdot u = v \cdot f,\ u \cdot s = s' \cdot v$).

Theorem 3.2 *If $\mathscr{C}$ is a pointed regular category, then the following conditions are equivalent:*

(i). $\mathscr{C}$ is protomodular;
(ii). The Short Five Lemma *holds, that is, given a commutative diagram, where the horizontal rows are short exact sequences*

$$
\begin{array}{ccccccccc}
0 & \longrightarrow & X & \xrightarrow{\ k\ } & A & \xrightarrow{\ f\ } & B & \longrightarrow & 0 \\
& & \downarrow t & & \downarrow u & & \downarrow v & & \\
0 & \longrightarrow & X' & \xrightarrow{\ k'\ } & A' & \xrightarrow{\ f'\ } & B' & \longrightarrow & 0
\end{array}
$$

if t and v are isomorphisms, u is an isomorphism as well.

(iii). The Split Short Five Lemma *holds, that is, given a commutative diagram* (3), *where the horizontal rows are split short exact sequences, if t and v are isomorphisms, u is an isomorphism as well.*

More information on protomodularity, and the proofs of these results, can be found in [3, 4].

3.2 Special Subcategories of Top$^{\mathbb{T}}$

We recall that a full subcategory $\mathscr{A}$ of a category $\mathscr{C}$ is said to be *reflective* if the inclusion functor $G: \mathscr{A} \to \mathscr{C}$ has a left adjoint. This is equivalent to giving, for each object C of $\mathscr{C}$, a *universal arrow* $\eta_C: C \to GA_C$ *from C to G*, i.e., a $\mathscr{C}$-morphism $\eta_C: C \to GA_C$, with A_C an object of $\mathscr{A}$, such that every such morphism $C \to GA'$ factors through η_C in a unique way. If the morphisms η_C are (regular) epimorphisms, one says that $\mathscr{A}$ is *(regular-)epireflective*.

3.2.1 Haus$^{\mathbb{T}}$ and TotDisc$^{\mathbb{T}}$ are Homological

Proposition 3.3 *The following full subcategories of* Top$^{\mathbb{T}}$ *are regular-epireflective subcategories:*

1. Haus$^{\mathbb{T}}$ *of Hausdorff topological algebras;*
2. TotDisc$^{\mathbb{T}}$ *of totally disconnected topological algebras.*

Proof Let A be a topological algebra.

1. Consider the quotient morphism $\eta_A: A \to A/\overline{\{0\}}$. Then $A/\overline{\{0\}}$ is Hausdorff by Proposition 2.10, and η_A has the required universal property: if $f: A \to B$, with B Hausdorff, then in its (regular epi,mono)-factorization $A \xrightarrow{e} f(A) \xrightarrow{m} B$, e is a quotient map and $f(A)$ is Hausdorff. Hence the kernel of e is a closed subalgebra of A containing $\overline{\{0\}}$. This implies that f factors through η_A, necessarily in a unique way because η_A is an epimorphism.

2. Consider the quotient morphism $\rho_A: A \to A/\Gamma(0)$. By Proposition 2.21, $\Gamma(0)$ is a closed normal subalgebra of A. If B is totally disconnected, given any morphism $f: A \to B$, $f(\Gamma(0))$, as a connected subset of B, must be $\{0\}$, and so f factors through $A/\Gamma(0)$. It remains to show that $A/\Gamma(0)$ is totally disconnected, that is, $\Gamma([0]) = \{[0]\}$. If we pullback $\rho_A: A \to A/\Gamma(0)$ along the inclusion $\Gamma([0]) \to A/\Gamma(0)$, we obtain a new short exact sequence

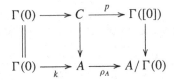

since regular epimorphisms are pullback stable and $p(\Gamma(0)) = \{0\}$. Then, by Proposition 2.23, C is connected because both $\Gamma(0)$ and $\Gamma([0])$ are. Therefore $C = \Gamma(0)$, and so $\Gamma([0]) = \{[0]\}$ as claimed.   □

Observing that any regular-epireflective subcategory of a homological category is homological (see [6] for details), we conclude:

Corollary 3.4 *The categories* $\mathsf{Haus}^{\mathbb{T}}$ *and* $\mathsf{TotDisc}^{\mathbb{T}}$ *are homological.*   □

3.2.2 $\mathsf{HComp}^{\mathbb{T}}$ and $\mathsf{HLocComp}^{\mathbb{T}}$ are Homological

Let $\mathsf{HComp}^{\mathbb{T}}$ and $\mathsf{HLocComp}^{\mathbb{T}}$ be the full subcategories of Hausdorff compact and of Hausdorff locally compact topological algebras, respectively. First of all, if $\mathscr{C}$ is any of these two subcategories and $f : A \to B$ belongs to $\mathscr{C}$, then its (regular epi, mono)-factorization in $\mathsf{Top}^{\mathbb{T}}$

$$A \xrightarrow{\;e\;} f(A) \xrightarrow{\;m\;} B$$

belongs to $\mathscr{C}$, since $f(A)$ is (locally) compact provided A is, and Hausdorff provided B is. Moreover, this gives again a (regular epi, mono)-factorization in $\mathscr{C}$: for every open surjection $e : A \to f(A)$ between Hausdorff (locally) compact topological algebras, its kernel is a Hausdorff (locally) compact topological algebra. Since $\mathscr{C}$ is closed under finite limits in $\mathsf{Top}^{\mathbb{T}}$, we may conclude that $\mathscr{C}$ is regular. To conclude that $\mathscr{C}$ is protomodular it is enough to observe that, given (2) in $\mathscr{C}$, it is a short exact sequence in $\mathscr{C}$ if and only if it is a short exact sequence in $\mathsf{Top}^{\mathbb{T}}$, therefore the Short Five Lemma holds in $\mathscr{C}$.

(We point out that $\mathsf{HComp}^{\mathbb{T}}$ is in fact semi-abelian, as shown in [4].)

4 Split Extensions: Semi-direct Products

4.1 *Semidirect Products of Groups*

In the category of groups, it is well known that split extensions of B by X correspond to actions $\xi : B \times X \to X$ of B on X. Moreover, each such action defines a particular split extension, usually called the *semidirect product* $X \rtimes_{\xi} B$ *of X and B with respect to the action* ξ, which is isomorphic to the original split extension. Let us recall these correspondences.

Given a split extension

$$X \xrightarrow{\;k\;} A \underset{f}{\overset{s}{\longleftrightarrow}} B \qquad (4)$$

where, for simplicity, we consider $k : X \to A$ an inclusion, define

$$\xi : B \times X \to X$$
$$(b, x) \mapsto \xi(b, x) = s(b) + x - s(b).$$

Then $\xi(e, x) = x$ and $\xi(b, x + x') = \xi(b, x) + \xi(b, x')$, $\xi(b + b', x) = \xi(b, \xi(b', x))$, that is ξ is an action of B on X as claimed. Given such a ξ, we define the *semidirect product of X and B* as the group

$$X \rtimes_\xi B = (X \times B, +_\xi), \quad \text{with } (x, b) +_\xi (x', b') = (x + \xi(b, x'), b + b').$$

Then

$$X \xrightarrow{\langle 1, 0 \rangle} X \rtimes_\xi B \underset{\pi_B}{\overset{\langle 0, 1 \rangle}{\rightleftarrows}} B$$

is a split extension and, moreover, the homomorphisms

$$\psi : A \to X \rtimes_\xi B \qquad \text{and} \quad \phi : X \rtimes_\xi B \to A$$
$$a \mapsto (a - sf(a), f(a)) \qquad\qquad (x, b) \mapsto x + s(b)$$

are inverse to each other and make the following diagram commute

$$
\begin{array}{ccc}
X & \xrightarrow{\ k\ } A & \underset{\ }{\overset{s}{\longleftarrow}} B \\
\| & \phi \big\Updownarrow \psi \ {\scriptstyle f} & \| \\
X & \underset{\langle 1,0 \rangle}{\longrightarrow} X \rtimes_\xi B & \underset{\pi_B}{\overset{\langle 0,1 \rangle}{\longleftarrow}} B
\end{array}
$$

We will say that the two split extensions are *isomorphic*.

The action ξ, external to the category of groups, is completely determined by an internal morphism, which we also denote by ξ, defined in the sequel. Consider the split extension

$$B \flat X \xrightarrow{\ k_0\ } X + B \underset{[0,1]}{\overset{\iota_B}{\rightleftarrows}} B$$

where $X + B$ is the coproduct of X and B in Grp, i.e. the free product of X and B, with coprojections ι_X and ι_B, and $[0, 1]$ is the unique morphism such that $[0, 1] \cdot \iota_X = 0$ and $[0, 1] \cdot \iota_B = 1_B$. In the following diagram the existence of a unique such $\xi : B \flat X \to X$ follows from the equalities $f \cdot [k, s] \cdot k_0 = [0, 1] \cdot k_0 = 0$.

$$
\begin{array}{ccc}
B \flat X & \xrightarrow{\ k_0\ } X + B & \underset{[0,1]}{\overset{\iota_B}{\longleftarrow}} B \\
{\scriptstyle \xi} \big\downarrow & \big\downarrow {\scriptstyle [k,s]} & \| \\
X & \xrightarrow{\ k\ } A & \underset{f}{\overset{s}{\longleftarrow}} B
\end{array}
\qquad (5)
$$

This morphism $\xi : B \flat X \to X$ is called the *internal action associated to the split extension* (4). This construction is part of a monad on Grp which allows for a categorical definition of semidirect product encompassing the one we described for groups. Considering that the study of this monad would go beyond our purpose here, we chose to describe directly semidirect products in semi-abelian varieties and in the corresponding categories of topological algebras. For the categorical study of semidirect products see [13], and for the proof that $\mathsf{Top}^{\mathbb{T}}$ has semidirect products see [4]; our description of semidirect products of semi-abelian algebras is based on [19] and [23].

4.2 Semidirect Products of Semi-abelian Algebras

Let $\mathscr{C}$ be a semi-abelian variety, and fix n, α_i, $i = 1, \ldots, n$, and θ satisfying conditions (SA). First we note that, given a split extension (4) in $\mathscr{C}$, as in groups it defines a morphism $\xi \colon B \flat X \to X$ as in diagram (5). Moreover, mimicking the maps ψ and ϕ above, we may consider the following diagram in Set

$$
\begin{array}{ccccc}
X & \xrightarrow{\ k\ } & A & \underset{f}{\overset{s}{\rightleftarrows}} & B \\
{\scriptstyle \psi_X}\downarrow & & {\scriptstyle \phi}\big\uparrow\big\downarrow{\scriptstyle \psi} & & \big\| \\
X^n & \underset{\langle 1,0\rangle}{\longrightarrow} & X^n \times B & \underset{\pi_B}{\overset{\langle 0,1\rangle}{\rightleftarrows}} & B,
\end{array}
\tag{6}
$$

where $\psi_X(x) = \underline{\alpha}(x, 0)$, $\psi(a) = (\underline{\alpha}(a, sf(a)), f(a))$, and $\phi(\underline{x}, b) = \theta(\underline{x}, s(b))$. Then the three squares in the diagram commute, i.e. $\psi \cdot k = \langle 1, 0\rangle \cdot \psi_X$, $\psi \cdot s = \langle 0, 1\rangle$, and $f \cdot \phi = \pi_B$. Moreover, $\phi \cdot \psi = 1_A$, and

$$(\psi \cdot \phi)(\underline{x}, b) = (\underline{x}, b) \text{ if and only if } \underline{\alpha}(\theta(\underline{x}, s(b)), s(b)) = \underline{x}.$$

(We leave the proof of this assertion as an exercise.)

Theorem 4.1 *Given a split extension (4) and $\xi \colon B \flat X \to X$ as in (5), consider*

$$Y = \{(\underline{x}, b) \in X^n \times B \mid \underline{\alpha}(\theta(\underline{x}, s(b)), s(b)) = \underline{x}\}.
\tag{7}$$

Then, in the following commutative diagram

$$
\begin{array}{ccccc}
X & \xrightarrow{\ k\ } & A & \underset{f}{\overset{s}{\rightleftarrows}} & B \\
\big\| & & {\scriptstyle \phi}\big\uparrow\big\downarrow{\scriptstyle \psi} & & \big\| \\
X & \underset{\langle \psi_X, 0\rangle}{\longrightarrow} & Y & \underset{\pi_B}{\overset{\langle 0,1\rangle}{\rightleftarrows}} & B,
\end{array}
\tag{8}
$$

the bottom row is a split extension and ψ and ϕ are isomorphisms of $\mathbb{T}$-algebras if, and only if, for each m-ary operation τ of the variety, the operation τ_Y is defined, for each $(\underline{x}_i, b_i)$, $i = 1, \ldots, m$, by:

$$(\xi^n \underline{\alpha}_{B \flat X} (\tau_{B \flat X} (\theta_{B \flat X} (\underline{x}_1, b_1), \ldots, \theta_{B \flat X} (\underline{x}_m, b_m)), \tau_{B \flat X} (\underline{b})), \tau_B (\underline{b})). \tag{9}$$

Proof From the definition of Y and our remark above it follows that ψ and ϕ are inverse to each other. It remains to show that the operations defined via (9) are the right ones to make the bottom row of (8) a split extension in $\mathsf{Set}^{\mathbb{T}}$. Given $(\underline{x}_i, b_i) \in Y$, for $i = 1, \ldots, m$, let

$$u = \tau_A (\theta_A (\underline{x}_1, s(b_1)), \ldots, \theta_A (\underline{x}_m, s(b_m))).$$

Then

$$\begin{aligned} \tau_Y ((\underline{x}_1, b_1), \ldots, (\underline{x}_m, b_m)) &= \psi \tau_A (\phi(\underline{x}_1, b_1), \ldots, \phi(\underline{x}_m, b_m)) = \psi(u) \\ &= (\underline{\alpha}_A (u, sf(u)), f(u)). \end{aligned}$$

Assuming that f and s are morphisms (and so they preserve the operations):

$$f(u) = \tau_B (\theta_B (f^n(\underline{x}_1), fs(b_1)), \ldots, \theta_B (f^n(\underline{x}_m), fs(b_m))),$$

and, since $f \cdot s = 1_B$ and X is the kernel of f,

$$f(u) = \tau_B (\theta_B (\underline{0}, b_1), \ldots, \theta_B (\underline{0}, b_m)) = \tau_B (\underline{b}).$$

Thus, $sf(u) = \tau_A (s^m (\underline{b}))$, and we obtain:

$$\begin{aligned} \tau_Y ((\underline{x}_1, b_1), \ldots, (\underline{x}_m, b_m)) &= (\underline{\alpha}_A (u, \tau_A (s^m (\underline{b}))), \tau_B (\underline{b})) \\ &= (\underline{\alpha}_A (\tau_A (\theta_A (\underline{x}_1, s(b_1)), \ldots, \theta_A (\underline{x}_m, s(b_m))), \tau_A (s^m (\underline{b}))), \tau_B (\underline{b})) \\ &= ([k, s]^n \underline{\alpha}_{B \flat X} (\tau_{B \flat X} (\theta_{B \flat X} (\underline{x}_1, b_1), \ldots, \theta_{B \flat X} (\underline{x}_m, b_m)), \tau_{B \flat X} (\underline{b})), \tau_B (\underline{b})). \end{aligned}$$

Since ξ is the restriction of $[k, s]$ to $B \flat X$, we finally obtain (9).

The proof that the operations τ_Y make Y a $\mathbb{T}$-algebra and the bottom row a split extension in $\mathsf{Set}^{\mathbb{T}}$ is straightforward. $\qquad\square$

Remark 4.2 We observe that:

1. When $\mathscr{C}$ is the variety of groups, then $n = 1$ and we recover the description of the classical semidirect product. For a general semi-abelian variety the algebra Y is the *categorical semidirect product induced by ξ*. Therefore Theorem 4.1 shows that *every split extension is isomorphic to a split extension given by a semidirect product*.
2. As we observed in Remarks 1.1, a semi-abelian theory may have more than one choice for n, $\underline{\alpha}$ and θ. Different choices will give distinct isomorphic descriptions of the semidirect product. We will present such an example in 4.2.2.

The Theorem above gives us a description of the semidirect product of X and B, for a given action ξ, as a subset Y of $X^n \times B$ equipped with the operations induced by the operations in X, B, and $B \flat X$ through the action ξ. Below we will analyse instances of this construction, but before that we characterize the setting where Y is the all product $X^n \times B$.

Corollary 4.3 *If $\mathscr{C}$ is a semi-abelian variety, with n, (α_i), θ satisfying (SA), then semi-direct products in $\mathscr{C}$ as described in (7) are all the $X^n \times B$ if, and only if, the α_i's verify the following extra condition, for $x_1, \ldots, x_n, y$,*

$$\alpha_i(\theta(x_1, \ldots, x_n, y), y) = x_i, \tag{SA3}$$

for every $i = 1, \ldots, n$.

Proof If this condition is satisfied, then the description (7) of the semidirect products gives immediately $Y = X^n \times B$.

To prove the converse, consider diagram (6) for the split extension below:

$$
\begin{array}{ccccc}
A & \xrightarrow{\langle 1,0\rangle} & A \times A & \underset{\pi_2}{\overset{\langle 1,1\rangle}{\rightleftarrows}} & A \\
\| & & \phi \Big\uparrow \Big\downarrow \psi & & \| \\
A & \xrightarrow{\langle \psi_A, 0\rangle} & A^n \times A & \underset{\pi_2}{\overset{\langle 0,1\rangle}{\rightleftarrows}} & A.
\end{array}
$$

Under our assumption, $\psi \cdot \phi = 1$, which translates into condition (SA3). $\square$

Theorem 4.4 *Let $\mathscr{C}$ be a semi-abelian variety.*

1. *If $\mathscr{C}$ is a variety of right Ω-loops, that is, if $\mathscr{C}$ has a unique constant 0 and, among the operations of $\mathscr{C}$, there are binary operations $+$ and $-$ satisfying conditions $(\Omega 1) - (\Omega 4)$ of 1.2.2, then the semidirect product $X \times_\xi B$ is $X \times B$.*
2. *Moreover, if the semidirect products are described as $X^n \times B$ as in Corollary 4.3, then $\mathscr{C}$ is a variety of right Ω-loops.*

Proof 1. If $\mathscr{C}$ is a variety of right Ω-loops, in its description as semi-abelian variety we may use $n = 1$, $\alpha = -$, and $\theta = +$. Then condition $(\Omega 4)$ is exactly condition (SA3) of the Corollary above, and therefore $Y = X \times B$ in (7).

2. Under these conditions we define $+$ and $-$ by:

$$x + y = \theta(\underline{\alpha}(x, 0), y), \quad x - y = \theta(\underline{\alpha}(x, y), 0).$$

Then $+$ and $-$ verify equations $(\Omega 1) - (\Omega 4)$:

$$x + 0 = \theta(\underline{\alpha}(x, 0), 0) = x,$$

$$0 + x = \theta(\underline{\alpha}(0, 0), x) \stackrel{(SA1)}{=} \theta(\underline{0}, x) \stackrel{(SA2)}{=} x,$$

$$(x - y) + y = \theta(\underline{\alpha}(x, y), 0) + y =$$
$$= \theta(\underline{\alpha}(\theta(\underline{\alpha}(x, y), 0), 0), y) \stackrel{(SA3)}{=} \theta(\underline{\alpha}(x, y), y) \stackrel{(SA2)}{=} x$$

$$(x + y) - y = \theta(\underline{\alpha}(x, 0), y) - y =$$
$$= \theta(\underline{\alpha}(\theta(\underline{\alpha}(x, 0), y), y), 0) \stackrel{(SA3)}{=} \theta(\underline{\alpha}(x, 0), 0) \stackrel{(SA2)}{=} x$$

□

We note that assertion 1 was first proved by E.B. Inyangala in his PhD thesis [25].

Remark 4.5 As shown in Corollary 4.3, condition (SA3) assures us of the existence of an isomorphism $\psi_X : X \to X^n$. If $n \geq 2$, this implies that all the non-trivial algebras of such variety are infinite.

4.2.1 When $X^2 \times B$ works Let $\mathscr{C}$ be the variety defined by:

- a unique constant 0,
- two binary operations α_1 and α_2,
- a ternary operation θ

satisfying conditions (SA1)–(SA3). By Corollary 4.3 we know that the semidirect product of X and B with respect to ξ is given by the split extension

$$X \xrightarrow{\langle \psi_X, 0 \rangle} X^2 \times B \underset{\pi_B}{\overset{\langle 0, 1 \rangle}{\rightleftarrows}} B$$

with the operations described in Theorem 4.1.

As an example of an algebra of $\mathscr{C}$, consider the set $\mathbb{R}^{\mathbb{N}}$ of real sequences. (We may in fact replace $\mathbb{R}$ by any non-trivial right Ω-loop.) We equip $\mathbb{R}^{\mathbb{N}}$ with the following operations:

$$\alpha_1(x, y) = (x_{2n-1} - y_{2n-1})_{n \in \mathbb{N}} = (x_1 - y_1, x_3 - y_3, \ldots),$$

$$\alpha_2(x, y) = (x_{2n} - y_{2n})_{n \in \mathbb{N}} = (x_2 - y_2, x_4 - y_4, \ldots),$$

$$\theta(x, y, z) = (x_1 + z_1, y_1 + z_2, x_2 + z_3, y_2 + z_4, \ldots).$$

We remark that, by the Theorem above, as an algebra in $\mathscr{C}$, $\mathbb{R}^{\mathbb{N}}$ is a right Ω-loop; the corresponding operations are, as expected:

$$x + y = (x_n + y_n)_{n \in \mathbb{N}},$$
$$x - y = (x_n - y_n)_{n \in \mathbb{N}}.$$

4.2.2 When there are Different Choices of n Let $\mathscr{C}$ be a variety with

- a unique constant 0,
- a binary subtraction σ,
- a binary sum ρ,
- a ternary operation ζ,

such that σ and ρ satisfy the usual group equations, and

$$\zeta(\sigma(x, y), \sigma(x, y), y) = x.$$

Then $\mathscr{C}$ is semi-abelian, and we have two different choices of operations satisfying conditions (SA):

(1) $n = 1, \alpha = \sigma$, and $\theta = \rho$;
(2) $n = 2, \alpha_1 = \alpha_2 = \sigma, \theta = \zeta$.

An example of such an algebra is the group of rational numbers $(\mathbb{Q}, +)$, with

$$\sigma(x, y) = x - y, \quad \rho(x, y) = x + y, \quad \zeta(x, y, z) = \frac{x + y + 2z}{2}.$$

In this variety, for each X, B, and action ξ, there are two ways of describing the semidirect product $X \rtimes_\xi B$: as in groups, using the first description, or as

$$X \xrightarrow{\langle \psi_X, 0 \rangle} X \times X \times B \underset{\pi_B}{\overset{\langle 0, 1 \rangle}{\rightleftarrows}} B$$

in case we use the second description. For $x \in X$,

$$\psi_X(x) = (\sigma(x, 0), \sigma(x, 0)) = (x, x),$$

that is, the inclusion $\langle \psi_X, 0 \rangle \colon X \to X \times X \times B$ is given by the diagonal map $\langle 1, 1, 0 \rangle$.

4.3 Semidirect Products of Topological Semi-abelian Algebras

In [4] it is shown that *topological semi-abelian algebras have semidirect products.* Here we show how they are built. They rely on the construction of semidirect products in the algebras, equipping them with a suitable topology.

A split extension in $\mathsf{Top}^{\mathbb{T}}$

$$X \xrightarrow{k} A \underset{f}{\overset{s}{\rightleftarrows}} B \tag{10}$$

induces once more an internal action $\xi : B\flat X \to X$ as in the diagram

$$
\begin{array}{ccccc}
B\flat X & \xrightarrow{\ k_0\ } & X + B & \underset{[0,1]}{\overset{\iota_B}{\rightleftarrows}} & B \\
{\scriptstyle \xi}\Big\downarrow & & {\scriptstyle [k,s]}\Big\downarrow & & \Big\| \\
X & \xrightarrow{\ k\ } & A & \underset{f}{\overset{s}{\rightleftarrows}} & B
\end{array}
\tag{11}
$$

where the entire diagram is built in $\mathsf{Top}^{\mathbb{T}}$. We point out that, applying the forgetful functor into $\mathsf{Set}^{\mathbb{T}}$, i.e., forgetting the topologies involved, the rows of the diagram are again split extensions. Then we can form the semidirect product $X \rtimes_\xi B$ in the variety and note that, equipping $X \rtimes_\xi B$ with the product topology (as a subspace of $X^n \times B$),

$$
\begin{aligned}
\psi : A &\to X \rtimes_\xi B \\
a &\mapsto (\underline{\alpha}(a, sf(a)), f(a))
\end{aligned}
$$

is continuous because its compositions with the projections, into X_i $(i = 1, \ldots, n)$ and B, are continuous; and

$$
\begin{aligned}
\phi : X \rtimes_\xi B &\to A \\
(\underline{x}, b) &\mapsto \theta(\underline{x}, s(b))
\end{aligned}
$$

is continuous because it is the composition of the continuous maps

$$
X \rtimes_\xi B \longrightarrow X^n \times B \xrightarrow{\ k^n \times s\ } A^n \times A \xrightarrow{\ \theta\ } A.
$$

Moreover, with this topology, in the diagram below all the homomorphisms are continuous

$$
\begin{array}{ccccc}
X & \xrightarrow{\ k\ } & A & \underset{f}{\overset{s}{\rightleftarrows}} & B \\
\Big\| & & {\scriptstyle \phi}\Big\uparrow\Big\downarrow{\scriptstyle \psi} & & \Big\| \\
X & \xrightarrow{\langle \psi_X, 0\rangle} & X \rtimes_\xi B & \underset{\pi_B}{\overset{\langle 0,1\rangle}{\rightleftarrows}} & B,
\end{array}
$$

and so this construction provides a split extension – *the semidirect product in* $\mathsf{Top}^{\mathbb{T}}$ – isomorphic to the former one.

5   Split Extensions: Classifiers

In this section we focus on another facet of split extensions of Grp, its categorical formulation, and its validity for topological groups.

5.1 Groups have Split Extension Classifiers

In the category of groups split extensions

$$X \xrightarrow{\ k\ } A \underset{f}{\overset{s}{\rightleftarrows}} B \tag{12}$$

correspond bijectively to homomorphisms $\varphi \colon B \to \mathrm{Aut}(X)$, where $\mathrm{Aut}(X)$ is the group of automorphisms of X. Indeed, given (12) and considering $X \subseteq A$ for simplicity, we define $\varphi \colon B \to \mathrm{Aut}(X)$ by $\varphi(b)(x) = s(b) + x - s(b)$. Then φ induces a morphism between the split extensions

$$
\begin{array}{ccccc}
X & \xrightarrow{\ k\ } & A & \overset{s}{\underset{f}{\rightleftarrows}} & B \\
\big\| & & \big\downarrow & & \big\downarrow{\scriptstyle\varphi} \\
X & \longrightarrow & \mathrm{Hol}(X) & \rightleftarrows & \mathrm{Aut}(X)
\end{array}
$$

where $\mathrm{Hol}(X)$ is the semidirect product of X and $\mathrm{Aut}(X)$ with respect to the evaluation action (that is, the classic holomorph of the group X).

Conversely, every homomorphism $\varphi \colon B \to \mathrm{Aut}(X)$ defines an action $\xi \colon B \times X \to X$ with $\xi(b, x) = \varphi(b)(x)$, and so a split extension

$$X \xrightarrow{\ \langle 1,0\rangle\ } X \rtimes_\xi B \underset{\pi_B}{\overset{\langle 0,1\rangle}{\rightleftarrows}} B$$

isomorphic to the former one. This property is shared by some semi-abelian varieties, but not all. One remarkable example is the variety of Lie algebras, as we will mention in the last section. Here we chose to concentrate on the case of groups. A detailed study of this property for different semi-abelian varieties can be found in [7].

Definition 5.1 If $\mathscr{C}$ is a pointed protomodular category, an object X is said to have a *split extension classifier* if there exists an object $\mathrm{Aut}(X)$ in $\mathscr{C}$ and a split extension

$$X \xrightarrow{\ \kappa\ } \mathrm{Hol}(X) \underset{\pi}{\overset{\iota}{\rightleftarrows}} \mathrm{Aut}(X)$$

such that, for each split extension

$$X \xrightarrow{\ k\ } A \underset{f}{\overset{s}{\rightleftarrows}} B$$

there exists a unique morphism $\varphi \colon B \to \mathrm{Aut}(X)$ such that the following diagram commutes

$$X \xrightarrow{\ k\ } A \underset{f}{\overset{s}{\rightleftarrows}} B$$

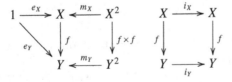

where $\tilde{\varphi}$ is determined by φ and 1_X. If every object X of $\mathscr{C}$ has split extension classifier, one says that $\mathscr{C}$ *has split extension classifiers* (or $\mathscr{C}$ *is action representative*; see [8, 9, 7]).

5.2   A Digression through Split Extension Classifiers for Internal Groups

Let $\mathscr{C}$ be a category with finite products. An internal group in $\mathscr{C}$ is a $\mathscr{C}$-object X together with morphisms $m_X \colon X^2 \to X$, $i_X \colon X \to X$, and $e_X \colon 1 \to X$ making the following diagrams commute,

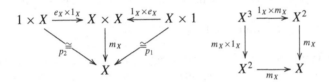

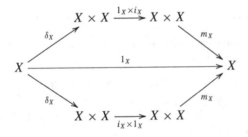

where $\delta_X \colon X \to X \times X$ is the diagonal map. Given two internal groups X, Y, an internal homomorphism $f \colon X \to Y$ is a $\mathscr{C}$-morphism such that (as expected, the commutativity of the second diagram follows for free). They form the category $\mathsf{Grp}(\mathscr{C})$ of internal groups in $\mathscr{C}$. When $\mathscr{C} = \mathsf{Set}$ (resp. Top) we recover the category of (topological) groups and (continuous) homomorphisms as $\mathsf{Grp}(\mathscr{C})$.

Remark 5.2 Clearly one can define *internal* $\mathbb{T}$-*algebra*, for any algebraic theory $\mathbb{T}$, in a similar way.

The other ingredient needed in this digression is a special adjunction. It is well known that, for every triple X, Y, Z of sets, maps $Z \times X \to Y$ correspond bijectively to maps $Z \to Y^X$. This bijection is in fact natural, that is, it induces a natural isomorphism, due to the fact that the Set-endofunctors $(\ ) \times X$ and $(\ )^X$ are adjoint. In a general category $\mathscr{C}$ with finite products, one says that an object X is *exponentiable* if the endofunctor $(\ ) \times X : \mathscr{C} \to \mathscr{C}$ has a right adjoint, which we will denote by $(\ )^X$:

$$
\mathscr{C} \underset{(\ )\times X}{\overset{(\ )^X}{\rightleftarrows}} \mathscr{C} \quad \top \tag{13}
$$

The counit of this adjunction is usually denoted by $\mathrm{ev}_X : Y^X \times X \to Y$ (or simply by ev). If every object of $\mathscr{C}$ is exponentiable, one says that $\mathscr{C}$ is a *cartesian closed category*.

Our aim here is to prove the following

Theorem 5.3 *If $\mathscr{C}$ is a finitely complete category and X is an exponentiable object of $\mathscr{C}$ equipped with an internal group structure, then X has a split extension classifier in $\mathsf{Grp}(\mathscr{C})$.*

The proof of this Theorem is based on the Lemma we prove next.

Lemma 5.4 *Let $\mathscr{C}$ be a category with finite products, and X an internal group in $\mathscr{C}$. If X is exponentiable in $\mathscr{C}$, then:*

1. *X^X has a natural structure of internal monoid, which, in case $\mathscr{C} = \mathsf{Set}$, is given by the composition of maps.*
2. *X^X has a distinct submonoid, $\mathrm{Hom}(X, X)$, which, in case $\mathscr{C} = \mathsf{Set}$, recovers the endomorphisms of the group X.*

Proof Throughout this proof we will use often the natural bijection

$$
(Z \times X \to X) \longleftrightarrow (Z \to X^X);
$$

given a morphism $Z \times X \to X$, we will call the corresponding one from Z to X^X its mate, and vice-versa.

1. The morphism $\mu_X : X^X \times X^X \to X^X$ we want to define is equivalently determined by its mate $X^X \times X^X \times X \to X$. Having in mind the particular case of sets, the latter morphism is defined by the composite

$$X^X \times X^X \times X \xrightarrow{1 \times ev} X^X \times X \xrightarrow{ev} X.$$

The unit $\eta_X : 1 \to X^X$ of the monoid is the mate of the identity $1 \times X \cong X \to X$.

We leave as an exercise to complete the proof that this defines an internal monoid structure in $\mathscr{C}$.

2. Let us first build the object $\mathrm{Hom}(X, X)$, which will be denoted simply by HX (to save space in our diagrams below). As stated in the Lemma, we know how it looks when $\mathscr{C} = \mathsf{Set}$. This means that we want to select those maps f that equalize two possible calculations, $f(x_1)f(x_2)$ and $f(x_1x_2)$, for $x_1, x_2 \in X$, which can be mimicked as:

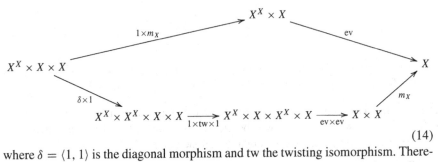

$$(14)$$

where $\delta = \langle 1, 1 \rangle$ is the diagonal morphism and tw the twisting isomorphism. Therefore, we consider the mates u, v of these composites and define HX as their equalizer

$$HX \xrightarrow{h} X^X \underset{v}{\overset{u}{\rightrightarrows}} X^{X \times X}.$$

In particular, diagram (14) commutes when we replace X^X by HX and consider the restriction of ev to HX. Then, from $u \cdot \eta_X = v \cdot \eta_X$ it follows that the unit factors through h, and to check that μ_X can be restricted to HX we verify that the composite

$$HX \times HX \xrightarrow{h \times h} X^X \times X^X \xrightarrow{\mu_X} X^X$$

equalizes u and v, that is, we check the commutativity of the following diagram.

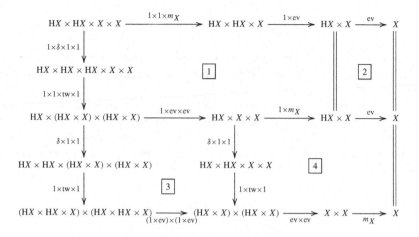

Diagrams $\boxed{2}$ and $\boxed{3}$ are trivially commutative, while commutativity of $\boxed{1}$ and $\boxed{4}$ follows from the commutativity of (14) when X^X is replaced by HX.   □

Proof of the Theorem Let HX be the internal monoid built in the Lemma. Then split extensions with kernel X will be classified by the internal group Aut(X) of *invertible elements* of HX, defined by the pullback diagram

$$
\begin{array}{ccc}
\mathrm{Aut}(X) & \xrightarrow{\;\;!\;\;} & 1 \\
\langle\varphi_1,\varphi_2\rangle \downarrow & & \downarrow \langle\eta_X,\eta_X\rangle \\
\mathrm{H}X \times \mathrm{H}X & \xrightarrow[\langle\mu_X,\mu_X^\circ\rangle]{} & \mathrm{H}X \times \mathrm{H}X,
\end{array}
$$

(where $\mu_X^\circ = \mu_X \cdot \mathrm{tw}$), with the monoid structure induced by HX, and the unit $\eta: 1 \to \mathrm{Aut}(X)$ and the inversion $\iota: \mathrm{Aut}(X) \to \mathrm{Aut}(X)$ given by the pullback universal property.

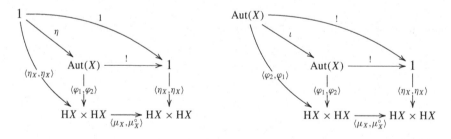

To show that μ_X defines a multiplication in Aut(X) needs extra calculation, which we leave to the reader.

It remains to show that, for every split extension (12), there is a unique morphism $\varphi: B \to \mathrm{Aut}(X)$ inducing a morphism of split extensions

$$X \xrightarrow{\ k\ } A \underset{f}{\overset{s}{\rightleftarrows}} B \qquad\qquad (15)$$

where ξ is the internal action induced by the evaluation

$$\mathrm{Aut}(X) \times X \xrightarrow{\langle \varphi_1, \varphi_2 \rangle \times 1_X} HX \times HX \times X \xrightarrow{\pi_1 \times 1_X} HX \times X \xrightarrow{\ \mathrm{ev}\ } X.$$

Now we outline the construction of the morphism $\varphi \colon B \to \mathrm{Aut}(X)$.

(a) In the case of groups (in **Set**), $f(s(b) + x - s(b)) = f(-s(b) + x + s(b)) = 0$; in the general case, one can check that the composites σ and ρ, given respectively by

$$B \times X \xrightarrow{\delta \times 1} B^2 \times X \xrightarrow{1 \times \mathrm{tw}} B \times X \times B \xrightarrow{s \times 1 \times s} A \times X \times A \xrightarrow{1 \times k \times i} A^3 \xrightarrow{m \times 1} A^2 \xrightarrow{m} A$$

$$B \times X \xrightarrow{\delta \times 1} B^2 \times X \xrightarrow{1 \times \mathrm{tw}} B \times X \times B \xrightarrow{s \times 1 \times s} A \times X \times A \xrightarrow{i \times k \times 1} A^3 \xrightarrow{m \times 1} A^2 \xrightarrow{m} A,$$

when composed with f give the zero morphism, and therefore they factor through X, defining morphisms $\sigma', \rho' \colon B \times X \to X$, which induce $\sigma'', \rho'' \colon B \to X^X$.

(b) Both σ'' and ρ'' factor through HX, since diagram (14) is commutative when we replace X^X by B and ev by σ' or ρ'.

(c) Finally one needs to check that the diagram

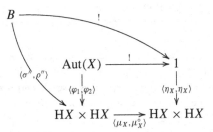

is commutative, obtaining then the claimed morphism $\varphi \colon B \to \mathrm{Aut}(X)$.

Now straightforward calculations show that φ induces diagram (15).   □

Corollary 5.5. *If $\mathscr{C}$ is a finitely complete and cartesian closed category, then* $\mathrm{Grp}(\mathscr{C})$ *has split extension classifiers.*   □

Although **Top** is not cartesian closed, the exponentiable topological spaces are characterized as the *core-compact* (or *quasi-locally compact*) spaces, that is, those spaces X such that, for each $x \in X$ and each neighbourhood V of x, there is a neighbourhood U relatively compact in V; by U *relatively compact in* V we mean that every open cover $(V_i)_{i \in I}$ of V has a finite subcover of U. Every locally compact space is core-compact, and the two conditions coincide when X is Hausdorff.

If X is core-compact and Y is any topological space, then the topology of Y^X is generated by the sets

$$\langle U, V \rangle = \{f \in Y^X ; \text{ every open cover of } f^{-1}(V) \text{ has a finite subcover of } U\},$$

where $U \in OX$ and $V \in OY$. When X is compact and Hausdorff, this topology is exactly the *compact-open topology*, that is, the topology generated by

$$\langle K, V \rangle = \{f \in Y^X ; f(K) \subseteq V\},$$

where K is a compact subset of X and $V \in OY$. (See [20] for details.)

Therefore from the Theorem we may conclude that

Corollary 5.6. *If X is a topological group whose topology is core-compact, then X has a split extension classifier.*   □

Recently it was shown that all topological groups have a split extension classifier. This will be the subject of the next section.

5.3   Topological Groups have Split Extensions Classifiers

In this section we outline the arguments used in [15] to show that Corollary 5.6 can be extended to all topological groups. These arguments are essentially topological. The only property of topological groups used is that every $T0$ topological group is a Tychonoff space, that is, it is a dense subspace of its Stone-Čech compactification, via the embedding $\beta_X : X \to \beta X$.

We recall that, although Top is not cartesian closed, it can be nicely embedded in a cartesian closed category, the category PsTop of pseudotopological spaces and continuous maps. A pseudotopological space is a set X endowed with an ultrafilter convergence structure, i.e. a relation $R \subseteq UX \times X$ such that the principal ultrafilter induced by x converges to x, for every $x \in X$. A map $f : X \to Y$ between pseudotopological spaces is said to be continuous if it preserves convergence. The convergence structure of topological spaces is clearly pseudotopological, and maps between topological spaces preserve convergence if, and only if, they are continuous; therefore we get a full embedding

$$\text{Top} \longrightarrow \text{PsTop}.$$

It is shown in [31] that, when exponentials exist in Top, they are built as in PsTop. Therefore, to conclude that we can perform in Top the construction described in the proof of Theorem 5.3 we only need to perform it in PsTop and then check that the pseudotopological space $\mathrm{Aut}(X)$ is in fact topological. The space $\mathrm{Aut}(X)$ is defined as a subspace of $HX \times HX$, which in turn is a subspace of $X^X \times X^X$. Our argument in Corollary 5.6 is that $\mathrm{Aut}(X)$ is topological provided that X^X is, but in fact it is enough to assure that the convergence structure in

$$\mathrm{Iso}(X) = \{f \colon X \to X \,;\, f \text{ is an homeomorphism}\}$$

is topological, since $\mathrm{Aut}(X)$ is clearly a subspace of $\mathrm{Iso}(X) \times \mathrm{Iso}(X)$.

The arguments of [15] can be stated as in the following theorem, whose proof we omit because it is rather technical (for details see [15]).

Theorem 5.7. *Let X be a topological space, and X^X the pseudotopological space obtained via the adjunction* (13). *Consider* $\mathrm{Iso}(X)$ *as a subspace of* X^X. *Then:*

1. *Denoting by RX the reflexion of X into* Top$_0$, *if* $\mathrm{Iso}(RX)$ *is topological, then so is* $\mathrm{Iso}(X)$.
2. *If X is a Tychonoff space, then* $\mathrm{Iso}(X)$ *is a subspace of* $\mathrm{Iso}(\beta X)$. *In particular, it is a topological space.*

Corollary 5.8. *Topological groups have split extension classifiers.*

Proof. The first assertion of the Theorem reduces the problem to $T0$ topological groups, which are immediately Tychonoff spaces (see [30; Section 19, Theorem 10]). Hence, by the second assertion we conclude that $\mathrm{Aut}(X)$, as a subspace of $\mathrm{Iso}(X) \times \mathrm{Iso}(X)$, and therefore also a subspace of $\mathrm{Iso}(\beta X) \times \mathrm{Iso}(\beta X)$, is topological. This also says that the topology of $\mathrm{Aut}(X)$ is inherited from the compact-open topology in $\beta X^{\beta X}$. □

6 Some Open Problems

6.1 Coproducts of Topological Algebras

As stated in Remark 1.7, coproducts of topological algebras are built as in the corresponding category of algebras, equipped with the largest topology making both the inclusions of the summands and the operations continuous. Although this topology is easily defined, it is very difficult to describe in a handy manner, reason why we pose the following open question.

> **? Given topological groups A and B, how can we describe the topology on the free product $C = A \amalg B$?**

We can pose a similar question for infinite coproducts of topological groups, or, more generally, for (infinite) coproducts of (semi-abelian) topological algebras.

So that the reader gets an idea of the nature of the obstacles, we list the properties that characterize a neighbourhood basis $\mathcal{U}$ for 0 (in a topological group X):

1. $\forall U \in \mathcal{U} \; \exists V \in \mathcal{U} \; : \; V + V \subseteq U$;
2. $\forall U \in \mathcal{U} \; \exists V \in \mathcal{U} \; : \; -V \subseteq U$;
3. $\forall U \in \mathcal{U} \; \forall x \in X \; \exists V \in \mathcal{U} \; : \; x + V \subseteq U$;
4. $\forall U \in \mathcal{U} \; \forall x \in X \; \exists V \in \mathcal{U} \; : \; x + V + x^{-1} \subseteq U$;
5. $\forall U, V \in \mathcal{U} \; \exists W \in \mathcal{U} \; : \; W \subseteq U \cap V$.

In the case of a coproduct, the (most) weird condition is (4). Indeed, since an element of C is of the form $a_1 b_1 \ldots a_n b_n$, with $a_i \in A$, $b_i \in B$, (4) says that, for every neighbourhood U of 0, and for every element $a_1 b_1 \ldots a_n b_n$ of X, there exists a neighbourhood V of 0 so that, for every element $a_1' b_1' \ldots a_k' b_k'$ of V,

$$a_1 b_1 \ldots a_n b_n a_1' b_1' \ldots a_k' b_k' b_n^{-1} a_n^{-1} \ldots b_1^{-1} a_1^{-1} \in U.$$

This problem – of not being able to describe this topology in a workable way – causes serious difficulties on the study of some categorical features of $\mathsf{Top}^{\mathbb{T}}$.

6.2   Split Extension Classifiers of Topological Algebras

Although it has been proven that topological groups have split extension classifiers, the known proof is quite complex and gives no clue on how to deal with the same problem for other (semi-abelian) topological algebras. Hence, a more categorical proof of this fact could be a great contribution to this study. In [7; Section 6] some steps towards this direction were made, but we reached a step where it was essential to know how to handle coproducts of topological groups. It is proved there that a sufficient condition for the existence of split extension classifiers for topological algebras is the *normal amalgamation property*, meaning that, for any pushout in $\mathsf{Top}^{\mathbb{T}}$

$$
\begin{array}{ccc}
X & \xrightarrow{\;k_1\;} & A_1 \\
{\scriptstyle k_2}\big\downarrow & & \big\downarrow{\scriptstyle f_1} \\
A_2 & \xrightarrow[\;f_2\;]{} & C
\end{array}
$$

with k_1, k_2 normal monomorphisms, f_1, f_2 are monic, and, moreover, $f_1 \cdot k_1 = f_2 \cdot k_2$ is a normal monomorphism.

> **? Is the normal amalgamation property valid in TopGrp?**

We can pose the same problem for $\mathrm{Top}^{\mathbb{T}}$, with $\mathbb{T}$ a semi-abelian theory.

6.3   Split Extension Classifiers: Topological Lie Algebras

Still regarding the existence of split extension classifiers, the case of topological Lie algebras is of particular importance. Indeed, like groups, Lie algebras have split extension classifiers. They are given by the Lie algebra of derivations. Moreover, as shown in [7], an adaptation of the proof of Theorem 5.3 shows that in the category of internal Lie algebras every exponentiable internal Lie algebra has a split extension classifier. Therefore, as for topological groups, we can conclude that:

Theorem 6.1.   *In the category of topological Lie algebras over a commutative ring R with unit, every core-compact topological Lie algebra has a split extension classifier.*

But the following problem still needs an answer.

> **? Does every topological Lie algebra (over a topological commutative ring R with unit) have a split extension classifier?**

As we have said before, the existing proof for topological groups [15] does not seem to give any clue for the solution of this problem.

6.4   Algebraic Coherence for Topological Groups

For an object B of a category $\mathscr{C}$, the *category* $\mathrm{Pt}_B(\mathscr{C})$ *of points over B in $\mathscr{C}$* has as objects triples $(A, f : A \to B, s : B \to A)$ where f is a split epimorphism with a chosen splitting s; morphisms $\alpha : (A, f, s) \to (A', f', s')$ are $\mathscr{C}$-morphisms $\alpha : A \to A'$ such that $f' \cdot \alpha = f, \alpha \cdot s = s'$.

Definition 6.2 A category $\mathscr{C}$ is said to be *algebraically coherent* [17] if, for every morphism $p : X \to Y$ in $\mathscr{C}$, the change-of-base functor

$$p^* : \mathrm{Pt}_Y(\mathscr{C}) \to \mathrm{Pt}_X(\mathscr{C})$$

is *coherent*, meaning that p^* preserves strong epimorphisms and the comparison morphism

$$p^*(A, f, s) \amalg p^*(A', f', s') \to p^*((A, f, s) \amalg (A', f', s'))$$

is a strong epimorphism.

As shown in [17, Proposition 3.13],

Proposition 6.3 *A homological category is algebraically coherent if, and only if, for every diagram of (vertical) split extensions of the form*

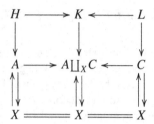

the induced arrow $H \coprod L \to K$ *is a strong epimorphism.*

> **?** Is TopGrp **algebraically coherent? And** HausGrp?

6.5   *Local Algebraic Cartesian Closedness for Topological Groups*

Finally, in case the answer to Question 6.4 is positive, one can ask

> **?** Is TopGrp **a locally algebraically cartesian closed category?**

By *locally algebraically cartesian closed category* (usually called (lacc)) [22] we mean a finitely complete category $\mathscr{C}$ where, for every $p \colon X \to Y$ in $\mathscr{C}$, p^* is a left adjoint.

Acknowledgements I thank both Andrea Montoli and the anonymous referee for their comments on a preliminary version of this text, and I thank Andrea Montoli for his suggestions towards the construction of the example of 3.1.3. I also want to express my gratitude to Francis Borceux for suggesting me, about two decades ago, to collaborate with him in this interesting subject.

References

1. Adámek, J., Herrlich, H., Strecker, G.E.: Abstract and Concrete Categories: The Joy of Cats. Pure and Applied Mathematics (New York). Wiley, New York (1990). Republished. In: Reprints in Theory and Applications of Categories, No. 17 (2006), pp. 1–507 (2006)
2. Barr, M.: Exact categories. In: Springer Lecture Notes in Math. vol. 236, pp. 1–120 (1971)
3. Borceux, F., Bourn, D.: Mal'cev, protomodular, homological and semi-abelian categories. Kluwer, Mathematics and its Applications (2004)
4. Borceux, F., Clementino, M.M.: Topological semi-abelian algebras. Adv. Math. **190**, 425–453 (2005)
5. Borceux, F., Clementino, M.M.: Topological protomodular algebras. Topology Appl. **153**, 3085–3100 (2006)
6. Borceux, F., Clementino, M.M., Gran, M., Sousa, L.: Protolocalizations of homological categories. J. Pure Appl. Algebra **212**, 1898–1927 (2008)
7. Borceux, F., Clementino, M.M., Montoli, A.: On the representability of actions for topological algebras. In: Categorical Methods in Algebra and Topology, vol. 46, pp. 41–66. Textos de Matemática, University of Coimbra, Coimbra (2014)
8. Borceux, F., Janelidze, G., Kelly, G.M.: On the representability of actions in a semi-abelian category. Theory Appl. Categ. **14**, 244–286 (2005)
9. Borceux, F., Janelidze, G., Kelly, G.M.: Internal object actions. Comment. Math. Univ. Carolin. **46**, 235–255 (2005)
10. Bourbaki, N.: Éléments de Mathématique. Topologie Générale. Hermann, Paris (1971)
11. Bourn, D.: Normalization, equivalence, kernel equivalence and affine categories. In: Lect. Notes Math., vol. 1488, pp. 43–62. Springer (1991)
12. Bourn, D.: From Groups to Categorial Algebra. Introduction to protomodular and Mal'tsev categories. Compact Textbooks in Mathematics, Birkhäuser/Springer (2017)
13. Bourn, B., Janelidze, G.: Protomodularity, descent, and semidirect products. Theory Appl. Categ. **4**, 37–46 (1998)
14. Bourn, B., Janelidze, G.: Characterization of protomodular varieties of universal algebra. Theory Appl. Categ. **11**, 143–147 (2002)
15. Cagliari, F., Clementino, M.M.: Topological groups have representable actions. Bull. Belg. Math. Soc. Simon Stevin **26**, 519–526 (2019)
16. Carboni, A., Kelly, G.M., Pedicchio, M.C.: Some remarks on Mal'tsev and Goursat categories. Appl. Categ. Structures **1**, 385–421 (1993)
17. Cigoli, A., Gray, J.R.A., Van der Linden, T.: Algebraically coherent categories. Theory Appl. Categ. **30**, 1864–1905 (2015)
18. Clementino, M.M.: Towards categorical behaviour of groups. CIM Bull. **2**, 7–12 (2007)
19. Clementino, M.M., Montoli, A., Sousa, L.: Semidirect products of (topological) semi-abelian algebras. J. Pure Appl. Algebra **219**, 183–197 (2015)
20. Escardó, M.H., Heckmann, R.: Topologies on spaces of continuous functions. Topol. Proc. **26**(2), 545–564 (2001–2002)
21. Gran, M.: An introduction to regular categories. In: Clementino, M.M., Facchini, A., Gran, M. (eds.) New Perspectives in Algebra, Topology and Categories, Coimbra Mathematical Texts 1, pp. 113–145. Springer Nature and University of Coimbra (2021)
22. Gray, J.R.A.: Algebraic exponentiation in general categories. Appl. Categ. Structures **20**, 543–567 (2012)
23. Gray, J.R.A., Martins-Ferreira, N.: On algebraic and more general categories whose split epimorphisms have underlying product projections. Appl. Categ. Structures **23**, 429–446 (2015)
24. Higgins, P.: Groups with multiple operators. Proc. Lond. Math. Soc. **6**(3), 366–416 (1956)
25. Inyangala, E.B.: Semidirect products and crossed modules in varieties of right Ω-loops. Theory Appl. Categ. **25**, 426–435 (2011)
26. Janelidze, G., Márki, L., Tholen, W.: Semi-abelian categories. J. Pure Appl. Algebra **168**, 367–386 (2002)

27. Johnstone, P.T.: A note on the semiabelian variety of Heyting semilattices. Fields Inst. Commun. **43**, 317–318 (2004)
28. Johnstone, P.T., Pedicchio, M.C.: Remarks on continuous Mal'cev algebras. In: Proceedings of the Eleventh International Conference of Topology (Trieste, 1993). Rend. Istit. Mat. Univ. Trieste, vol. 25, no. 1–2, pp. 277–297 (1994)
29. Mac Lane, S.: Categories for the Working Mathematician, 2nd edn. Springer, New York (1998)
30. Pontryagin, L.S.: Topological groups. Translated from the second Russian edition by. Arlen Brown Gordon and Breach Science Publishers Inc, New York-London-Paris (1966)
31. Schwarz, F.: Product compatible reflectors and exponentiability. In: Categorical Topology (Toledo, Ohio, 1983). Sigma Ser. Pure Math., vol. 5, pp. 505–522. Heldermann, Berlin (1984)
32. Ursini, A.: Sulle varietà di algebre con una buona teoria degli ideali. (Italian) Boll. Un. Mat. Ital. **6**, 90–95 (1972)

Chapter 3
Commutative Monoids, Noncommutative Rings and Modules

Alberto Facchini

Abstract These are the notes of a non-standard course of Algebra. It deals with elementary theory of commutative monoids and non-commutative rings. Most of what is taught in a master course of Commutative Algebra holds not only for commutative rings, but more generally for any commutative monoid, which shows that the additive group structure on a commutative ring has little importance.

In the rest of the notes of the course presented here, we introduce the basic notions of non-commutative rings and their modules, stressing the difference with what happens in the case of commutative rings.

Keywords Commutative monoid · Preordered abelian group · Associative ring · Module over a ring

Math. Subj. Classification: 06F05 · 16-01 · 16Dxx · 20Mxx

Introduction

These are the notes of a course I gave in Louvain-la-Neuve in September 2018. It is a non-standard course of Algebra. It contains some topics that are not usually taught in master courses in Mathematics. The first topic is the elementary theory of commutative monoids. It is very standard to teach a course of Commutative Algebra, teaching commutative rings and modules over them (localization at prime ideals, and so on). But most things taught in those courses hold not only for commutative rings, but more generally for any commutative monoid. This occurs from the most elementary things (prime ideals, localizations, spectrum of the ring), to more "advanced" topics (valuations, Krull domains/monoids, divisorial ideals, class group). In other words, the additive group structure on a commutative ring is of little consequence.

A. Facchini (✉)
Dipartimento di Matematica "Tullio Levi-Civita", Università di Padova, Via Trieste 63, 35121 Padova, Italy
e-mail: facchini@math.unipd.it

© The Author(s), under exclusive license to Springer Nature Switzerland AG 2021
M. M. Clementino et al. (eds.), *New Perspectives in Algebra, Topology and Categories*, Coimbra Mathematical Texts 1,
https://doi.org/10.1007/978-3-030-84319-9_3

The promoter of this idea was Chouinard [5]. In this topic, what I present is very easy, but not so much known among mathematicians.

Then I pass to a quick introduction to the theory of non-commutative rings, their modules, and Grothendieck group. My main aim, as far as non-commutative rings and their modules are concerned, is to stress the points where their properties differ from those of modules over commutative rings. The path I follow explaining the various topics is also partially non-standard, and relies on my personal taste.

I don't give most proofs. The interested reader can find them in several text books. For examples, for further notions about commutative monoids, one can see the books [6] and [13]. For non-commutative rings the best text books are [2] and [14]. My books [7] and [8] are also a possible reference.

1 Commutative Monoids

One of the structures in which we can come across most frequently in Algebra is the structure of monoid.

1.1 Commutative Monoids and Their Morphisms

An *(additive) monoid M* is a set with an operation (addition)

$$+: M \times M \to M, \qquad (x, y) \mapsto x + y,$$

which is associative (that is, $x + (y + z) = (x + y) + z$ for every $x, y, z \in M$) and has a *zero element*, usually denoted by 0, that is, an element $0 \in M$ such that $x + 0 = 0 + x = x$ for every $x \in M$. In these notes, all the monoids we will consider will be commutative, that is, $x + y = y + x$ for every $x, y \in M$. In other words, "monoid" and "commutative monoid" will have the same meaning for us.

A *monoid morphism* is a mapping f of a monoid M into a monoid N such that $f(0) = 0$ and $f(x + y) = f(x) + f(y)$ for every $x, y \in M$. The composite mapping of two monoid morphisms is a monoid morphism. Thus we have a category of commutative monoids, which we will denote by CMon.

Monomorphisms in the category CMon (that is, the morphisms $f: M \to N$ such that, for every pair $g, h: P \to M$ of monoid morphisms, $fg = fh$ implies $g = h$) are exactly the monoid morphisms that are injective mappings. This is not the case for epimorphisms: not all epimorphisms in CMon (that is, the morphisms $f: M \to N$ such that, for every pair $g, h: N \to P$ of monoid morphisms, $gf = hf$ implies $g = h$) are necessarily onto mappings. It is sufficient to consider the inclusion of the monoid $\mathbb{N}_0$ of non-negative integers into the additive monoid $\mathbb{Z}$ of integers, which is a non-surjective epimorphism.

A subset N of a commutative additive monoid M is a *submonoid* of M if it is closed under the addition of M and contains the zero element of M. For a monoid M, the set of all elements $a \in M$ with an opposite in M will be denoted by $U(M)$, that is, $U(M) := \{ x \in M \mid$ there exists $y \in M$ with $x + y = 0 \}$. If such an element $y \in M$ exists, it is unique, is denoted by $-x$, and is called the *opposite* of x. The subset $U(M)$ turns out to be a submonoid of M, and is an abelian group, often (improperly) called the *group of units* of M. The monoid M is *reduced* if $U(M) = \{0\}$, that is, if $x + y = 0$ implies $x = y = 0$ for every $x, y \in M$.

1.2 Preorders

A *preorder* on a set A is a relation on A that is reflexive and transitive. We will denote by Preord the category of all preordered sets. Its objects are the pairs (A, ρ), where A is a set and ρ is a preorder on A. The morphisms $f : (A, \rho) \to (A', \rho')$ in Preord are the mappings f of A into A' such that $a\rho b$ implies $f(a)\rho' f(b)$ for all $a, b \in A$. As usual, when there is no danger of confusion, that is, when the preorder is clear from the context, we will denote the preordered set (A, ρ) simply by A.

The main examples of preordered sets (A, ρ) are those in which the preorder ρ is a partial order (i.e., ρ is antisymmetric) or an equivalence relation (i.e., ρ is symmetric). The full subcategories of Preord whose objects are all preordered sets (A, ρ) with ρ a partial order (an equivalence relation) will be denoted by ParOrd (Equiv, respectively).

Proposition 1.1 *Let A be a set. There is a one-to-one correspondence between the set of all preorders ρ on A and the set of all pairs $(\sim, \leq)$, where $\sim$ is an equivalence relation on A and $\leq$ is a partial order on the quotient set $A/\sim$. The correspondence associates with every preorder ρ on A the pair $(\simeq_\rho, \leq_\rho)$, where $\simeq_\rho$ is the equivalence relation defined, for every $a, b \in A$, by $a \simeq_\rho b$ if $a\rho b$ and $b\rho a$, and $\leq_\rho$ is the partial order on $A/\simeq_\rho$ defined, for every $a, b \in A$, by $[a]_{\simeq_\rho} \leq_\rho [b]_{\simeq_\rho}$ if $a\rho b$. Conversely, for any pair $(\sim, \leq)$ with $\sim$ an equivalence relation on A and $\leq$ a partial order on $A/\sim$, the corresponding preorder $\rho_{(\sim,\leq)}$ on A is defined, for every $a, b \in A$, by $a\rho_{(\sim,\leq)}b$ if $[a]_\sim \leq [b]_\sim$.*

The objects of Preord that are objects in both the full subcategories ParOrd and Equiv are the objects of the form $(A, =)$, where $=$ denotes the equality relation on A. The pair (Equiv, ParOrd) is a pretorsion theory in Preord in the sense of [9].

The category of finite preordered sets is isomorphic to the category of finite topological spaces, the full subcategory of Top whose objects are the topological spaces with only finitely many points. (If X is a finite topological space, the corresponding preorder $\leq$ on X is defined by $x \leq y$ if and only if x belongs to the closure of the subset $\{y\}$ of X. Every closed set in a finite topological space X is a union of closures of points.)

More generally, the category of preordered sets is isomorphic to the category of Alexandrov topological spaces, the full subcategory of Top whose objects are the topological spaces whose topology is an Alexandrov topology. A topology is *Alexandrov* if the intersection of any family of open subsets is an open set (equivalently, if the union of any family of closed subsets is a closed subset).

If M is a commutative additive monoid, a preorder $\leq$ on M is *translation-invariant* if, for every $x, y, z \in M$, $x \leq y$ implies $x + z \leq y + z$. There is a natural translation-invariant preorder on any commutative additive monoid M, called the *algebraic preorder* on M, defined, for all $x, y \in M$, by $x \leq y$ if there exists $z \in M$ such that $x + z = y$. If x is an element of a monoid M and $n \geq 0$, we can inductively define the *n-th multiple* nx of x setting $0x := 0$ and $nx := (n-1)x + x$. An element u of a commutative monoid M is an *order-unit* if for every $x \in M$ there exists an integer $n \geq 0$ such that $x \leq nu$. For example, let M be the monoid $\mathbb{N}_0^n$ of all n-tuples of non-negative integers. The algebraic preorder on M is the component-wise order, that is, $(x_1, \ldots, x_n) \leq (y_1, \ldots, y_n)$ if and only if $x_i \leq y_i$ for every $i = 1, \ldots, n$, and an element $(u_1, \ldots, u_n)$ of $\mathbb{N}_0^n$ is an order-unit if and only if $u_i > 0$ for every $i = 1, \ldots, n$.

A submonoid N of a monoid M is said to be *divisor-closed* if $x \in M$, $y \in N$, and $x \leq y$ in the algebraic preorder $\leq$ of M, implies $x \in N$. The term "divisor-closed" becomes clear if we move on to the multiplicative notation. More precisely, if the operation in the commutative monoid M is denoted as multiplication instead of addition, then the algebraic preorder on M is the relation $\mid$ (divides), and a submonoid N of M is divisor-closed if, for every element $y \in N$, it contains all divisors of y in M. The group of units $U(M)$ of an arbitrary commutative monoid M is a divisor-closed submonoid of M contained in all divisor-closed submonoids of M.

Let X be a subset of a monoid M. Let $\mathcal{F}$ be the family of all submonoids of M that contain X. The family $\mathcal{F}$ is always non-empty, because $M \in \mathcal{F}$. The intersection of all the submonoids in $\mathcal{F}$ is the smallest submonoid of M that contains X. It is called the submonoid of M *generated* by X and is denoted by $[X]$. It is easily seen that if $X = \emptyset$, then $[X] = \{0\}$, the zero submonoid of M. If $X \neq \emptyset$, then $[X] = \{ x_1 + \cdots + x_n \mid n \geq 0 \text{ and } x_i \in X \text{ for } i = 1, \ldots, n \}$ (sums of finitely many elements of X, possibly with repetitions). Conventionally, the sum of zero elements of M, that is, the sum of no element of M, is the zero element of M.

A subset X of a monoid M is a *set of generators* of M if $[X] = M$. A monoid M is *finitely generated* if it has a finite set of generators, and *cyclic* if it has a set of generators with one element.

1.3 Congruences

If $f : M \to N$ is a monoid morphism, the *kernel pair* of f is the equivalence relation $\sim_f$ on the set M defined, for every $x, y \in M$, by $x \sim_f y$ if $f(x) = f(y)$.

A *congruence* on a monoid M is an equivalence relation $\sim$ on the set M such that $x \sim y$ and $z \sim w$ implies $x + z \sim y + w$ for every $x, y, z, w \in M$. Equivalently, an equivalence relation $\sim$ on a monoid M is a congruence if $x \sim y$ implies $x + z \sim$

$y + z$ for every $x, y, z \in M$. It is easily verified that the kernel pair $\sim_f$ of any monoid morphism $f \colon M \to N$ is a congruence on the monoid M.

If M is a monoid and $\sim$ is a congruence on M, the *factor monoid* $M/\sim$ is the set of all *congruence classes* $[x]_\sim := \{\, y \in M \mid y \sim x \,\}$, where x ranges in M, with the addition inherited from that of M:

$$[x]_\sim + [y]_\sim := [x + y]_\sim \quad \text{for every } x, y \in M.$$

This operation on $M/\sim$ is well defined, as is easily verified. It is the unique operation on the quotient set $M/\sim$ which makes the *canonical projection* $\pi \colon M \to M/\sim$, defined by $\pi(x) = [x]_\sim$ for every $x \in M$, a monoid morphism. Every congruence on a monoid is the kernel pair of a monoid morphism.

A subset P of $M \times M$ is a *set of generators* for a congruence $\sim$ of the monoid M if the intersection of all congruences of M that contain P is $\sim$. A congruence $\sim$ on a monoid M is *finitely generated* if it has a finite set of generators.

An element x of a monoid M is said to be *idempotent* if $x + x = x$. A monoid M is *archimedean* if for every pair (x, y) of elements of M with $y \nleq 0$ there exists a positive integer n such that $x \leq ny$. Equivalently, this means that M is either $\{0\}$ or has exactly two divisor closed submonoids. More generally, for any x, y in a commutative monoid M, define $x \asymp y$ if there exist positive integers n and m such that $x \leq ny$ and $y \leq mx$. It is easy to prove that $\asymp$ is the smallest congruence on M such that every element in the quotient monoid $M/\asymp$ is idempotent. The equivalence classes of M modulo $\asymp$ are additively closed subsets of M, called the *archimedean components* of M.

Here is another important example of congruence. Recall that, for any monoid M, $U(M)$ denotes the abelian additive group of all the elements of M with an opposite in M. The relation $\sim$ on M, defined, for every $x, y \in M$, by $x \sim y$ if there exists $z \in U(M)$ with $x = y + z$, turns out to be a congruence on M. The congruence class $[x]_\sim$ is the *coset* $x + U(M) := \{\, x + z \mid z \in U(M) \,\}$. We will denote by M_{red} the factor monoid $M/\sim$. The monoid M_{red} is always a reduced monoid, i.e., does not have non-zero elements with an opposite element. Thus every commutative monoid M is an extension of the reduced monoid M_{red} by the abelian group $U(M)$. Again, we have a pretorsion theory in the category CMon. The torsion class is the class of abelian groups. The torsionfree class is the class of all reduced commutative monoids.

As a further example of congruence, define an equivalence $\sim$ on any commutative monoid M, setting, for every $x, y \in M$, $x \equiv y$ if there exists $z \in M$ with $x + z = y + z$. It is easily seen that $\equiv$ is a congruence on M, called the *stable congruence*, and that the factor monoid $M/\equiv$ is a cancellative monoid. Recall that a monoid N is *cancellative* if $x + z = y + z$ implies $x = y$ for every $x, y, z \in N$. Hence $\equiv$ is the smallest congruence on M with $M/\equiv$ cancellative.

1.4 The Additive Monoid $\mathbb{N}_0$ of Natural Numbers

Consider the additive monoid $\mathbb{N}_0$ whose elements are the natural numbers $0, 1, 2, \ldots$
Fix k and n in $\mathbb{N}_0$ with $n \geq 1$, and define the relation $\sim_{k,n}$ on $\mathbb{N}_0$ setting, for every $x, y \in \mathbb{N}_0$,

$$x \sim_{k,n} y \text{ if } \begin{cases} x = y \\ \text{or} \\ x \geq k, \ y \geq k \text{ and } x \equiv y \pmod{n}. \end{cases}$$

Here $x \equiv y \pmod{n}$ means that x and y are integers congruent modulo n, that is, n divides $x - y$ in $\mathbb{Z}$. It is easily verified that $\sim_{k,n}$ is a congruence on $\mathbb{N}_0$. In the factor monoid

$$\mathbb{N}_0/\!\sim_{k,n} \ = \ \{[x]_{\sim_{k,n}} \mid x \in \mathbb{N}_0\},$$

the elements are $[0]_{\sim_{k,n}}, [1]_{\sim_{k,n}}, \ldots, [k+n-1]_{\sim_{k,n}}$. They are pairwise distinct elements. Therefore $\mathbb{N}_0/\!\sim_{k,n}$ is a monoid with exactly $k+n$ elements. Notice that

$$[0]_{\sim_{k,n}} = \{0\},$$
$$[1]_{\sim_{k,n}} = \{1\},$$
$$[2]_{\sim_{k,n}} = \{2\},$$
$$\vdots$$
$$[k-2]_{\sim_{k,n}} = \{k-2\},$$
$$[k-1]_{\sim_{k,n}} = \{k-1\},$$
$$[k]_{\sim_{k,n}} = \{k, k+n, k+2n, k+3n, \ldots\},$$
$$[k+1]_{\sim_{k,n}} = \{k+1, k+1+n, k+1+2n, k+1+3n, \ldots\},$$
$$\vdots$$
$$[k+n-2]_{\sim_{k,n}} = \{k+n-2, k+n-2+n, k+n-2+2n, \ldots\},$$
$$[k+n-1]_{\sim_{k,n}} = \{k+n-1, k+n-1+n, k+n-1+2n, \ldots\}.$$

1.5 Congruences in the Monoid $\mathbb{N}_0$

In the additive monoid $\mathbb{N}_0$, the congruences are exactly the equality $=$ and the congruences $\sim_{k,n}$, where $k \geq 0$ and $n \geq 1$. To see it, notice that if $\equiv$ is a congruence on $\mathbb{N}_0$ different from the equality, then there are natural numbers $a < b$ with $a \equiv b$. Let k be the smallest $a \in \mathbb{N}_0$ such that $a \equiv b$ for some $b \neq a$, let b_0 be the smallest natural number $b > k$ with $k \equiv b$, and set $n := b_0 - k$. The congruence $\sim_{k,n}$ is the principal congruence generated by the relation $(k, k+n)$.

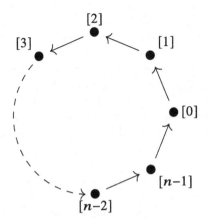

Fig. 1 The cyclic group $\mathbb{Z}/n\mathbb{Z}$

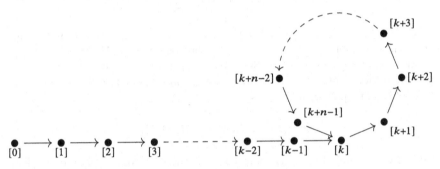

Fig. 2 The cyclic monoid $\mathbb{N}_0/\sim_{k,n}$

The monoid $\mathbb{N}_0$ is cyclic generated by 1. The monoids $\mathbb{N}_0/\sim_{k,n}$ are cyclic generated by $[1]_{\sim_{k,n}}$. Conversely, every cyclic monoid is isomorphic to either $\mathbb{N}_0$ or $\mathbb{N}_0/\sim_{k,n}$ for some $k, n \in \mathbb{N}_0, n \geq 1$.

Recall that finite cyclic groups are isomorphic to $\mathbb{Z}/n\mathbb{Z}$ for some n and that the most natural representation of $\mathbb{Z}/n\mathbb{Z}$ is that in Fig. 1. Finite cyclic *monoids* have a slightly different behavior. The representation of $\mathbb{N}_0/\sim_{k,n}$, analogous to that of $\mathbb{Z}/n\mathbb{Z}$ in Fig. 1, is that in Fig. 2, i.e., $\mathbb{N}_0/\sim_{k,n}$ consists of a cycle of length n with a tail of length k that begins in $[0]_{\sim_{k,n}}$.

1.6 Prime Ideals and Localizations

An *ideal* of a commutative monoid M is a subset I of M such that $x \in I$ and $y \in M$ imply $x + y \in I$. A *prime ideal* of a commutative monoid M is a subset P of M such that $M \setminus P$ is a divisor-closed submonoid of M. That is, P is a proper subset of M and, for any $x, y \in M$, one has $x + y \in P$ if and only if either $x \in P$ or $y \in P$.

The union of any family of prime ideals of a commutative monoid M is a prime ideal, so that the set $\mathrm{Spec}(M)$ of all prime ideals of M, partially ordered by set inclusion, is a complete lattice whose greatest element is the prime ideal $M \setminus U(M)$ and whose least element is the empty ideal $\emptyset$. In particular, a commutative monoid has one prime ideal if and only if M is an abelian group. The spectrum $\mathrm{Spec}(M)$ of a commutative monoid M is a commutative monoid. The monoid structure in $\mathrm{Spec}(M)$ is given by the union $\bigcup$ of prime ideals, and the zero is the empty ideal. The spectrum $\mathrm{Spec}(M)$ is also equipped with a topology, where a basis of open sets is given by the sets

$$D(a) := \{\, P \in \mathrm{Spec}(M) \mid a \notin P \,\}, \quad a \in M.$$

For any monoid morphism $f \colon M \to N$, there is a continuous map

$$f^* \colon \mathrm{Spec}(N) \to \mathrm{Spec}(M), \quad Q \mapsto f^{-1}(Q)$$

(notice that the inverse image of a prime ideal via a monoid morphism is a prime ideal). The operation $\bigcup$ on $\mathrm{Spec}(M)$ and the topology of $\mathrm{Spec}(M)$ are compatible, i.e., the mapping $\bigcup \colon \mathrm{Spec}(M) \times \mathrm{Spec}(M) \to \mathrm{Spec}(M)$ is continuous [16]. Thus $\mathrm{Spec}(M)$ is a topological monoid. For any commutative monoid M, there is a natural isomorphism of topological monoids

$$\mathrm{Spec}(M) \cong \mathrm{Hom}_{\mathrm{CMon}}(M, \{0, 1\})$$

[16]. Here the monoid $\{0, 1\}$ is endowed with multiplication and is a topological monoid with respect to the topology in which the open subsets are $\emptyset$, $\{1\}$ and $\{0, 1\}$. For instance, the topological monoids $\mathrm{Spec}(\mathbb{N})$ and $\{0, 1\}$ are isomorphic. Observe that the set $\mathrm{Hom}_{\mathrm{CMon}}(M, \{0, 1\})$ is contained in $\{0, 1\}^M$. The topology on $\mathrm{Hom}_{\mathrm{CMon}}(M, \{0, 1\})$ is the subspace topology induced by the product topology on $\{0, 1\}^M$.

One has that $\mathrm{Spec}(M) \cong \mathrm{Spec}(M/\asymp)$, where $\asymp$ is the smallest congruence on M for which every element in the quotient monoid $M/\asymp$ is idempotent (see 1.3). More precisely, the canonical morphism $q \colon M \to M/\asymp$ induces an isomorphism of topological monoids $q^* \colon \mathrm{Spec}(M/\asymp) \to \mathrm{Spec}(M)$. See [16].

There is a relation between monoids and semilattices. A *join-semilattice* (or *upper semilattice*) is a partially ordered set in which every nonempty finite subset has a least upper bound. Dually, a *meet-semilattice* (or *lower semilattice*) is a partially ordered set in which every nonempty finite subset has a greatest lower bound. A *semilattice with 1* is a meet-semilattice with a greatest element 1. A morphism of semilattices with 1 is a mapping that respects the greatest lower bound of two elements and the greatest elements 1. If M is a commutative additive monoid in which $2x = x$ for all $x \in M$, that is, every element is *idempotent*, one defines $y \le x$ if $x + y = y$. In this way, M becomes a semilattice with 1. Conversely, if L is a semilattice with 1, then L is a monoid with respect to the operation $\wedge$. The category of monoids satisfying the identity $2x = x$ turns out to be equivalent to the category of semilattices with 1.

There is a notion of tensor product of commutative monoids, and one finds the isomorphism $M/\asymp \cong M \otimes \{0, 1\}$. For all finitely generated semilattices L with 1, there is an isomorphism ev: $L \to \mathrm{Hom}(\mathrm{Hom}(L, \{0, 1\}), \{0, 1\})$, defined by $\mathrm{ev}(x)(f) = f(x)$ for every $x \in L$ and $f \in \mathrm{Hom}(L, \{0, 1\})$.

The localization of a commutative monoid M at a prime ideal P is similar to that of commutative rings. If P is a prime ideal of M, consider the cartesian product $M \times (M \setminus P)$, that is, the set of all pairs (x, s) with $x, s \in M$ and $s \notin P$. Define an equivalence relation $\equiv$ on $M \times (M \setminus P)$ setting $(x, s) \equiv (x', s')$ if there exists an element $t \in M \setminus P$ such that $x + s' + t = x' + s + t$. Let $x - s$ denote the equivalence class of (x, s) modulo the equivalence relation $\equiv$ (notice here that the minus sign in $x - s$ is just suggestive notation). The *localization* M_P of M at P is the monoid whose elements are all $x - s$ with $x \in M$ and $s \in M \setminus P$, and in which the addition is defined by

$$(x - s) + (x' - s') = (x + x') - (s + s').$$

There is a canonical morphism $f: M \to M_P$, defined by $f(x) = x - 0$ for every $x \in M$.

For instance, we have already seen that every monoid M has a unique least prime ideal $\emptyset$ and a unique greatest prime ideal $\overline{P} := M \setminus U(M)$. The localization $M_\emptyset$ of M at its empty prime ideal $\emptyset$ is an abelian group, which is usually called the *Grothendieck group* of M, or the *group of differences*, or the *enveloping group* of M, and denoted by $G(M)$. If M is cancellative, $M_P \subseteq M_\emptyset$ for every prime ideal P of M (more precisely, there is an embedding of monoids $M_P \to M_\emptyset$ for each prime P). The localization $M_{\overline{P}}$ of M at $\overline{P} := M \setminus U(M)$ is isomorphic to M.

Proposition 1.2 *Let M be a commutative monoid and P a prime ideal. For every prime ideal Q of M contained in P, set $Q_P := \{ x - y \in M_P \mid x \in Q, y \in M \setminus P \}$. Then the prime ideals of M_P are in one-to-one correspondence ($Q \leftrightarrow Q_P$) with the prime ideals of M contained in P.*

And now we present an operation that does not have an analogue for commutative rings. For every prime ideal P of a commutative monoid M, the monoid $(M_P)_{\mathrm{red}} = M_P/U(M_P)$ is called the *reduced localization* of M at P. If $x, x' \in M$ and $s, s' \in M \setminus P$, then $x - s + U(M_P) = x' - s' + U(M_P)$ in $(M_P)_{\mathrm{red}}$ if and only if there exist elements $t, t' \in M \setminus P$ such that $x + t = x' + t'$.

For every prime ideal P, there is a canonical morphism $\varphi: M \to (M_P)_{\mathrm{red}}$, defined by $\varphi(x) = x - 0 + U(M_P)$, which is surjective. Its kernel pair is the congruence $\sim_P$ on M defined, for every $x, y \in M$, by $x \sim_P y$ if there exist $z, t \in M \setminus P$ such that $x + z = y + t$. Hence we could have equivalently defined the reduced localization $(M_P)_{\mathrm{red}}$ of a commutative monoid M at a prime ideal P as the factor monoid $M/\sim_P$. For instance, the largest prime ideal of a commutative monoid M is $M \setminus U(M)$, and the smallest one is $\emptyset$. We leave to the reader to show that the reduced localization of M at the prime ideal $M \setminus U(M)$ is M_{red}, and the reduced localization of M at the prime ideal $\emptyset$ is the trivial monoid with one element.

Proposition 1.3 *Let M be a commutative monoid and $\pi : M \to M_{\text{red}} = M/U(M)$, $\pi : x \mapsto x + U(M)$, the canonical projection. Then $\pi^* \colon \operatorname{Spec}(M_{\text{red}}) \to \operatorname{Spec}(M)$ is a homeomorphism.*

The proofs of all these results are easy. Possible references for the results presented here about commutative monoids are [10] and [13].

2 Preordered Groups, Positive Cones

A structure often useful to describe factorizations of elements in an integral domain or direct-sum decompositions in particular classes of modules is the structure of preordered abelian group.

If G is an abelian group, a translation-invariant preorder $\leq$ on G is completely determined by the set of elements $x \in G$ with $x \geq 0$, because for any $x, y \in G$, we have that $x \leq y$ if and only if $y - x \geq 0$. (To see this, notice that $x \leq y$ implies $0 = x + (-x) \leq y + (-x)$, and conversely $y - x \geq 0$, that is, $0 \leq y - x$, implies $x = 0 + x \leq (y - x) + x = y$.) More precisely:

Lemma 2.1 *There is a one-to-one correspondence between the set of all submonoids of an abelian group G and the set of all translation-invariant preorders on G. This correspondence associates with every translation-invariant preorder $\leq$ on G the positive cone $G^+ := \{ x \in G \mid 0 \leq x \}$. Conversely, if M is a submonoid of G, the corresponding preorder $\leq_M$ on G is defined, for every $x, y \in G$, by $x \leq_M y$ if $y - x \in M$.*

A *preordered abelian group* $(G, +, \leq)$ is an abelian group $(G, +)$ with a translation-invariant preorder $\leq$ on G. Equivalently, a preordered abelian group can be defined as a pair (G, C), where G is an abelian group and C is a submonoid of G. Preordered abelian groups form a category in which the morphisms $f \colon (G, +, \leq) \to (H, +, \leq)$ are the group morphisms $f \colon G \to H$ for which $x \leq y$ implies $f(x) \leq f(y)$ for every $x, y \in G$ (equivalently, such that $f(G^+) \subseteq H^+$). For a very nice introduction about preordered abelian groups, a very nice reference is a chapter in the book [12], where most of the proofs about preordered groups we present here are given.

A *partially ordered abelian group* is a preordered abelian group $(G, +, \leq)$ in which $\leq$ is a partial order, that is, the preorder $\leq$ is antisymmetric.

A submonoid of an abelian group G is sometimes called a *cone* in G. A reduced submonoid of an abelian group G is sometimes called a *strict cone* in G. Thus a strict cone is a submonoid C with the property that $x \in C$ and $-x \in C$ imply $x = 0$.

It is easily seen that, for a preordered abelian group $(G, +, \leq)$, $\leq$ is a partial order if and only if the positive cone G^+ of G is a reduced submonoid of G. Thus the one-to-one correspondence of the previous lemma induces a one-to-one correspondence between the set of all reduced submonoids of the abelian group G and the set of all translation-invariant partial orders on G.

In the category of preordered abelian groups there is also a pretorsion theory very similar to the pretorsion theory we met in Sect. 1.2. The torsionfree objects are now the partially ordered abelian groups. The torsion objects are the preordered abelian groups for which the preorder is an equivalence relation, as follows.

Recall that for any preorder $\leq$ on a set S, the *equivalence relation* $\simeq_{\leq}$ *associated with* $\leq$ is defined, for all x, $y \in S$, by $x \simeq_{\leq} y$ if $x \leq y$ and $y \leq x$ (Proposition 1.1). In the case of preordered abelian groups, we have the following.

Proposition 2.2 *Let* G *be a preordered abelian group. Set* $H := \{ x \in G \mid x \leq 0$ *and* $0 \leq x \}$. *Then:*

(a) *H is a subgroup of G.*
(b) *Define a relation* $\preceq$ *on* G/H *by* $x + H \preceq y + H$ *if* $x \leq y$, *for every* x, $y \in G$. *This definition is independent of the choice of the representatives x and y of $x + H$ and $y + H$, that is, the relation $\preceq$ on G/H is well defined.*
(c) *The relation $\preceq$ defined in (b) is a partial order on G/H, and G/H, with this partial order, turns out to be a partially ordered group.*

Conversely, if G is an abelian group, H is a subgroup of G and $\preceq$ is a translation-invariant partial order on G/H, the relation $\leq$ on G, defined by $x \leq y$ if $x + H \preceq y + H$, is a translation-invariant preorder on G. There is a canonical one-to-one correspondence between the set of all translation-invariant preorders on G and the set of all pairs $(H, \preceq)$ with H a subgroup of G and $\preceq$ a translation-invariant partial order on G/H.

Proposition 2.2 is the analogue of Proposition 1.1 for abelian groups. The pretorsion theory is therefore the following. For any preordered abelian group G, the torsionfree quotient of G is G/H with the induced partial order. The torsion subobject is G endowed with the equivalence relation for which two elements g, $g' \in G$ are equivalent if and only if $g - g' \in H$.

Notice that, in Proposition 2.2, if G^+ is the positive cone of the preordered group G, then the positive cone of the corresponding partial group G/H is $G^+/H = (G^+)_{\mathrm{red}}$. Therefore the pretorsion theory on the category of preordered abelian groups corresponds to the pretorsion in the category of commutative monoids CMon, in which the torsion objects are abelian groups, and torsionfree objects are reduced commutative monoids.

For any commutative monoid M, endow the Grothendieck group $G(M)$ of M with the structure of preordered group given by $G(M)^+ = \{[m] \mid m \in M\}$, where $[m]$ is the image of $m \in M$ under the canonical map $\psi_M \colon M \to G(M)$. Every monoid morphism $\varphi \colon M \to N$ induces a morphism of preordered groups $G(\varphi) \colon G(M) \to G(N)$. Hence G is a functor of CMon into the category of preordered abelian groups. For every monoid morphism $\varphi \colon M \to N$, there is a commutative diagram

$$
\begin{array}{ccc}
M & \xrightarrow{\ \varphi\ } & N \\
\psi_M \downarrow & & \downarrow \psi_N \\
G(M) & \xrightarrow{\ G(\varphi)\ } & G(N)
\end{array}
$$

It follows that if F is the forgetful functor of the category of preordered abelian groups into the category CMon that sends a preordered abelian group $(G, +, \leq)$ to the commutative monoid $(G, +)$, then ψ is a natural transformation of the identity functor CMon $\rightarrow$ CMon into the composite functor FG: CMon $\rightarrow$ CMon. The functor FG is the functor "localization at the empty prime ideal $\emptyset$".

For the proofs and related results we refer the reader to the paper [3].

3 Some Set Theory

Some students ask me what the difference is between sets and classes. This will be needed in the sequel. For instance, in Lemma 4.1 we will deal with a monoid $V(\mathscr{C})$ that is large in the sense that it can be a class and not a set. To this end, we need some notions of axiomatic set theory.

3.1 ZFC

The most popular and accepted form of axiomatic set theory is ZFC, the *Zermelo-Fraenkel set theory* with the axiom of choice. It has a single primitive ontological notion, the notion of set. That is, it treats only sets (and not classes): all individuals in the universe of discourse are sets. Sets are denoted with lower case letters. The only binary relations are equality and set membership, denoted by $\in$. Thus the formula $x \in y$ indicates that x and y are sets and that x *belongs to* y (or x *is an element* of y, or x *is a member* of y). We can only use the logical symbols ($\neg, \wedge, \vee, \rightarrow, \leftrightarrow, \forall, \exists$), = (equality), parentheses, lower case letters (variable symbols) and the symbol $\in$. (One must follow the rules studied in any course of mathematical logic to get well-formed formulas!) Here is a list of the axioms of ZFC. Notice that the axioms are formulas, to which we have added some comments for clarity.

1. Axiom of extensionality. Two sets are equal if they have the same elements, that is, a set is determined by its elements:

$$\forall x \forall y (\forall z (z \in x \leftrightarrow z \in y) \Rightarrow x = y).$$

2. Axiom of regularity. Every non-empty set x contains an element y such that x and y are disjoint sets.

$$\forall x [\exists a (a \in x) \Rightarrow \exists y (y \in x \wedge \neg \exists z (z \in y \wedge z \in x))].$$

3. Axiom schema of specification (also called the axiom schema of separation). If z is a set, and ϕ is a property that the elements x of z can have or not have,

then there exists a subset y of z containing the elements x of z which satisfy the property ϕ.

$$\forall z \forall w_1 \ldots w_n \exists y \forall x (x \in y \Leftrightarrow (x \in z \wedge \phi)).$$

Here ϕ is a formula in the language of ZFC in the variables

$$x, y, z, w_1, \ldots, w_n$$

with free variables among $x, z, w_1, \ldots, w_n$ and y not free in ϕ.

4. Axiom of pairing. If x and y are sets, then there exists a set whose elements are exactly x and y.

$$\forall x \forall y \exists z \forall w (w \in z \Leftrightarrow w = x \vee w = y).$$

5. Union axiom. For any set x there is a set whose elements are exactly the elements of the elements of x:

$$\forall x \, \exists y \, \forall z (z \in y \Leftrightarrow \exists w (z \in w \wedge w \in x)).$$

6. Axiom schema of collection. If ϕ is a formula in the language of ZFC with free variables among $x, y, z, w_1, \ldots, w_n$ and with a non-free variable w, then

$$\forall z \, \forall w_1, \ldots, w_n ((\forall x \in z \exists! y \phi) \Rightarrow \exists w \forall x \in z \exists y \in w \, \phi).$$

Here $\exists! y$ means "there exists a unique y such that...". The axiom essentially says that if $f : z \to z'$ is a function, then the image of f is set. A function $f : z \to z'$ is a triple (f, z, z') of sets, where $f \subseteq z \times z'$ and for every $x \in z$ there exists a unique $y \in z'$ with $(x, y) \in f$.

7. Axiom of infinity. The axiom essentially states that there exists a set with infinitely many members.

$$\exists x \, (\emptyset \in x \wedge \forall y (y \in x \Rightarrow y \cup \{y\} \in x)).$$

8. Axiom of power set. For any set x, there is a set y whose elements are exactly the subsets of x.

$$\forall x \exists y \forall z (z \in y \Leftrightarrow (\forall q (q \in z \Rightarrow q \in x))).$$

9. Axiom of choice. For any set X, every equivalence relation on X has a set of representatives.

The axiom of choice AC is independent from the other axioms of ZFC, and ZFC is independent from the continuum hypothesis $2^{\aleph_0} = \aleph_1$.

3.2 Grothendieck's Universes

The idea is: fix a set, which we call a *universe*, big enough because we put in it all what we need, but not too big because we want it to be a set and not a class. The formal definition is the following:

A *universe* is a set U satisfying the following properties:

(a) $X \in Y \in U \to X \in U$.
(b) $X, Y \in U \to \{X, Y\} \in U$.
(c) $X, Y \in U \to X \times Y \in U$.
(d) $X \in U \to \mathcal{P}(X) \in U$.
(e) $X \in U \to \bigcup_{Y \in X} Y \in U$.
(f) The set ω of natural numbers is an element of U.
(g) If $X \in U$ and $f : X \to U$ is a mapping, then $\{ f(Y) \mid Y \in X \} \in U$.

Important: the axioms of ZFC do not guarantee the existence of a universe. Following Grothendieck, we adjoin a further axiom to the axioms of ZFC:

Axiom of Universes: Every set is a member of a universe.

Given any universe U, if the axioms of ZFC are satisfied by the class of all sets with the relation $\in$, then they are also satisfied by the set of all sets belonging to U with the relation $\in$ between them. Hence we can argue remaining in the universe U, which we suppose fixed once for all. In the universe, we find all what we need, and if we do not find it, we can always adjoin it to the universe thanks to the Axiom of universes. In other words, we decide to work in a set that we possibly expand.

But the problem remains: it is not possible to deal with the category Set of all sets in our universe, in this universe in expansion.

3.3 NBG

For the notion of class, we must introduce NBG.

The *Von Neumann-Bernays-Gödel set theory* (NBG) is a conservative extension of ZFC. The ontology of NBG includes proper classes. The members of both sets and proper classes are sets. Classes cannot be members. "Conservative extension" means that a statement in the language of ZFC is provable in NBG if and only if it is provable in ZFC, that is, any theorem in NBG which speaks only about sets is a theorem in ZFC. In NBG, quantified variables in the defining formula can range only over sets.

Let us try to be more precise. The characteristic of NBG is the distinction between proper classes and sets. NBG is a two-sorted theory, that is, two types of variables are used in NBG. Lower case letters will denote variables ranging over sets, and upper case letters will denote variables ranging over classes. The atomic sentences $a \in b$ and $a \in A$ are defined for a, b sets and A a class, but $A \in a$ or $A \in B$ are not defined for any two classes A, B. Equality can have the form $a = b$ or $A = B$. $a = A$ stands for $\forall x(x \in a \leftrightarrow x \in A)$ and is an abuse of notation.

NBG can also be presented as a one-sorted theory of classes, with sets being those classes that are members of at least one other class. That is, NBG can be presented as a system having only one type of variables (class variables) with a unary relation $\mathcal{M}(A)$ ($\mathcal{M}$ stands for the German word Menge, set), and $\mathcal{M}(A)$ indicates that A is a set. Thus $\mathcal{M}(A) \leftrightarrow \exists B(A \in B)$. Notice that NBG admits the class V of all sets, but it does not admit the class of all classes or the set of all sets.

Here is a list of the axioms of NBG. Notice that the first five ones coincide with five axioms of ZFC and deal only with sets, not classes.

1. Axiom of extensionality. Two sets are equal if they have the same elements:

$$\forall a \forall b (\forall z (z \in a \Leftrightarrow z \in b) \Rightarrow a = b).$$

2. Axiom of pairing. If x and y are sets, then there exists a set whose elements are exactly x and y.
3. Union axiom. For any set x there is a set whose elements are exactly the elements of the elements of x.
4. Axiom of power set. For any set x, there is a set y whose elements are exactly the subsets of x.
5. Axiom of infinity.

$$\exists x \left(\emptyset \in x \wedge \forall y (y \in x \Rightarrow y \cup \{y\} \in x) \right).$$

The remaining axioms are primarily concerned with classes rather than sets.
6. Axiom of extensionality for classes. Two classes are equal if they have the same elements:
$$A = B \Leftrightarrow \forall x (x \in A \leftrightarrow x \in B).$$

7. Axiom of regularity for classes. Every non-empty class A contains an element disjoint from A.

$$\exists x(x \in A) \Rightarrow \exists y(y \in A \wedge \neg \exists z(z \in y \wedge z \in A)).$$

Finally, the last two axioms are particular to NBG:
8. Axiom of limitation of size: For any class A, there exists a set a such that $a = A$ if and only if there is no bijection between A and the class V of all sets.

This is really a powerful axiom. By this axiom, every proper class is equipotent to the class V of all sets. Moreover, the axiom of choice for classes holds, because the class of ordinals is not a set, so that there is a bijection between the ordinals and any proper class, and any class can be well ordered. Equivalently, if A is any class and $\sim$ is an equivalence relation on A, a class of representatives exists.

9. Class comprehension schema: For any formula ϕ containing no quantifiers over classes (it may contain class and set parameters), there exists a class A such that $\forall x(x \in A \leftrightarrow \phi(x))$.

It can be proved that NBG can be finitely axiomatized. What is important for us, is that in NBG, which is a conservative extension of ZFC, we can deal with classes and have the axiom of choice for classes. Thus every category has a skeleton, we have a class of representatives for any equivalence relation on any class, and we can define an equivalence between two categories either as a functor with a quasi-inverse or as a fully faithful essentially surjective functor.

4 The Monoid $V(\mathscr{C})$, Discrete Valuations, Krull Monoids

4.1 The Monoid $V(\mathscr{C})$

We will denote by $\mathrm{Ob}\,\mathscr{C}$ the class of objects of any category $\mathscr{C}$. Recall that a *terminal object* in a category $\mathscr{C}$ is an object T of $\mathscr{C}$ with the property that, for every $A \in \mathrm{Ob}(\mathscr{C})$, there is a unique morphism $A \to T$ in $\mathscr{C}$. Similarly, I is called an *initial object* of $\mathscr{C}$ if for every $A \in \mathrm{Ob}(\mathscr{C})$ there is exactly one morphism $I \to A$. Finally, an object Z of $\mathscr{C}$ is called a *null object* (or a *zero object*) if it is both initial and terminal. Thus an object I is initial if and only if $\mathrm{Hom}_{\mathscr{C}}(I, A)$ has cardinality 1 for every object A, and T is terminal if and only if $\mathrm{Hom}_{\mathscr{C}}(A, T)$ has cardinality 1 for every A. Obviously, an object is an initial object in a category $\mathscr{C}$ if and only if it is a terminal object in the dual category $\mathscr{C}^{\mathrm{op}}$.

Let $\mathscr{C}$ be a category and let 0 be a zero object of $\mathscr{C}$. Then there exist exactly one morphism $A \to 0$ and exactly one morphism $0 \to B$ for every pair A, B of objects. Their composite morphism $A \to B$ is called the *zero morphism* of A into B. In fact, it is easily seen that, in a category $\mathscr{C}$ with a zero object, there is a unique zero morphism $A \to B$ for every pair A, B of objects of $\mathscr{C}$. (One must prove that if 0, $0'$ are two zero objects, then the composite morphism $A \to 0 \to B$ is equal to the composite morphism $A \to 0' \to B$.)

Let $\mathscr{C}$ be a category. For every object A of $\mathscr{C}$, let $\mathrm{Iso}(A)$ denote the *isomorphism class* of A, that is, the class of all objects of $\mathscr{C}$ isomorphic to A. The class $\mathrm{Iso}(A)$ is a subclass of the class $\mathrm{Ob}(\mathscr{C})$ of all objects of $\mathscr{C}$, and the isomorphism classes $\mathrm{Iso}(A)$ form a partition of $\mathrm{Ob}(\mathscr{C})$. Let $V(\mathscr{C})$ denote a *skeleton* of $\mathscr{C}$, that is, a class of representatives of the objects of $\mathscr{C}$ modulo isomorphism. Notice that $V(\mathscr{C})$ exists by the axiom of choice for classes (see Sect. 3.3, Axiom 8). For every object A in $\mathscr{C}$, there is a unique object $\langle A \rangle$ in $V(\mathscr{C})$ isomorphic to A. Thus there is a mapping $\mathscr{C} \to V(\mathscr{C})$, $A \mapsto \langle A \rangle$, that associates with every object A of $\mathscr{C}$ the unique object $\langle A \rangle$ in $V(\mathscr{C})$ isomorphic to A. Assume that a product $A \times B$ exists in $\mathscr{C}$ for every pair A, B of objects of $\mathscr{C}$. Define an addition $+$ in $V(\mathscr{C})$ by $A + B := \langle A \times B \rangle$ for every A, $B \in V(\mathscr{C})$. In this way we get a monoid that is *large*, in the sense that it is a class and not a set when the category $\mathscr{C}$ is not skeletally small:

Lemma 4.1 *Let $\mathscr{C}$ be a category with a terminal object and in which a product $A \times B$ exists for every pair A, B of objects of $\mathscr{C}$. Then $V(\mathscr{C})$ is a large reduced commutative monoid.*

Notice that if $\mathscr{C}$ is an arbitrary category, so that the product $A \times B$ does not necessarily exist for any pair A, B of objects of $\mathscr{C}$, then the skeleton $V(\mathscr{C})$ turns out to be a class in which the operation induced by product is only partially defined, that is, it is a mapping $+: S \to V(\mathscr{C})$ for a subclass S of $V(\mathscr{C}) \times V(\mathscr{C})$.

4.2 Discrete Valuations, Krull Monoids

Let M be a monoid. A *discrete valuation* on a monoid M is a non-zero monoid morphism $v: M \to \mathbb{N}_0$. Here $\mathbb{N}_0$ is the additive monoid of nonnegative integers. Every discrete valuation $M \to \mathbb{N}_0$ induces a non-zero group morphism $G(M) \to \mathbb{Z}$ that maps $\psi_M(M)$ into $\mathbb{N}_0$. Here $\psi_M: M \to G(M)$ is the canonical map that sends each $x \in M$ to $x - 0$. Conversely, every non-zero group morphism $f: G(M) \to \mathbb{Z}$ with $f(\psi_M(M)) \subseteq \mathbb{N}_0$ induces a discrete valuation $M \to \mathbb{N}_0$. Thus discrete valuations can be also seen as those non-zero group morphisms $G(M) \to \mathbb{Z}$ that map $\psi_M(M)$ into $\mathbb{N}_0$, i.e., non-zero morphisms of preordered groups, where $G(M)$ is the preordered group whose positive cone is the image $\psi_M(M)$ of M in $G(M)$, and $\mathbb{Z}$ is endowed with its usual linear order.

A monoid morphism $f: M \to M'$ is called a *divisor morphism* if, for every $x, y \in M$, $f(x) \leq f(y)$ implies $x \leq y$. Here $\leq$ denotes the algebraic preorder. A monoid M is a *Krull monoid* if there exists a divisor morphism of M into a free commutative monoid. Equivalently, a monoid M is a Krull monoid if and only if there exists a set $\{\, v_i \mid i \in I \,\}$ of monoid morphisms $v_i: M \to \mathbb{N}_0$ such that: (1) if $x, y \in M$ and $v_i(x) \leq v_i(y)$ for every $i \in I$, then $x \leq y$; (2) for every $x \in M$, the set $\{\, i \in I \mid v_i(x) \neq 0 \,\}$ is finite.

Our main application of Krull monoids will be to the *reduced* monoid $V(\mathscr{C})$. We leave to the reader the proof of the following elementary Lemma.

Lemma 4.2 *A commutative monoid M is a Krull monoid if and only if the reduced monoid M_{red} is a Krull monoid.*

Reduced Krull monoids are characterized among Krull monoids in the next elementary Lemma.

Lemma 4.3 *Let $f: M \to F$ be a divisor morphism of a commutative monoid M into a free commutative monoid F. The following conditions are equivalent:*

(a) *The monoid M is reduced and cancellative.*
(b) *The monoid M is reduced.*
(c) *The morphism f is injective.*

Proposition 4.4 *Let M be an additive, cancellative, commutative monoid with Grothendieck group $G(M)$. The following conditions are equivalent:*

(a) *M is a Krull monoid.*
(b) *There exists a set* $\{\, v_i \mid i \in I \,\}$ *of non-zero group morphisms*

$$v_i : G(M) \to \mathbb{Z}$$

  such that: (1) $M = \{\, x \in G(M) \mid v_i(x) \geq 0 \text{ for every } i \in I \,\}$; *and* (2) *for every* $x \in G(M)$, *the set* $\{\, i \in I \mid v_i(x) \neq 0 \,\}$ *is finite.*
(c) *There exist an abelian group* G, *a set* I *and a subgroup* H *of the free abelian group* $\mathbb{Z}^{(I)}$ *such that* $M \cong G \oplus (H \cap \mathbb{N}_0^{(I)})$.

For the proofs, see [8] and [11].

5 Modules

We will always suppose in these notes that our rings R are associative rings with an identity 1_R (unless explicitly stated, like in the next paragraph). Ring morphisms are assumed to preserve identities.

5.1 Left Modules

Let R be a ring. It is possible to define left modules over the ring R in two equivalent ways. For every abelian group M, we denote by $\mathrm{End}(M)$ the endomorphism ring of M.

Definition 5.1 A *left R-module* (or *left module over the ring R*) is a triple $(M, +, \cdot)$, where $(M, +)$ is an additive abelian group and $\cdot : R \times M \to M$, $\cdot : (r, m) \mapsto rm$, is a mapping, called *left scalar multiplication*, with the following properties for every $r, r' \in R$, and every $m, m' \in M$:

(i) $r(r'm) = (rr')m$;
(ii) $(r + r')m = rm + r'm$;
(iii) $r(m + m') = rm + rm'$;
(iv) $1_R m = m$.

Definition 5.2 A *left R-module* (or *left module over the ring R*) is a triple $(M, +, \lambda)$, where $(M, +)$ is an additive abelian group and $\lambda : R \to \mathrm{End}(M)$ is a ring morphism of R into the ring $\mathrm{End}(M)$ of all endomorphisms of the abelian group $(M, +)$.

These two definitions are equivalent in the following sense. Assume that $(M, +, \cdot)$ is a module defined as in Definition 5.1. Let $\lambda : R \to \mathrm{End}(M)$ be the mapping defined by $\lambda(r)(m) = rm$ for every $r \in R$, $m \in M$. Then λ is a ring morphism of R into the ring $\mathrm{End}(M)$ of all endomorphisms of the abelian group $(M, +)$.

To see it, we must check four conditions: that $\lambda(r) \in \text{End}(M)$ for every $r \in R$, $\lambda(r + r') = \lambda(r) + \lambda(r')$, $\lambda(rr') = \lambda(r)\lambda(r')$, $\lambda(1_R) = 1_{\text{End}(M)}$. These four conditions follow from properties (iii), (ii), (i), (iv) of Definition 5.1 respectively. Thus $(M, +, \lambda)$ becomes a left module as defined in Definition 5.2.

Conversely, let $(M, +, \lambda)$ be a module as in Definition 5.2. Define a scalar multiplication $\cdot : R \times M \to M$ setting $\cdot : (r, m) \mapsto rm := \lambda(r)(m)$ for every $r \in R, m \in M$. Then from the fact that λ maps R into $\text{End}(M)$ and respects addition, multiplication and the identity, we get the four properties (iii), (ii), (i), (iv) of Definition 5.1 respectively, that is, $(M, +, \cdot)$ is a left module in the sense of Definition 5.1.

Thus the two definitions of a left module are logically equivalent, and we will use both, depending on the convenience.

Definition 5.3 Let R be a ring and let M, N be left R-modules. A *module morphism* (or *module homomorphism*) of M into N is a mapping $f : M \to N$ such that, for every $x, y \in M$ and every $r \in R$, $f(x + y) = f(x) + f(y)$ and $f(rx) = rf(x)$.

We can be very precise and describe the logical equivalence of the two definitions 5.1 and 5.2 of left R-modules in categorical terms. Define a category $R\text{-Mod}_1$ in which: the objects are all modules $(M, +, \cdot)$ defined as in Definition 5.1; the morphisms $f : (M, +, \cdot) \to (M', +, \cdot)$ in $R\text{-Mod}_1$ are the module morphisms as defined in Definition 5.3. Composition in $R\text{-Mod}_1$ is the composition of mappings. Similarly, we can define another category $R\text{-Mod}_2$ whose objects are all modules $(M, +, \lambda)$ defined as in Definition 5.2. A morphism $f : (M, +, \lambda_M) \to (M', +, \lambda'_{M'})$ in $R\text{-Mod}_2$ is a group morphism $f : (M, +) \to (M', +)$ such that the diagram

$$
\begin{array}{ccc}
M & \xrightarrow{\ f\ } & M' \\
{\scriptstyle \lambda_M(r)}\downarrow & & \downarrow{\scriptstyle \lambda_{M'}(r)} \\
M & \xrightarrow[\ f\ ]{} & M'
\end{array}
$$

is commutative for every $r \in M$, that is, such that $f \circ \lambda_M(r) = \lambda_{M'}(r) \circ f$ for every $r \in R$. Composition in $R\text{-Mod}_2$ is the composition of mappings. Then the assignment $(M, +, \cdot) \mapsto (M, +, \lambda)$ can be extended to a functor $F : R\text{-Mod}_1 \to R\text{-Mod}_2$, and the assignment $(M, +, \lambda) \mapsto (M, +, \cdot)$ can be extended to a functor $G : R\text{-Mod}_2 \to R\text{-Mod}_1$. These two functors F and G are one the inverse of the other, so that the categories $R\text{-Mod}_1$ and $R\text{-Mod}_2$ turn out to be isomorphic.

When R is a division ring D, left D-modules are usually called *left vector spaces* over the division ring D.

5.2 Right Modules

Let us pass to define *right* modules. The definition is similar to that of left modules, but the scalars act on the right instead of on the left.

Definition 5.4 A *right R-module* (or *right module over the ring R*) is a triple $(M, +, \cdot)$, where $(M, +)$ is an additive abelian group and $\cdot \colon M \times R \to M$, $\cdot \colon (m, r) \mapsto mr$, is a mapping, called *right scalar multiplication*, with the following properties for every $r, r' \in R$, and every $m, m' \in M$:

(i) $(mr)r' = m(rr')$;
(ii) $m(r + r') = mr + mr'$;
(iii) $(m + m')r = mr + m'r$;
(iv) $m1_R = m$.

For a second equivalent definition, analogous to that of Definition 5.2, we need the notion of ring anti-homomorphism.

Definition 5.5 Let R and S be rings. A *ring anti-homomorphism* $f \colon R \to S$ is a mapping of the set R into the set S such that:

(i) $f(r + r') = f(r) + f(r')$ for every $r, r' \in R$;
(ii) $f(rr') = f(r')f(r)$ for every $r, r' \in R$;
(iii) $f(1_R) = 1_S$.

Example 5.6 Let k be a field, n be a positive integer, and $\mathbb{M}_n(k)$ be the ring of $n \times n$ matrices with entries in k. The *transposition* $t \colon \mathbb{M}_n(k) \to \mathbb{M}_n(k)$ defined by $A \mapsto A^t$ (where A^t is the transpose of A) is a ring *anti-isomorphism*, that is, a ring anti-homomorphism that is also a bijective mapping.

Example 5.7 If $(R, +, \cdot)$ is a ring, its *opposite ring* is the ring $(R, +, \circ)$, where $\circ \colon R \times R \to R$ is a new operation on the set R defined by $r \circ r' = r' \cdot r$. Usually, if R is a ring, its opposite ring is denoted by R^{op}. It is easily see that R^{op} is a ring for every ring R. Then the identity mapping $\iota_R \colon R \to R$, defined by $r \in R \mapsto r$, viewed as a mapping $R \to R^{\mathrm{op}}$, is an anti-isomorphism of R onto R^{op}.

Definition 5.8 Let R be a ring. A *right R-module* $(M, +, \rho)$ is an abelian group $(M, +)$ together with a ring anti-homomorphism $\rho \colon R \to \mathrm{End}(M)$.

For right modules it is also easy to see that the two Definitons 5.4 and 5.8 give the same structures, or, if we want to be more precise, that the two corresponding categories are isomorphic. Both for right modules and for left modules we will not distinguish between the two possible definitions. We will consider the category R-Mod of all left R-modules and we will use both definitions with left scalar multiplication or with the ring morphism $R \to \mathrm{End}(M)$. Similarly, on the other side, we will consider the category Mod-R of all right R-modules and we will use both definitions with right scalar multiplication or the ring anti-homomorphism $R \to \mathrm{End}(M)$, as it is more convenient.

Remark 5.9 It is clear that ring anti-homomorphisms

$$R \to \mathrm{End}(M)$$

and ring homomorphisms

$$R^{\mathrm{op}} \to \mathrm{End}(M)$$

coincide. Therefore right R-modules and left R^{op}-modules are exactly the same thing. Similarly, left R-modules coincide with right R^{op}-modules. Also, notice that if the ring R is commutative, a mapping

$$R \to \mathrm{End}(M)$$

is a ring homomorphism if and only if it is a ring anti-homomorphism. It follows that right modules and left modules coincide over a commutative ring R. If we want to be more precise, we can use the categorical language, and say that there is an isomorphism of categories between the category of all right R-modules and the category of all left R^{op}-modules. Similarly, for R commutative, the category of all right R-modules and the category of all left R-modules are isomorphic.

If M is a right R-module, we will usually denote it by M_R, and if M is a left R-module, we will denote it by $_R M$. That is, we will write the ring R of "scalars" on the side on which it acts.

If $f: M_R \to N_R$ is a module morphism, then f is a monomorphism in the category Mod-R if and only if f is injective, it is an epimorphism if and only if it is surjective, and it is an isomorphism if and only if it is bijective.

5.3   Abelian Groups = $\mathbb{Z}$-modules

For any ring R, there is a unique ring morphism $\mathbb{Z} \to R$, that is, $\mathbb{Z}$ is an initial object in the category of rings.

In particular, let $(G, +)$ be a non-zero abelian group and $\mathrm{End}(G)$ its endomorphism ring. As we have just said, there is a unique ring homomorphism $\lambda: \mathbb{Z} \longrightarrow \mathrm{End}(G)$. Equivalently, there is a unique left $\mathbb{Z}$-module structure on any abelian group G. The scalar multiplication $\cdot: \mathbb{Z} \times M \to M$ of M is given by $nx =$ "n-th multiple of x in the additive group M" for every $n \in \mathbb{Z}$, $x \in M$. That is,

$$nx = \begin{cases} \underbrace{x + \cdots + x}_{n \text{ times}} & \text{if } n > 0 \\ 0_M & \text{if } n = 0 \\ \underbrace{(-x) + (-x) + \cdots + (-x)}_{-n \text{ times}} & \text{if } n < 0 \end{cases}$$

Thus, left $\mathbb{Z}$-modules and abelian groups coincide. If we want to be more precise, the category Ab of all abelian groups is isomorphic to the category $\mathbb{Z}$-Mod of all left $\mathbb{Z}$-modules, an isomorphism $F: \mathbb{Z}$-Mod $\to$ Ab being the forgetful functor F.

5.4 Is Left Better Than Right?

The definition of left R-modules, which correspond to ring homomorphisms, seems more natural than that of right R-modules, corresponding to the less natural notion of ring anti-homomorphism. The reason of this lies in the fact that we are used to write mappings on the left, and not on the right. To be more precise, let A and B be sets and assume that we have a mapping $f : A \to B$. Then we use to denote the image of an element $a \in A$ by $f(a)$. Also, if $f : A \to B$ and $g : B \to C$ are two mappings, we denote their composite mapping by $g \circ f$, which is the mapping that sends an element $a \in A$ to $(g \circ f)(a) = g(f(a))$. The choice of this notation was arbitrary, and we could write mappings on the right. For a mapping $f : A \to B$, it is possible to denote the image of an element $a \in A$ by $(a)f$, with the mapping f on the right of the elements a on which f acts. In this case, if $f : A \to B$ and $g : B \to C$ are two mappings, it is more natural to denote the composite mapping by $f \circ g$, because it sends the element $a \in A$ to $((a)f)g$. Notice that, in some settings, mappings *are* denoted on the right. For instance, in group theory, it is common to denote an action g, for instance, conjugation, on an element a in the form a^g. Here g is written as an exponent, that is, on the right of the elements a on which it acts.

If, for any reason, we write mappings on the right, then right R-modules correspond to ring homomorphisms $R \to \mathrm{End}(M)$, and left R-modules correspond to ring anti-homomorphisms of R into $\mathrm{End}(M)$. If, in the ring $\mathrm{End}(M)$ of all endomorphisms of an abelian group M, we write endomorphisms on the right, then the ring of all endomorphisms of M with endomorphisms written on the right is $\mathrm{End}(M)^{\mathrm{op}}$.

From now on, we will always write, as usual, mappings on the left, and most modules we will consider will be right modules M_R.

5.5 Two Exercises

(1) Let M be a right R-module, $x, y \in M$, $r, s \in R$. Show that:

    (i) $0_M r = 0_M$.
    (ii) $x 0_R = 0_M$.
    (iii) $(-x)r = -(xr)$ and $x(-r) = -(xr)$. We will denote the element $(-x)r = x(-r) = -(xr)$ by $-xr$.
    (iv) $(x - y)r = xr - yr$ and $x(r - s) = xr - xs$.

[Recall that in any additive group G, one writes $a - b$ to denote the sum of a and the opposite of b. That is, $a - b := a + (-b)$. Notice that in this exercise we only use properties (i), (ii) and (iii) of Definition 5.4.]

(2) Let R be a ring with identity 1_R, $(M, +)$ an additive abelian group and $R \times M \to M$, $(r, m) \mapsto rm$, a mapping that satisfies properties (i), (ii) and (iii) of Definition 5.1. Let M_0 be the set of all $x \in M$ with $1_R x = 0_M$, and M_1 be the set of all $x \in M$ with $1_R x = x$. Show that:

(i) M_0 and M_1 are subgroups of M.
(ii) M is the direct sum of M_0 and M_1 as abelian groups.
(iii) $rx = 0_M$ for every $r \in R$ and $x \in M_0$.
(iv) M_1 is a left R-module with respect to the left scalar multiplication induced by the left scalar multiplication on M.

Sometimes "modules" are defined as the algebraic structures satisfying properties (i), (ii) and (iii) of Definition 5.1, and those satisfying property (iv) are called "unitary modules". Thus every "non-unitary module" M is the direct sum of a "module" M_0 on which R acts trivially and a "unitary module" M_1. A non-unitary left R-module M can be also described as an abelian group M with a "ring morphism" $R \to \mathrm{End}(M)$ that does not necessarily map 1_R to $1_{\mathrm{End}(M)}$.

6 Representations/Modules/Actions of Other Algebraic Structures

In this section I will present my personal point of view on the organization of algebraic structures and their representations (modules).

6.1 k-algebras

Let k be a commutative ring with identity. A (not necessarily associative) k-algebra is any unitary k-module M with a k-bilinear mapping $(x, y) \mapsto xy$ of $M \times M$ into M (equivalently, a k-linear mapping $M \otimes_K M \to M$). Thus all algebra axioms are satisfied except at most for associativity of multiplication. Here we are following Bourbaki's terminology [4]. The content of this part of these notes is essentially taken from [1, Section 2]. It is possible to construct the *opposite* M^{op} of any such algebra M by defining multiplication in M^{op} via $(x, y) \mapsto yx$.

If M, M' are k-algebras, a k-algebra morphism $\varphi \colon M \to M'$ is any k-linear mapping such that $\varphi(xy) = \varphi(x)\varphi(y)$ for every $x, y \in M$. A *derivation* of a k-algebra M is any k-linear mapping $D \colon M \to M$ such that $D(xy) = (D(x))y + x(D(y))$ for every $x, y \in M$.

If M is any k-algebra, its endomorphisms form a (not necessarily commutative) monoid, that is, a semigroup with a two-sided identity, with respect to composition of mappings $\circ$, and its derivations form a Lie k-algebra $\mathrm{Der}(M)$ with respect to the operation $[D, D'] = D \circ D' - D' \circ D$ for every $D, D' \in \mathrm{Der}(M)$. The definition of Lie k-algebra will be given at the beginning of Sect. 6.2.

The main example of associative k-algebra is, for any k-module A_k, the *endomorphism ring* $\mathrm{End}(A_k)$ of A_k. If M is any (not necessarily associative) k-algebra and $x \in M$, the mapping $\lambda_x \colon M_k \to M_k$, defined by $\lambda_x(y) = xy$ for every $y \in M$, is an element of the associative ring $\mathrm{End}(M_k)$.

For any (not necessarily associative) k-algebra M, there is a canonical mapping λ of M into the associative k-algebra $\mathrm{End}_k(M)$, defined by $\lambda \colon x \mapsto \lambda_x$ for every $x \in M$. This mapping λ is a k-algebra morphism if and only if M is associative.

Thus, for any k-algebra M, it is natural to define a left M-module as we did in Sect. 5, that is, as a k-module A_k with a k-algebra morphism $\lambda \colon M \to \mathrm{End}(A_k)$. In fact, consider the natural isomorphism

$$\mathrm{Hom}_k(X_k \otimes_k Y_k, Z_k) \cong \mathrm{Hom}_k(X_k, \mathrm{Hom}_k(Y_k, Z_k)) \qquad X_k, Y_k, Z_k \ k\text{-modules.}$$

For a fixed k-algebra M, any k-bilinear mapping $\mu \colon M_k \times A_k \to A_k$ (any *left scalar multiplication*) can be equivalently described by a k-algebra morphism $\lambda \colon M \to \mathrm{End}(A_k)$, where $\mathrm{End}(A_k)$ is the k-algebra of all endomorphisms of the k-module A_k.

Similarly, we can define *right M-modules* as k-modules A_k with a k-algebra anti-homomorphism $\rho \colon M \to \mathrm{End}(A_k)$. Again, a mapping $M \to M'$ is a k-algebra anti-homomorphism if and only if it is a k-algebra morphism $M^{\mathrm{op}} \to M'$. It follows that right M-modules coincide with left M^{op}-modules. Similarly, left M-modules coincide with right M^{op}-modules.

If the k-algebra M is commutative, then a mapping $M \to M'$ is a k-algebra anti-homomorphism if and only if it is a k-algebra homomorphism $M \to M'$, so right M-modules coincide with left M-modules whenever M is commutative.

6.2 Lie k-algebras

Let k be a commutative ring with identity. A *Lie k-algebra L* is a k-algebra for which, denoting the k-bilinear mapping of $L \times L$ into L by $(x, y) \mapsto [x, y]$, one has:

(1) *(Alternativity)* $[x, x] = 0$ for every $x \in L$.
(2) *(Jacobi identity)* $[x, [y, z]] + [y, [z, x]] + [z, [x, y]] = 0$ for every $x \in L$.

The main example of Lie k-algebra is, for any k-algebra M, the *Lie k-algebra of derivations* $\mathrm{Der}_k(M)$ of the k-algebra M. If M is any k-algebra and D, D' are two derivations of M, then the composite mapping DD' is not a derivation of M in general, but $DD' - D'D$ is, as we have already remarked in Sect. 7.1. Thus, for any k-algebra M, we can define the Lie k-algebra $\mathrm{Der}_k(M)$ as the subset of $\mathrm{End}_k(M)$ consisting of all derivations of M with multiplication $[D, D'] := DD' - D'D$ for every $D, D' \in \mathrm{Der}_k(M)$.

A well known second example of Lie k-algebra is the following. Let L be any associative k-algebra. Define $[x, y] := xy - yx$ for every $x, y \in L$. This is a k-bilinear mapping $L \times L \to L$, and L, with this multiplication, turns out to be a Lie k-algebra.

As a third example, let A be any k-module and L the associative k-algebra $L := \mathrm{End}_k(A)$ of all k-endomorphisms of A. Then L with the operation $[-, -]$ defined as in the previous paragraph, is a Lie k-algebra, denoted by $\mathfrak{gl}(A)$.

For any Lie k-algebra M and any element $x \in M$, the mapping $\lambda_x \colon M \to M$, defined by $\lambda_x = [x, -]$, is a derivation of the Lie algebra M, that is, it is an element of the Lie k-algebra $\mathrm{Der}_k(M)$, usually called the *adjoint* of x, or the *inner derivation* defined by x, and usually denoted by $\mathrm{ad}_M x$ instead of λ_x.

For every Lie k-algebra M, there is a canonical Lie k-algebra morphism $\mathrm{ad}\colon M \to \mathrm{Der}_k(M)$, defined by $\mathrm{ad}\colon x \mapsto \mathrm{ad}_M x$ for every $x \in M$.

It is possible to define *left M-modules* for any Lie k-algebra M. Let M be any Lie k-algebra. A *left M-module* is a k-module A with a Lie k-algebra morphism $\lambda\colon M \to \mathfrak{gl}(A)$. Similarly, we can define *right M-modules* as k-modules A with a k-algebra anti-homomorphism $\rho\colon M \to \mathfrak{gl}(A)$. But any Lie k-algebra M is isomorphic to its opposite algebra M^{op} via the isomorphism $M \to M^{\mathrm{op}}, x \mapsto -x$. It follows that the category of all right M-modules is canonically isomorphic to the category of all left M-modules for any Lie k-algebra M. Therefore it is useless to introduce both right and left modules, it is sufficient to introduce left M-modules and call them simply "M-modules". This cannot be done for associative k-algebras, because for an associative k-algebra M the structure of right M-modules can be very different from that of its left M-modules. For instance, it is easy to construct examples of associative k-algebras that are right noetherian, but not left noetherian, e. g. the $\mathbb{Z}$-algebra of triangular 2×2-matrices $\begin{pmatrix} \mathbb{Q} & 0 \\ \mathbb{Q} & \mathbb{Z} \end{pmatrix}$. Over such an associative k-algebra, the structure of the category of right modules is very different from that of left modules.

6.3 Monoids

In Sect. 1 we have considered commutative additive monoids. In this Subsect. 6.3, we will consider *multiplicative* monoids, *not necessarily commutative*. Thus a *monoid* will be a semigroup with a two-sided identity, that is, an element $1_M \in M$ such that $1_M x = x 1_M = x$ for every $x \in M$. The main example of monoid is, for any set X, the monoid X^X of all mappings $X \to X$. In this monoid, multiplication is composition of mappings. If M is any monoid and $x \in M$, we have the mapping $\lambda_x\colon M \to M$, that is, a morphism in the category of sets, defined by $\lambda_x(y) = xy$ for every $y \in M$. This λ_x is an element of the monoid M^M. We have a canonical injective monoid morphism $\lambda\colon M \to M^M, \lambda\colon x \mapsto \lambda_x$.

Correspondingly, we can define "left M-modules", now called *left M-sets*, for any monoid M. A *left M-set* is any set X with a monoid morphism $\lambda\colon M \to X^X$. Similarly, we can define *right M-sets* as sets X with a monoid anti-homomorphism $\rho\colon M \to X^X$. Again, right M-sets coincide with left M^{op}-sets, and left M-sets coincide with right M^{op}-sets. If the monoid M is commutative, right M-sets coincide with left M-sets.

The concept of monoid is somehow pervasive in Category Theory, essentially because composition of morphisms is required to be associative and the requirement of identity morphisms. Hence given any fixed monoid M, one can consider any object A of any category $\mathscr{C}$ (for instance another monoid A or a vector space A) with a monoid morphism $M \to \mathrm{End}_{\mathscr{C}}(A)$. That is, a monoid M has representations in any category $\mathscr{C}$.

6.4 Monoids with Zero

A *monoid with zero* is a multiplicative monoid with an element $0_M \in M$ such that $0_M x = x 0_M = 0_M$ for every $x \in M$. The zero element in a monoid, when it exists, is unique. By definition, a morphism of monoids with zero must respect multiplication, send the identity to the identity and send zero to zero.

One of the main examples of monoid with zero is the endomorphism monoid of any object in the category of pointed sets. The *category* Set_* *of pointed sets* has as objects the pairs (X, x_0), where X is a non-empty set and x_0 is a selected element of X, called the *base point* of X. A morphism $(X, x_0) \to (X', x_0')$ in Set_* is any mapping $f \colon X \to X'$ such that $f(x_0) = x_0'$. For any pointed set (X, x_0), the endomorphism monoid $\mathrm{End}_{\mathsf{Set}_*}(X, x_0)$ of (X, x_0) in the category Set_* is a monoid with zero. The zero in this monoid is the mapping $X \to X$ that sends all the elements of X to x_0.

If M is any monoid with zero 0_M and $x \in M$, we have the morphism

$$\lambda_x \colon (M, 0_M) \to (M, 0_M)$$

in the category Set_*, defined by $\lambda_x(y) = xy$ for every $y \in M$. There is a canonical injective morphism of monoids with zero $\lambda \colon (M, 0_M) \to \mathrm{End}_{\mathsf{Set}_*}(M, 0_M)$, $\lambda \colon x \mapsto \lambda_x$.

Correspondingly, define *left M-sets* for any monoid $(M, 0_M)$ with zero, as follows. A *left M-set* is any pointed set (X, x_0) with a morphism of monoids with zero

$$\lambda \colon (M, 0_M) \to \mathrm{End}_{\mathsf{Set}_*}(X, x_0).$$

Similarly, define *right M-sets* as pointed sets (X, x_0) with an anti-homomorphism of monoids with zero $\rho \colon (M, 0_M) \to \mathrm{End}_{\mathsf{Set}_*}(X, x_0)$. Clearly, right M-sets coincide with left M^{op}-sets, left M-sets coincide with right M^{op}-sets, and, for M commutative, right M-sets coincide with left M-sets.

6.5 Near-Rings

A similar situation occurs for near-rings, where a *near-ring* is a ring $(R, +, \cdot)$ for which the group $(R, +)$ is not necessarily abelian and for which multiplication on the right distributes over addition, i.e., $(x + y)z = xz + yz$, but multiplication on the left does not necessarily distribute over addition. The main example is the near-ring G^G of all mappings $G \to G$ for a group G. Hence a left module over a near-ring R must be defined as a group H with a near-ring morphism $R \to H^H$.

6.6 Groups and the Cayley Representation

A group is a special type of monoid, so that everything that we've said about monoids applies to groups. One of the main examples of group is, for any set X, the group $\mathrm{Sym}(X)$ of all bijections $X \to X$. If G is any group and $x \in G$, the mapping $\lambda_x : G \to G$ considered in Sect. 6.3 is a bijection. We have a canonical Cayley representation $\lambda : G \to \mathrm{Sym}(G)$, $\lambda : x \mapsto \lambda_x$. The mapping λ is an injective group morphism.

Correspondingly, we have "left G-sets". A *left G-set* is any set X with a group morphism $\lambda : G \to \mathrm{Sym}(X)$. Similarly, *right G-sets* are sets X with a group anti-homomorphism $\rho : G \to \mathrm{Sym}(X)$. But any group G is isomorphic to its opposite group G^{op} via the isomorphism $G \to G^{\mathrm{op}}$, $x \mapsto x^{-1}$. This is the mother of all symmetries in groups. Hence the category of all right G-sets is canonically isomorphic to the category of all left G-sets for all groups G, and it is useless to introduce both right and left G-sets.

Since groups are monoids, the categorical interpretation of left M-sets in Sect. 6.3 applies directly to left G-sets. Given any group G, we can construct the category $\mathscr{C}$ with a unique object $*$ and with endomorphism monoid $\mathrm{End}_{\mathscr{C}}(*) := G$. The functors of this category $\mathscr{C}$ into the category Set of sets correspond to G-sets and the natural transformations between two functors $\mathscr{C} \to \mathsf{Set}$ correspond to G-set morphisms.

6.7 Groups G and Action of G on G via Inner Automorphisms

There is another very natural action of a group G onto itself, different from that in the previous subsection. For any group G, we can construct its *automorphism group* $\mathrm{Aut}(G)$. If G is any group and $x \in G$, the mapping $\alpha_x : G \to G$, defined by $\alpha_x(y) = xyx^{-1}$ for every $y \in G$, is the *inner automorphism* of G given by conjugation by x. There is a canonical group morphism $\alpha : G \to \mathrm{Aut}(G)$, defined by $\alpha : x \mapsto \alpha_x$ for every $x \in G$.

Correspondingly, we have "left G-groups". For a fixed group G, a *left G-group* is any group H with a group morphism $\alpha : G \to \mathrm{Aut}(H)$. Similarly, we can define *right G-groups* as groups H with a group anti-homomorphism $\beta : G \to \mathrm{Aut}(H)$. As we have said above, any group G is isomorphic to its opposite group G^{op}. Hence the category of all right G-groups is canonically isomorphic to the category of all left G-groups for all groups G, and it is therefore useless to introduce both right G-groups and left G-groups.

As in Sect. 6.6 for G-sets, given any group G, we can construct the category $\mathscr{C}$ with a unique object $*$ and with endomorphism monoid $\mathrm{End}_{\mathscr{C}}(*) := G$. The functors of this category $\mathscr{C}$ into the category Grp of groups correspond to G-groups and the natural transformations between two functors $\mathscr{C} \to \mathsf{Grp}$ correspond to G-group morphisms.

The notion of G-group H is classical. Sometimes G is called an *operator group* on H [17, Definition 8.1].

A G-*group morphism* $f \colon (H, \varphi) \to (H', \varphi')$ is any group morphism $f \colon H \to H'$ such that $f(gh) = gf(h)$ for every $g \in G, h \in H$. We will denote by $\mathrm{Hom}_G(H, H')$ the set of all G-group morphisms of H into H'. G-groups form a category G-Grp. The category Grp of groups and the category 1-Grp, where 1 is the trivial group (that is, the group with one element), are isomorphic categories. This is the analogue of the fact that the category Ab of abelian groups and the category $\mathbb{Z}$-Mod of modules over the ring $\mathbb{Z}$ of integers are isomorphic categories, because 1 and $\mathbb{Z}$ are the initial objects in the category of groups and the category of rings, respectively.

Similarly we can present representations $G \to \mathrm{GL}_n(k)$ of a group G. Here k is a field. In general we can represent a group G fixing any category $\mathscr{C}$, an object C of $\mathscr{C}$ and a group morphism $G \to \mathrm{Aut}_{\mathscr{C}}(C)$.

7 Free Modules

7.1 Definition and First Properties of Free Modules

Let M_R be a right module over an (associative) ring R with identity. A *set X of generators* of M_R is a subset X of M_R such that if N is a submodule of M_R that contains X, then $N = M_R$. For instance, the empty set $X = \emptyset$ generates the zero module. If $X \neq \emptyset$ and $X \subseteq M_R$, then X is a set of generators of M_R if and only if, for every element $x \in M_R$, there exist $n \geq 1, x_1, \ldots, x_n \in X$ and $r_1, \ldots, r_n \in R$ such that $x = x_1 \cdot r_1 + \ldots + x_n \cdot r_n$.

Let us see now what a *free* set of generators is.

Definition 7.1 Let X be a set of generators of a right R-module M_R. The set X is called a *free* set of generators if, for every $n \geq 1, x_1, \ldots, x_n$ distinct elements of X and $r_1, \ldots, r_n$ in R, one has that $x_1 \cdot r_1 + \ldots + x_n \cdot r_n = 0$ implies $r_1 = \ldots = r_n = 0_R$.

Notice that every module M_R has sets X of generators, for instance $X = M_R$. Not every module has free sets of generators. For instance, the $\mathbb{Z}$-module $\mathbb{Z}/n\mathbb{Z}$ does not have a free set of generators for $n \geq 2$.

Definition 7.2 A right R-module M_R is said to be *free* if it has a free set of generators.

Let M_R be a right R-module and X a subset of M_R. We know that X is a set of generators of M_R if and only if every element of M_R can be written as a linear combination of elements of X. It is easily seen that X is a free set of generators of M_R if and only if every element of M_R can be written as a linear combination of distinct elements of X in a unique way; that is, $x_1 \cdot r_1 + \ldots + x_n \cdot r_n = x_1 \cdot r_1' + \ldots + x_n \cdot r_n'$ with $x_1, \ldots, x_n$ n distinct elements of X implies $r_1 = r_1', \ldots, r_n = r_n'$.

Example 7.3 Let R be a ring and X be an arbitrary set. Let $R^{(X)}$ be the set of all mappings $f: X \to R$ such that $f(x) = 0$ for almost all $x \in X$; that is, a mapping $f: X \to R$ is in $R^{(X)}$ if and only if there exists a finite subset F of X with $f(x) = 0$ for every $x \in X \setminus F$. Then $R^{(X)}$ is an abelian group with respect to the operation $+$ defined by

$$(f + g)(x) = f(x) + g(x)$$

for every $f, g \in R_R^{(X)}$ and every $x \in X$. The abelian group $R^{(X)}$ becomes a free right R-module $R_R^{(X)}$ with respect to the right scalar multiplication defined by

$$(fr)(x) = f(x)r$$

for every $f \in R_R^{(X)}, x \in X$ and $r \in R$.

For every fixed $x_0 \in X$, let $\delta_{x_0}: X \to R$ be the mapping defined by

$$x \mapsto \begin{cases} 1_R \ if \ x = x_0, \\ 0_R \ if \ x \neq x_0. \end{cases}$$

It is easy to see that $\Delta := \{ \delta_{x_0} \mid x_0 \in X \}$ is a free set of generators for $R_R^{(X)}$. The module $R_R^{(X)}$ is isomorphic to the direct sum of $|X|$ copies of the module R_R.

We also have the same on the left. The abelian group $R^{(X)}$ is a free left R-module $_R R^{(X)}$ with respect to the left scalar multiplication defined by

$$(rf)(x) = r(f(x))$$

for every $f \in {}_R R^{(X)}, x \in X$ and $r \in R$. In this case also, the set Δ is a free set of generators.

Proposition 7.4 (Universal Property of Free Modules). *Let M_R be a free right R-module, X a free set of generators for M_R and $\varepsilon: X \to M_R$ the embedding of X into M_R. Then, for every right R-module M'_R and every mapping $f: X \to M'_R$, there exists a unique right R-module morphism $\tilde{f}: M_R \to M'_R$ making the diagram*

commute, that is, such that $\tilde{f} \circ \varepsilon = f$.

We have the functors $F: \mathsf{Set} \to \mathsf{Mod}\text{-}R$, where, for every set X, $F(X)$ is the free module $R_R^{(X)}$, and the forgetful functor $U: \mathsf{Mod}\text{-}R \to \mathsf{Set}$. Proposition 7.4 says that F is a left adjoint of U, that is, $\mathrm{Hom}_R(R_R^{(X)}, M'_R) \cong M'^X$ for every set X and module M'_R.

Corollary 7.5 *If M_R is a free right module with free set X of generators, then $M_R \cong R_R^{(X)}$.*

When R is a division ring, every right R-module, that is, every right vector space over the division ring R, is free. For this, we need a:

7.2 Crash Course of Linear Algebra over Non-commutative Division Rings

Let us briefly recall some elementary notions of linear algebra. The reader is definitely an expert on the elementary theory of vector spaces over a field k: vector spaces over k (they are exactly what we have called k-modules), linear transformations (they are exactly what we have called k-module morphisms), the concept of set of generators, linear combinations, linear independence and bases. The reader knows that any two bases of a vector space over k have the same cardinality, and that this cardinality is called the dimension of the vector space. He knows that if we have a linear transformation f between two vector spaces V and W over k of finite dimensions n and m respectively, and we fix an ordered basis for V and an ordered basis for W, we can associate with f a $m \times n$ matrix with entries in k. He knows the rank of a linear transformation f (it is the dimension of the image of f), bilinear mappings, the determinant of a square matrix, its minimal polynomial, the characteristic polynomial, eigenvectors and eigenvalues and so on. Assume now that k is not a field, but a division ring, and consider right vector spaces over k, that is, right k-modules. It is very easy to see that all the previous concepts hold for right vector spaces over a division ring, until when bilinear mappings enter the picture. Bilinearity is a notion concerning modules over commutative rings, because, for a bilinear mapping $\beta \colon {}_kV \times {}_kW \to {}_kU$, we have that $\beta(\lambda v, \mu w)$ must be equal to both $\lambda\beta(v, \mu w) = \lambda\mu\beta(v, w)$ and $\mu\beta(\lambda v, w) = \mu\lambda\beta(v, w)$.

Thus, over an arbitrary division ring k we still have linear transformations (they are right k-module morphisms), sets of generators (again, we have already defined them for modules over arbitrary rings), linear combinations (that is, expressions of the form $\sum_{i=1}^{n} v_i\lambda_i$, where the elements v_i belong to a right vector space V_k and the scalars λ_i are in the division ring k), linear independence (a subset X of V_k is linearly independent if and only if it is a free set of generators for the subspace of V_k it generates), bases (i.e., free sets of generators). Any two bases of a right vector space over a division ring k have the same cardinality (same proof as in the case of a commutative k), and this cardinality is called the *dimension* of the right vector space. If we have a linear transformation f between two right vector spaces V_k and W_k, $\{v_1, \ldots, v_n\}$ is an ordered basis of V_k and $\{w_1, \ldots, w_m\}$ is an ordered basis of W_k, we can associate with f the $m \times n$ matrix $A_f = (\lambda_{i,j})_{i,j}$, where $f(v_j) = \sum_{i=1}^{n} w_i\lambda_{i,j}$. In this case also, if $v = \sum_{j=1}^{n} v_j a_j$ is an arbitrary element of

V_k and $\begin{pmatrix} a_1 \\ \vdots \\ a_n \end{pmatrix}$ is the $n \times 1$ matrix whose entries are the coefficients of v as a linear

combination of $v_1, \ldots, v_n$, then the $m \times 1$ matrix $A_f \begin{pmatrix} a_1 \\ \vdots \\ a_n \end{pmatrix}$ is the matrix whose

entries are the coefficients of $f(v)$ as a linear combination of $w_1, \ldots, w_m$. Notice
that if $f \colon V_k \to W_k$ and $g \colon W_k \to Y_k$, then $A_{g \circ f} = A_g A_f$.

The rank of a linear transformation f is the dimension of the image of f when the
division ring k is non-commutative also. The difficulties in the non-commutative case
appear when bilinear mappings and determinant, which are multilinear mappings,
are introduced. There are notions of right determinant and left determinant, due to
Dieudonné, one is linear on columns and the other on rows, but they are not easy
to handle. Consequently, it becomes impossible to deal (at least easily) with the
minimal polynomial, the characteristic polynomial, eigenvectors and eigenvalues.
But, until the appearance of bilinear mappings and determinant, the passage from
the commutative case to the non-commutative one is very smooth.

As we have already said, every module over a division ring, that is, every right
vector space and every left vector space over a division ring, is free. The converse is
also true: if R is a ring over which every right R-module is free, then R is a division
ring.

7.3 Rank of a Free Module

Let us go back to the study of free modules over arbitrary rings. Recall that $|X|$
denotes the cardinality of a set X.

Corollary 7.6 *If M_R and N_R are free right R-modules with free sets of generators
X and Y respectively, and $|X| = |Y|$, then $M_R \cong N_R$.*

If M_R is a free module, the cardinality of any free set of generators of M_R is called
a *rank* of the free module M_R.

Proposition 7.7 *Let M_R be a free right R-module. If M_R is finitely generated, then
every free set of generators of M_R is finite.*

Corollary 7.8 *Let R be a ring and let M_R be a free right R-module. If X is an
infinite free set of generators of M_R, then every free set of generators of M_R has
cardinality $|X|$.*

Hence, the rank of a free module with an infinite free set of generators is uniquely
defined (Corollary 7.8), while the only thing we can say about a finitely generated
free module is that every free set of generators is finite (Proposition 7.7).

8 IBN Rings

Let R be a ring (with identity). Let $\mathcal{F}_{\mathrm{fg}}$ be the full subcategory of Mod-R whose objects are all finitely generated free right R-modules. This subcategory has a zero object: the zero module, free of rank zero. We can proceed like in Sect. 4.1, constructing the monoid $V(\mathcal{C})$ for $\mathcal{C} = \mathcal{F}_{\mathrm{fg}}$. The set $\{ R_R^n \mid n \geq 0 \text{ an integer} \}$ contains a skeleton $V(\mathcal{F}_{\mathrm{fg}})$: every finitely generated free right R-module is isomorphic to R_R^n for some n, possibly a finite number of natural numbers n (see Example 8.2). Hence we have a reduced commutative monoid $V(\mathcal{F}_{\mathrm{fg}})$ with the operation induced by the direct sum $\oplus$. Equivalently, we can define $V(\mathcal{F}_{\mathrm{fg}})$ as the quotient monoid $\mathbb{N}_0/\sim$, where $\sim$ is the congruence on the additive monoid $\mathbb{N}_0$ defined, for every $n, m \in \mathbb{N}_0$, by $n \sim m$ if $R_R^n \cong R_R^m$. Thus $\sim$ depends on the fixed ring R.

Of course, $\sim$ is a congruence on $\mathbb{N}_0$, and therefore, as we said in Sect. 1.5, the congruence $\sim$ must be either the equality $=$ or one of the congruences $\sim_{k,n}$, where $k \geq 0$ and $n \geq 1$, for a unique pair (k, n). For arbitrarily fixed integers $k, n \geq 1$, it is possible to construct rings for which the associated congruence $\sim$ is exactly the congruence $\sim_{k,n}$.

A ring R is IBN or has IBN (invariant basis number) if the congruence $\sim$ is the equality $=$, that is, if for every $n, m \geq 0$, $R_R^n \cong R_R^m$ implies $n = m$. Equivalently, a ring R is IBN if and only if $V(\mathcal{F}_{\mathrm{fg}})$ is isomorphic to the additive monoid $\mathbb{N}_0$. For instance, division rings have IBN. Notice that R_R^0 has one element, and R_R^n has at least two elements for $n \geq 1$ and $R \neq 0$. This has two consequences: (1) A ring R has IBN if and only if, for every $n, m \geq 1$, $R_R^n \cong R_R^m$ implies $n = m$. (2) If $\sim_{k,n}$ is the congruence associated with a ring $R \neq 0$ as above, then necessarily $k \geq 1$.

Exercise 8.1 (1) Show that having IBN is a left/right symmetric condition, that is, a ring R has IBN if and only if $_R R^n \cong {_R R^m}$ implies $n = m$ for every $n, m \geq 0$.

(2) Show that a ring R has IBN if and only if for every $n, m \geq 1$, $A \in \mathbb{M}_{n \times m}(R)$, $B \in \mathbb{M}_{m \times n}(R)$, $AB = 1_n$, $BA = 1_m$ imply $n = m$. Here $\mathbb{M}_{n \times m}(R)$ denotes the set of all $n \times m$ matrices with entries in R.

(3) Show that if there exists a ring morphism $\varphi \colon R \to S$ and the ring S has IBN, then R has IBN as well.

(4) Show that if R is a ring, I is a two-sided ideal in R and the quotient ring R/I has IBN, then R has IBN as well. (Here notice the special case of $I = R$. In this case R/I is the zero ring, which is not an IBN ring.)

(5) Show that every non-zero commutative ring has IBN.

[Hint for (1): The functor Hom$(-, R)$ induces a duality between the category of finitely generated free right R-modules and the category of finitely generated free left R-modules. Hint for (3): Apply (2). If $A \in \mathbb{M}_{n \times m}(R)$, $B \in \mathbb{M}_{m \times n}(R)$ and we apply the morphism φ to the entries of A and B, we get two matrices in $\mathbb{M}_{n \times m}(S)$ and $\mathbb{M}_{m \times n}(S)$ such that…]

Example 8.2 Here is an example of a ring $R \neq 0$ with $R_R \cong R_R \oplus R_R$, so that in particular the ring R has not IBN. Let k be a field. Let V_k be a vector space over k of infinite dimension. Then $V_k \oplus V_k \cong V_k$. Let $R := \text{End}(V_k)$ be the endo-morphism ring of V_k, so that $_R V_k$ is a R-k-bimodule. Thus there is a covari-ant additive functor $\text{Hom}(_R V_k, -) \colon \text{Mod-}k \to \text{Mod-}R$. Applying this functor to the right k-module isomorphism $V_k \oplus V_k \cong V_k$, we get a right R-module isomor-phism $\text{Hom}(_R V_k, V_k) \oplus \text{Hom}(_R V_k, V_k) \cong \text{Hom}(_R V_k, V_k)$, that is, an isomorphism $R_R \oplus R_R \cong R_R$. This is an isomorphism between two free right R-modules of rank 2 and 1 respectively. Therefore R is not an IBN ring. Notice that, for this ring R, we have that $R_R \cong R_R^n$ for every $n \geq 1$. Thus $R_R^n \cong R_R^m$ for every $n, m \geq 1$. Hence, for this ring R, the monoid $V(\mathcal{F}_{\text{fg}})$ is a monoid with two elements. It is isomorphic to the multiplicative monoid $\{0, 1\}$ with two elements. The congruence associated with the ring R as at the beginning of this Sect. 8 is $\sim_{1,1}$.

9   Simple Modules, Semisimple Modules

A *simple* right module is a non-zero right module M_R whose submodules are only M_R and 0. Thus a simple module has exactly two submodules.

Lemma 9.1 *A right module M_R is simple if and only it is isomorphic to R_R/I for some maximal right ideal I of R.*

Lemma 9.2 (Schur's Lemma) *The endomorphism ring of a simple module is a division ring.*

A module M_R is *semisimple* if every submodule of M_R is a direct summand of M_R.

Remark 9.3 (1) Every simple module is semisimple.
(2) If R is a division ring, every R-module is semisimple.
(3) A module M_R is semisimple if and only if every short exact sequence with M_R in the middle, that is, every short exact sequence of the form $0 \to A_R \to M_R \to C_R \to 0$, splits.
(4) Submodules and homomorphic images of semisimple modules are semisimple modules.

Definition 9.4 Let M_R be a right R-module. The *socle* $\text{soc}(M_R)$ of M_R is the sum of all simple submodules of M_R.

Thus $\text{soc}(M) = 0$ if and only if M has no simple submodules.

Theorem 9.5 *The following conditions are equivalent for a right R-module M:*

(i)   *M is a sum of simple submodules, that is, M is equal to its socle.*
(ii) *M is a direct sum of simple submodules.*
(iii) *M is semisimple.*

10   Projective Modules

Definition 10.1 Let R be a ring. A right R-module P_R is *projective* if for every epimorphism $f : M_R \to N_R$ and every morphism $g : P_R \to N_R$, there exists a morphism $h : P_R \to M_R$ with $f \circ h = g$.

The situation in the previous definition is described by the following commutative diagram, in which the row is exact:

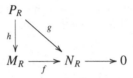

Lemma 10.2   (i) *Every free module is projective.*
(ii) *Every direct summand of a projective module is projective.*
(iii) *Every direct sum of projective modules is projective.*

Proposition 10.3 *The following conditions are equivalent for a right R-module P_R:*

(i) *The module P_R is projective.*
(ii) *Every short exact sequence of the form $0 \to M_R \to N_R \to P_R \to 0$ splits.*
(iii) *The module P_R is isomorphic to a direct summand of a free module.*

Corollary 10.4 *A module P_R is a finitely generated projective module if and only if it is isomorphic to a direct summand of R_R^n for some $n \geq 0$.*

A ring R is *semisimple artinian* if it is right artinian and has no non-zero nilpotent right ideal. To be more precise, we should call such a ring a *right semisimple artinian* ring, because it is defined relatively to the structure of the right module R_R and to right ideals. Also, we should define *left semisimple artinian* rings symmetrically. But as a consequence of the Artin-Wedderburn Theorem 10.6, it will follow that a ring is right semisimple artinian if and only if it is left semisimple artinian, so that a reference to the side is useless. In order not to have a too heavy terminology, we call the rings just defined semisimple artinian, without any reference to the side.

Theorem 10.5 *The following conditions are equivalent for a ring R.*

(i) *Every right R-module is projective.*
(ii) *Every short exact sequence of right R-modules splits.*
(iii) *Every right R-module is semisimple.*
(iv) *The module R_R is semisimple.*
(v) *The ring R is semisimple artinian.*

10.1   The Ring of n × n Matrices over a Division Ring

Let us describe the structure of the ring of all $n \times n$ matrices with entries in a division ring.

Let D be a division ring, $n \geq 1$ an integer, and $R := \mathbb{M}_n(D)$ be the ring of all $n \times n$ matrices with entries in D. It is possible to prove that R is a *simple* ring, that is, its two-sided ideals are only the trivial ones. For every $i, j = 1, 2, \ldots, n$, let $E_{i,j}$ be the matrix with the (i, j) entry equal to 1, and 0 in all the other entries. Notice that the elements $E_{i,i}$ *are idempotents of* R, $E_{1,1} + \cdots + E_{n,n} = 1$ and $E_{i,i}E_{j,j} = 0$ for $i \neq j$. Also, $E_{i,i}R$ is the set of all $n \times n$ matrices with entries in D, that are 0 on all rows except for the i-th row, and with arbitrary entries in D on the i-th row, and

$$R_R = E_{1,1}R \oplus E_{2,2}R \oplus \cdots \oplus E_{n,n}R$$

The modules $E_{i,i}R$ *are all pairwise isomorphic.* For instance, an isomorphism $E_{1,1}R \to E_{i,i}R$ is given by left multiplication by the matrix $E_{i,1}$. Moreover, *the module* $E_{1,1}R$ *is simple.* Thus $R_R = E_{1,1}R \oplus \cdots \oplus E_{n,n}R$ *is a direct sum of n simple isomorphic modules,* in particular R *is a semisimple artinian ring.*

Matrix transposition $t\colon A \mapsto A^t$ is a ring isomorphism

$$t\colon \mathbb{M}_n(D) \to (\mathbb{M}_n(D^{\mathrm{op}}))^{\mathrm{op}}.$$

Therefore R is isomorphic to the opposite ring of $\mathbb{M}_n(D^{\mathrm{op}})$, where D^{op} is also a division ring. Thus all properties we have seen on the right also hold on the left. Also, the category R-Mod, which is equivalent to the category Mod-R^{op}, is equivalent to the category $\mathbb{M}_n(D^{\mathrm{op}})$-Mod.

We have that the left ideal $RE_{i,i}$ *is the set of all* $n \times n$ *matrices with entries in* D, *that are* 0 *on all columns except for the i-th column, and with arbitrary entries in* D *on the i-th column.* Therefore

$$_RR = RE_{1,1} \oplus RE_{2,2} \oplus \cdots \oplus RE_{n,n},$$

and *the left ideals* $RE_{1,1}, \ldots, RE_{n,n}$ *are isomorphic simple modules.*

It is also possible to prove that *every simple right* R-*module is isomorphic to* $E_{1,1}R$, *and the endomorphism ring* $\mathrm{End}(E_{1,1}R)$ *of the simple module* $E_{1,1}R$ *is isomorphic to the division ring* D.

Now if M_R is any right R-module, then M_R is semisimple by Theorem 10.5. Hence M_R is a direct sum of simple submodules. But all simple right R-modules are isomorphic to $E_{1,1}R$. Thus we have seen that *every right* R-*module is isomorphic to a direct sum* $E_{1,1}R^{(X)}$ *for some set* X. *It is possible to prove that the cardinality of such a set* X *is uniquely determined.*

Notice that $\mathrm{Hom}(E_{1,1}R, E_{1,1}R)$ is an abelian group that cannot be endowed with a right R-module structure or a left R-module structure. For instance, assume that the division ring D is a finite field with q elements and $n = 2$. It is easy to show that $\mathrm{Hom}(E_{1,1}R, E_{1,1}R) \cong E_{1,1}RE_{1,1} \cong D$, hence has q elements in this case. But every right R-module is isomorphic to a direct sum of copies of $E_{1,1}R$, which has q^2 elements. Hence every finite right R-module has q^{2t} elements for some non-negative integer t. Thus no right R-module can have q elements. This proves that $\mathrm{Hom}(E_{1,1}R, E_{1,1}R)$ cannot be endowed with a right R-module structure. Similarly, it cannot be endowed with a left R-module structure.

As we have just said above, a matrix ring with entries in a division ring is a semisimple artinian ring. This is true for any finite direct product of such matrix rings:

Let $t, n_1, \ldots, n_t \geq 1$ be integers and $D_1, \ldots, D_t$ division rings. Then the ring $R := M_{n_1}(D_1) \times \cdots \times M_{n_t}(D_t)$ is a semisimple artinian ring.

It is interesting that the converse of this result also holds:

Theorem 10.6 (Artin-Wedderburn) *A ring R is semisimple artinian if and only if there exist integers $t \geq 0$, $n_1, \ldots, n_t \geq 1$ and division rings $D_1, \ldots, D_t$ such that*

$$R \cong M_{n_1}(D_1) \times \cdots \times M_{n_t}(D_t). \tag{11.i}$$

Moreover, if R is semisimple artinian, the integers $t, n_1, \ldots, n_t$ in the decomposition (11.i) are uniquely determined by R and $D_1, \ldots, D_t$ are determined by R up to ring isomorphism.

11　Superfluous Submodules and Radical of a Module

A submodule N of a module M_R is *superfluous* (or *small*, or *inessential*) in M_R if, for every submodule L of M_R, $N + L = M_R$ implies $L = M_R$. To denote that N is superfluous in M_R, we will write $N \leq_s M_R$.

Here are the main elementary properties of superfluous submodules:

(i) If $K \leq N \leq M_R$, then $N \leq_s M$ if and only if $K \leq_s M$ and $N/K \leq_s M/K$.

(ii) $K \leq_s M_R$ and $M_R \leq N_R$ imply $K \leq_s N_R$.

(iii) If $N, N' \leq M_R$, then $N + N' \leq_s M$ if and only if $N \leq_s M$ and $N' \leq_s M$.

(iv) The zero submodule is always a superfluous submodule of any module M_R, when $M_R = 0$ also.

(v) If $f : M \to M'$ is an R-module morphism and $N \leq_s M$, then $f(N) \leq_s M'$.

(vi) Assume $K_1 \leq M_1 \leq M, K_2 \leq M_2 \leq M$ and $M = M_1 \oplus M_2$. Then $K_1 \oplus K_2 \leq_s M_1 \oplus M_2$ if and only if $K_1 \leq_s M_1$ and $K_2 \leq_s M_2$.

We will say that an epimorphism $g : M_R \to N_R$ is *superfluous* if $\ker g$ is a superfluous submodule of M_R.

The *radical* $\mathrm{rad}(M_R)$ of a module M_R is the intersection of all maximal submodules of M_R. Note the duality with the definition of socle, which is the sum of all simple (=minimal) submodules of M_R.

Here are some elementary properties of the radical $\mathrm{rad}(M_R)$ of a module M_R:

(1) The submodule $\mathrm{rad}(M_R)$ is the sum of all superfluous submodules of M_R.
(2) $\mathrm{rad}(M_R/\mathrm{rad}(M_R)) = 0$.
(3) If $f\colon M_R \to M'_R$ is a morphism of R-modules, then

$$f(\mathrm{rad}(M_R)) \le \mathrm{rad}(M'_R).$$

12 The Jacobson Radical of a Ring

The radical of the right R-module R_R is called the *Jacobson radical* of the ring R. It is denoted by $J(R)$. Thus $J(R) := \mathrm{rad}(R_R)$ is the intersection of all maximal right ideals of R, but it is possible to show that $\mathrm{rad}(R_R) = \mathrm{rad}(_R R)$ for any ring R, so that $J(R)$ is also the intersection of all maximal left ideals of R. Therefore $J(R)$ is a two-sided ideal of R.

For every right R-module M_R, the *right annihilator* $\mathrm{r.\,ann}_R(M_R)$ of M_R is the set of all $r \in R$ such that $Mr = 0$. The right annihilator of any right R-module is a two-sided ideal of R. If $x \in M_R$, the *right annihilator* $\mathrm{r.\,ann}_R(x)$ of x is the set of all $r \in R$ such that $xr = 0$. The right annihilator of an element x of M_R is a right ideal of R.

Proposition 12.1 *The Jacobson radical $J(R)$ of any ring R is the intersection of the right annihilators $\mathrm{r.\,ann}_R(S_R)$ of all simple right R-modules S_R.*

13 Injective Modules

Fix two modules M_R and N_R. There are a covariant functor

$$\mathrm{Hom}(M_R, -)\colon \mathsf{Mod}\text{-}R \to \mathsf{Ab}$$

and a contravariant functor

$$\mathrm{Hom}(-, N_R)\colon \mathsf{Mod}\text{-}R \to \mathsf{Ab}.$$

These functors Hom are "left exact", in the sense that, for every fixed module M_R, if $0 \to N'_R \to N_R \to N''_R$ is exact, then so is $0 \to \mathrm{Hom}(M_R, N'_R) \to \mathrm{Hom}(M_R, N_R) \to \mathrm{Hom}(M_R, N''_R)$, and, for every fixed module N_R, if $M'_R \to M_R \to M''_R \to 0$ is exact, then so is $0 \to \mathrm{Hom}(M''_R, N_R) \to \mathrm{Hom}(M_R, N_R) \to \mathrm{Hom}(M'_R, N_R)$.

In general, the functors $\mathrm{Hom}(M_R, -)$ and $\mathrm{Hom}(-, N_R)$ are not "exact", that is, it is not always true that, for every fixed module M_R, if $0 \to N'_R \to N_R \to N''_R \to 0$ is a short exact sequence, then $0 \to \mathrm{Hom}(M_R, N'_R) \to \mathrm{Hom}(M_R, N_R) \to \mathrm{Hom}(M_R, N''_R) \to 0$ is necessarily exact, and, for every fixed module N_R, if $0 \to M'_R \to M_R \to M''_R \to 0$ is exact, then $0 \to \mathrm{Hom}(M''_R, N_R) \to \mathrm{Hom}(M_R, N_R) \to$

$\mathrm{Hom}(M'_R, N_R) \to 0$ is necessarily exact. It is easily seen that a module M_R is projective if and only if the functor $\mathrm{Hom}(M_R, -)$ is exact, that is, for every exact sequence $0 \to N'_R \to N_R \to N''_R \to 0$, the sequence of abelian groups $0 \to \mathrm{Hom}(M_R, N'_R) \to \mathrm{Hom}(M_R, N_R) \to \mathrm{Hom}(M_R, N''_R) \to 0$ is exact.

The proof of the following result is easy.

Proposition 13.1 *The following conditions are equivalent for an R-module E_R:*

(i) *The functor $\mathrm{Hom}(-, E_R)$: Mod-$R \to$ Ab is exact, that is, for every exact sequence $0 \to M'_R \to M_R \to M''_R \to 0$ of right R-modules, the sequence of abelian groups $0 \to \mathrm{Hom}(M''_R, E_R) \to \mathrm{Hom}(M_R, E_R) \to \mathrm{Hom}(M'_R, E_R) \to 0$ is exact.*

(ii) *For every monomorphism $M'_R \to M_R$ of right R-modules,*

$$\mathrm{Hom}(M_R, E_R) \to \mathrm{Hom}(M'_R, E_R)$$

is an epimorphism of abelian groups.

(iii) *For every submodule M'_R of a right R-module M_R, every morphism*

$$M'_R \to E_R$$

can be extended to a morphism $M_R \to E_R$.

(iv) *For every monomorphism $f : M'_R \to M_R$ and every morphism*

$$g : M'_R \to E_R,$$

there exists a morphism $h : M_R \to E_R$ with $h \circ f = g$.

A module E_R is *injective* if it satisfies the equivalent conditions of Proposition 13.1.

Condition (iv) is described by the following commutative diagram, in which the row is exact:

$$
\begin{array}{ccc}
0 \longrightarrow M'_R & \overset{f}{\longrightarrow} & M_R \\
 & {\scriptstyle g}\searrow & \downarrow {\scriptstyle h} \\
 & & E_R
\end{array}
$$

Thus we have that:

(1) A module M_R is projective if and only if every short exact sequence of the form $0 \to A_R \to B_R \to M_R \to 0$ splits.

(2) A module M_R is injective if and only if every short exact sequence of the form $0 \to M_R \to B_R \to C_R \to 0$ splits.

(3) A module M_R is semisimple if and only if every short exact sequence of the form $0 \to A_R \to M_R \to C_R \to 0$ splits.

Proposition 13.2 (Baer's criterion). *A right module E over a ring R is injective if and only if for every right ideal I of R, every morphism $\sigma: I \to E$ can be extended to a morphism $\sigma^*: R \to E$.*

Definition 13.3 An additive abelian group G is *divisible* if $nG = G$ for every nonzero integer n (equivalently, for every positive integer n). Thus G is divisible if and only if, for every $g \in G$ and every $n > 0$, there exists $h \in G$ such that $nh = g$.

For instance, the abelian group $\mathbb{Z}$ is not divisible, and the abelian group $\mathbb{Q}$ is divisible. Homomorphic images of divisible abelian groups are divisible. It is possible to prove that every divisible abelian group is a direct sum of copies of $\mathbb{Q}$ and Prüfer groups $\mathbb{Z}(p^\infty)$.

Proposition 13.4 *A $\mathbb{Z}$-module G is injective if and only if it is a divisible abelian group.*

Exercise 13.5 Show that an abelian group is divisible if and only if it is a homomorphic image of $\mathbb{Q}^{(X)}$ for some set X.

Proposition 13.6 *Direct summands of injective modules are injective.*

Theorem 13.7 *Every right R-module can be embedded in an injective right R-module.*

Corollary 13.8 *The following conditions are equivalent for a right R-module E_R:*

(i) *The module E_R is injective.*
(ii) *Every short exact sequence that begins with E_R splits, that is, every short exact sequence of right R-modules of the form $0 \to E_R \to B_R \to C_R \to 0$ splits.*
(iii) *The module E_R is a direct summand of every module of which it is a submodule.*

14 Projective Covers

Every module is a homomorphic image of a projective module, because every module M_R is a homomorphic image of the free module $R^{(M_R)}$. Now we look for the smallest possible representation of M_R as a homomorphic image of a projective module.

Definition 14.1 (Projective cover). A projective cover of a module M_R is a pair (P_R, p) where P_R is a projective right R-module and $p: P \to M$ is a superfluous epimorphism (that is, an epimorphism $p: P \to M$ with ker p a superfluous submodule of P).

Theorem 14.2   (1) (Fundamental lemma of projective covers) *Let (P, p) be a projective cover of a right R-module M. If Q is a projective module and $q: Q \to M$ is an epimorphism, then Q has a direct-sum decomposition $Q = P' \oplus P''$ where $P' \cong P$, $P'' \subseteq \ker(q)$ and $(P', q|_{P'}: P' \to M)$ is a projective cover.*

(2) (Uniqueness of projective covers up to isomorphism) *Projective covers, when they exist, are unique up to isomorphism in the following sense. If $(P, p), (Q, q)$ are any two projective covers of a right R-module M, there is an isomorphism $h \colon Q \to P$ such that $p \circ h = q$.*

15 Injective Envelopes

A submodule N of a module M_R is *essential* (or *large*) in M_R if, for every submodule L of M_R, $N \cap L = 0$ implies $L = 0$. In this case, we will write $N \leq_e M_R$.

Exercise 15.1 Show that

(a) If $K \leq N \leq M_R$, then $K \leq_e M$ if and only if $K \leq_e N$ and $N \leq_e M$.
(b) If $N, N' \leq M_R$, then $N \cap N' \leq_e M$ if and only if $N \leq_e M$ and $N' \leq_e M$.
(c) The submodule M is always essential in M_R, when $M_R = 0$ also.
(d) If $f \colon M \to M'$ is a morphism of R-modules and $N' \leq_e M'$, then $f^{-1}(N') \leq_e M$.
(e) A submodule N of an R-module M is essential in M if and only if for every $x \in M, x \neq 0$, there exists $r \in R$ with $xr \in N$ and $xr \neq 0$.
(f) Assume $N_1 \leq M_1 \leq M$, $N_2 \leq M_2 \leq M$ and $M = M_1 \oplus M_2$. Show that $N_1 \oplus N_2 \leq_e M_1 \oplus M_2$ if and only if $N_1 \leq_e M_1$ and $N_2 \leq_e M_2$.

A monomorphism $f \colon N_R \to M_R$ is said to be *essential* if its image $f(N_R)$ is an essential submodule of M_R.

Exercise 15.2 (a) Show that a monomorphism $f \colon N \to M$ is essential if and only if for every module L and every morphism $g \colon M \to L$, if gf is injective, then g is injective.

(b) Let $f \colon N \to M$ and $g \colon M \to P$ be two monomorphisms. Show that the composite mapping gf is an essential monomorphism if and only if both f and g are essential monomorphisms.

Let M_R be a right R-module. An *extension* of M_R is a pair (N_R, f), where N_R is a right R-module and $f \colon M_R \to N_R$ is a monomorphism. An *essential extension* of M_R is an extension (N_R, f) where $f \colon M_R \to N_R$ is an essential monomorphism. An extension (N_R, f) of M_R is *proper* if f is not an isomorphism.

Proposition 15.3 *A module M_R is injective if and only if it does not have proper essential extensions.*

Definition 15.4 An *injective envelope* of a module M_R is a pair (E_R, i), where E_R is an injective right R-module and $i \colon M_R \to E_R$ is an essential monomorphism. Equivalently, (E_R, i) is an essential extension of M_R with E_R an injective module.

For example, if i is the inclusion of $\mathbb{Z}_{\mathbb{Z}}$ into $\mathbb{Q}_{\mathbb{Z}}$, then $(\mathbb{Q}_{\mathbb{Z}}, i)$ is an injective envelope of $\mathbb{Z}_{\mathbb{Z}}$. Dualizing the proof of the Fundamental lemma of projective covers, we get the following

Theorem 15.5 (Fundamental lemma of injective envelopes). *Let (E, i) be an injective envelope of a right R-module M. If F is an injective module and $j \colon M \to F$ is a monomorphism, then F has a direct-sum decomposition $F = F' \oplus F''$ where $F' \cong E$, $j(M) \subseteq F'$ and if $j' \colon M_R \to F'$ is the mapping obtained from j restricting the codomain to F', then (F', j') is an injective envelope of M.*

Theorem 15.6 *Every right R-module has an injective envelope, which is unique up to isomorphism in the following sense: if (E, i) and (E', i') are both injective envelopes of M, then there exists an isomorphism $h \colon E \to E'$ such that $hi = i'$.*

Theorem 15.7 *The following conditions are equivalent for an extension (E, ε) of a right R-module M:*

1. (a) (E, ε) *is an injective envelope of M, that is, an essential injective extension of M.*
2. (b) (E, ε) *is a maximal essential extension of M.*
3. (c) (E, ε) *is a minimal injective extension of M.*

16 The Monoid $V(R)$

Our main example of monoid $V(\mathscr{C})$ is when the category $\mathscr{C}$ is the full subcategory proj-R of Mod-R whose objects are all finitely generated projective right R-modules. We will denote such a monoid $V(\text{proj-}R)$ by $V(R)$. Thus $V(R)$ is a set of representatives of all finitely generated projective right R-modules up to isomorphism. Notice that $V(R)$ is a set, because every finitely generated projective R-module is isomorphic to a direct summand of the module $R_R^{(\aleph_0)}$. For any finitely generated projective right R-module P_R, the unique module in $V(R)$ isomorphic to P_R will be denoted by $\langle P_R \rangle$. Thus we have a mapping $\langle - \rangle \colon \text{Ob(proj-}R) \to V(R)$, with the property that, for every $P_R, Q_R \in \text{Ob(proj-}R)$, $\langle P_R \rangle = \langle Q_R \rangle$ if and only if $P_R \cong Q_R$. The set $V(R)$ becomes a reduced commutative monoid with respect to the addition defined by $\langle P_R \rangle + \langle Q_R \rangle = \langle P_R \oplus Q_R \rangle$ for every $\langle P_R \rangle, \langle Q_R \rangle \in V(R)$. The element $\langle R_R \rangle$ of the monoid $V(R)$ is an order-unit in $V(R)$.

For instance, if R is a semisimple artinian ring, finitely generated (projective) modules are direct sums of simple modules in a unique way up to isomorphism, and there are only finitely many simple modules up to isomorphism. Thus $V(R)$ is a finitely generated free monoid in this case. More precisely, for R a semisimple artinian ring, we have that $V(R) \cong \mathbb{N}_0^n$, where n is the number of simple right R-modules up to isomorphism.

We have defined $V(R)$ using the category proj-R, that is, *right* R-modules. Let us show that if we had taken as $\mathscr{C}$ the full subcategory R-proj of R-Mod whose objects are all finitely generated projective *left* R-modules, we would have got essentially the same object, that is, we would have got isomorphic monoids.

Proposition 16.1 *The functor* $\mathrm{Hom}(-, R)\colon \mathrm{proj}\text{-}R \to R\text{-proj}$ *is a duality, that is, an equivalence between the category* $\mathrm{proj}\text{-}R$ *and the dual category* $(R\text{-proj})^{\mathrm{op}}$ *of the category* $R\text{-proj}$.

Proof The functor $\mathrm{Hom}(-, R)$ is additive, hence preserves direct summands and finite direct sums and sends R_R to $\mathrm{Hom}(R_R, R) \cong {}_R R$.

It immediately follows that the two monoids $V(\mathrm{proj}\text{-}R)$ and $V(R\text{-proj})$ are isomorphic via the isomorphism defined by $\langle P_R \rangle \mapsto \langle {}_R\mathrm{Hom}(P_R, R_R)\rangle$ for every $\langle P_R \rangle \in V(R)$. In other words, if, instead of finitely generated projective right R-modules, we use finitely generated projective left R-modules, we essentially get the same monoid $V(R)$. Also notice that the categories $\mathrm{proj}\text{-}R$ and $R^{\mathrm{op}}\text{-proj}$, where R^{op} denotes the opposite ring of R, are isomorphic. Thus $V(R) \cong V(R^{\mathrm{op}})$.

A *right hereditary ring* is a ring in which every right ideal is projective. Similarly for left hereditary. There exist right hereditary rings that are not left hereditary. A *hereditary ring* is a ring that is both right hereditary and left hereditary. Hereditary commutative integral domains are called *Dedekind* domains. For instance, principal ideal domains are Dedekind domains.

Theorem 16.2 *Let R be a right hereditary ring. Then every submodule of a free right R-module is isomorphic to a direct sum of right ideals of R.*

In particular, every (finitely generated) projective right module over a right hereditary ring is isomorphic to a direct sum of (finitely many) right ideals of R.

As an example, we now compute the monoid $V(R)$ for a Dedekind domain R. Let R be a Dedekind domain. By Theorem 16.2, every finitely generated projective R-module is isomorphic to a direct sum $I_1 \oplus I_2 \oplus \cdots \oplus I_m$ of $m \geq 0$ non-zero ideals $I_1, I_2, \ldots, I_m$ of R. Moreover, two direct sums $I_1 \oplus I_2 \oplus \cdots \oplus I_m$ and $I_1' \oplus I_2' \oplus \cdots \oplus I_{m'}'$ of non-zero ideals I_i, I_j' are isomorphic if and only if $m = m'$ and $I_1 I_2 \ldots I_m \cong I_1' I_2' \ldots I_m'$ [15, Lemma 7.6]. Now every Dedekind domain is noetherian, so that the divisorial fractional ideals of R are the non-zero finitely generated R-submodules of the field of fractions K of R, and the product $I * J$ in the commutative monoid $D(R)$ of all divisorial fractional ideals of R coincides with the usual product IJ for any two ideals I, J of R. As every Dedekind domain is a Krull domain, the monoid $D(R)$ is a group. Therefore the class group $\mathrm{Cl}(R)$ of R is the factor group of the multiplicative group $D(R)$ modulo the subgroup $\mathrm{Prin}(R)$ of non-zero principal fractional ideals. Equivalently, $\mathrm{Cl}(R)$ is the multiplicative group of all isomorphism classes of non-zero ideals of the Dedekind domain R. If we map a non-zero element $\langle A_R \rangle$ of $V(R)$, with $A_R \cong I_1 \oplus I_2 \oplus \cdots \oplus I_m$ and $I_1, I_2, \ldots, I_m$ non-zero ideals of R, to the pair $(m, I_1 I_2 \ldots I_m)$, we get an isomorphism of the monoid of non-zero elements of $V(R)$ onto the direct product $\mathbb{N} \times \mathrm{Cl}(R)$ of the additive monoid $\mathbb{N}$ of positive integers and the multiplicatively group $\mathrm{Cl}(R)$. Thus $V(R)$ turns out to be isomorphic to the monoid $M := (\mathbb{N} \times \mathrm{Cl}(R)) \cup \{0\}$, that is, to the direct product $\mathbb{N} \times \mathrm{Cl}(R)$ to which a zero element is adjoined.

We will now show that the monoids $V(R)$ describe the behavior, as far as direct sums are concerned, not only of projective modules, but of *any* module or *any set of* modules. If M_R is a right module over a ring R, let add(M_R) be the full subcategory of Mod-R whose objects are all modules isomorphic to direct summands of direct sums M^n of finitely many copies of M. For example, proj-$R = $ add(R_R).

We can construct a monoid $V(\text{add}(M_R))$ in a way similar to that in which we have constructed the monoid $V(R)$. The monoid $V(\text{add}(M_R))$ is the monoid $V(\mathscr{C})$ constructed in Sect. 4 when $\mathscr{C}$ is the full subcategory add(M_R) of Mod-R. More precisely, we replace R_R with M_R in the construction of $V(R)$. That is, we fix a set $V(\text{add}(M_R))$ of representatives of the modules in add(M_R) up to isomorphism. Notice that $V(\text{add}(M_R))$ is a set, because every module in add(M_R) is isomorphic to a direct summand of a direct sum of countably many copies of M_R. For a module N_R in add(M_R), denote by $\langle N_R \rangle$ the unique module in $V(\text{add}(M_R))$ isomorphic to N_R. Then $V(\text{add}(M_R))$ becomes a commutative reduced monoid with respect to the addition defined by $\langle N_R \rangle + \langle N'_R \rangle = \langle N_R \oplus N'_R \rangle$ for all $\langle N_R \rangle, \langle N'_R \rangle \in V(\text{add}(M_R))$. The element $\langle M_R \rangle$ is an order-unit in $V(\text{add}(M_R))$. Clearly, the commutative monoid with order-unit $(V(\text{add } M_R), \langle M_R \rangle)$ is the algebraic object that describes the behavior of all direct-sum decompositions of the module M_R.

Given a ring S, let Proj-S denote the full subcategory of Mod-S whose objects are all projective right S-modules. If M_S is a right S-module, let Add(M_S) denote the full subcategory of Mod-S whose objects are all modules isomorphic to direct summands of direct sums of copies of M. Let M_S be a right S-module and let $E = \text{End}(M_S)$ be its endomorphism ring, so that $_E M_S$ is a bimodule.

Theorem 16.3 *The functors*

$$\text{Hom}_S(M, -)\colon \text{Mod-}S \to \text{Mod-}E \quad \text{and} \quad - \otimes_E M\colon \text{Mod-}E \to \text{Mod-}S$$

induce an equivalence between the full subcategory add(M_S) *of* Mod-S *and the full subcategory* proj-E *of* Mod-E. *In particular, the monoids with order-unit*

$$(V(\text{add}(M_S)), \langle M_S \rangle) \quad \text{and} \quad (V(E), \langle E_E \rangle)$$

are isomorphic. Moreover, if M_S is finitely generated, they induce an equivalence between the full subcategory Add(M_S) *of* Mod-S *and the full subcategory* Proj-E *of* Mod-E. $\qquad\qquad\square$

The Grothendieck group $G(V(R))$ of the monoid $V(R)$ is usually denoted by $K_0(R)$. We conclude with three examples.

Example 16.4 *(1)* Suppose that the ring R is a division ring, or more generally a local ring, that is, a ring with a unique maximal right ideal. Over such a ring every projective module is free of unique rank (local rings are IBN). Therefore proj-$R = \mathscr{F}_{\text{fg}}$ and $V(R) \cong \mathbb{N}_0$, so $K_0(R) \cong \mathbb{Z}$.

(2) For an arbitrary field F and arbitrarily fixed integers $k \geq 0$ and $n \geq 1$, it is possible to construct associative F-algebras R (called Leavitt algebras) over which every finitely generated projective module is free and for which the congruence

$\sim$ of $\mathbb{N}_0$ defined, for every $n, m \in \mathbb{N}_0$, by $n \sim m$ if $R_R^n \cong R_R^m$, is exactly the congruence $\sim_{k,n}$. See Sect. 8. Therefore, for such rings R, one has proj-$R = \mathcal{F}_{\mathrm{fg}}$ and $V(R) \cong \mathbb{N}_0/\sim_{k,n}$. The Grothendieck group $G(M)$ of the monoid $M = \mathbb{N}_0/\sim_{k,n}$ is the cyclic group $G(M) = \mathbb{Z}/n\mathbb{Z}$, and the canonical mapping $M \to G(M)$ is the mapping $\mathbb{N}_0/\sim_{k,n} \to \mathbb{Z}/n\mathbb{Z}$, $[t]_{\sim_{k,n}} \mapsto t + n\mathbb{Z}$ for every integer $t \geq 0$.

Example 16.5 A ring R is *semilocal* if $R/J(R)$ is semisimple artinian. It is possible to prove that if R is semilocal, then $V(R)$ is a finitely generated reduced Krull monoid [8, Corollary 3.30]. If M_R is an artinian right module over an arbitrary ring R, then the endomorphism ring $E := \mathrm{End}(M_R)$ is a semilocal ring, so that $V(\mathrm{add}(M_R)) \cong V(E)$ is a finitely generated reduced Krull monoid [8, p. 107].

Acknowledgements The author is partially supported by Ministero dell'Istruzione, dell'Università e della Ricerca (Progetto di ricerca di rilevante interesse nazionale "Categories, Algebras: Ring-Theoretical and Homological Approaches (CARTHA)"), Fondazione Cariverona (Research project "Reducing complexity in algebra, logic, combinatorics - REDCOM" within the framework of the programme "Ricerca Scientifica di Eccellenza 2018"), and Dipartimento di Matematica "Tullio Levi-Civita" of Università di Padova (Research program DOR1828909 "Anelli e categorie di moduli").

References

1. Altun-Özarslan, M., Facchini, A.: The Krull-Schmidt-Remak-Azumaya Theorem for G-groups. In: Leroy, A., Lomp, Ch., López-Permouth, S., Oggier, F. (eds.) Rings, Modules and Codes, pp. 25–38. American Mathematical Society, Providence (2019)
2. Anderson, F.W., Fuller, K.R.: Rings and Categories of Modules, 2nd edn. Springer, New York (1992). https://doi.org/10.1007/978-1-4612-4418-9
3. Ara, P., Facchini, A.: Direct sum decompositions of modules, almost trace ideals, and pullbacks of monoids. Forum Math. **18**, 365–389 (2006)
4. Bourbaki, N.: Éléments de mathématique. Fasc. XXVI, Groupes et algèbres de Lie. Chapitre I: Algèbres de Lie, Seconde édition. Hermann, Paris (1971)
5. Chouinard, L.G., II.: Krull semigroups and divisor class groups. Canad. J. Math. **33**, 1459–1468 (1981)
6. Clifford, A.H., Preston, G.B.: The Algebraic Theory of Semigroups, vol. I. American Mathematical Society, Providence (1961)
7. Facchini, A.: Module Theory. Endomorphism Rings and Direct Sum Decompositions in Some Classes of Modules. Birkhäuser Verlag, Basel (1998, reprinted in 2010). https://doi.org/10.1007/978-3-0348-0303-8
8. Facchini, A.: Semilocal Categories and Modules with Semilocal Endomorphism Rings. Birkhäuser/Springer, Cham (2019). https://doi.org/10.1007/978-3-030-23284-910.1007/978-3-030-23284-9
9. Facchini, A., Finocchiaro, C.A.: Pretorsion theories, stable category and preordered sets. Ann. Mat. Pura Appl. **199**, 1073–1089 (2020)
10. Facchini, A., Halter-Koch, F.: Projective modules and divisor homomorphisms. J. Algebra Appl. **2**, 435–449 (2003)
11. Facchini, A., Herbera, D.: Projective modules over semilocal rings. In: Huynh, D.V., Jain, S.K., López-Permouth, S.R. (eds.) Algebra and Its Applications, pp. 181–198. American Mathematical Society, Providence (2000)

12. Goodearl, K.R.: Von Neumann Regular Rings, 2nd edn. Robert E. Krieger Publishing Co. Inc., Malabar (1991)
13. Halter-Koch, F.: Ideal Systems. An Introduction to Multiplicative Ideal Theory. Marcel Dekker, New York (1998)
14. Lam, T.Y.: A First Course in Noncommutative Rings, 2nd edn. Springer, New York (2001). https://doi.org/10.1007/978-1-4419-8616-0
15. Passman, D.S.: A Course in Ring Theory. AMS Chelsea Publishing, American Mathematical Society, Providence (2004)
16. Pirashvili, I.: On the spectrum of monoids and semilattices. J. Pure Appl. Algebra **217**, 901–906 (2013)
17. Suzuki, M.: Group Theory I. Springer, Heidelberg (1982)

Chapter 4
An Introduction to Regular Categories

Marino Gran

Abstract This paper provides a short introduction to the notion of regular category and its use in categorical algebra. We first prove some of its basic properties, and consider some fundamental algebraic examples. We then analyse the algebraic properties of the categories satisfying the additional Mal'tsev axiom, and then the weaker Goursat axiom. These latter contexts can be seen as the categorical counterparts of the properties of 2-permutability and of 3-permutability of congruences in universal algebra. Mal'tsev and Goursat categories have been intensively studied in the last years: we present here some of their basic properties, which are useful to read more advanced texts in categorical algebra.

Keywords Regular category · Mal'tsev category · Goursat category · Variety of algebras · Mal'tsev conditions

Math. Subj. Classification 18E08 · 18E13 · 18C05 · 08B05

Introduction

In categorical algebra some structural properties of varieties of universal algebras are investigated by replacing the arguments involving elements of an algebraic structure and its operations with other ones using relations and commutative diagrams. A typical example is provided by the study of *Mal'tsev categories* [11], which can be seen as the categorical counterpart of *Mal'tsev varieties* (in the sense of [37]), also called 2-permutable varieties in the literature. Instead of requiring the existence, in the algebraic theory of the variety, of a ternary term $p(x, y, z)$ verifying the identities $p(x, y, y) = x$ and $p(x, x, y) = y$, one asks that any internal reflexive relation in the category is an equivalence relation. This categorical property, with its many equivalent formulations, has turned out to be strong enough to establish, in the

M. Gran (✉)

Institut de Recherche en Mathématique et Physique, Université catholique de Louvain, Chemin du Cyclotron 2 bte L7.01.02, 1348 Louvain-la-Neuve, Belgium

e-mail: marino.gran@uclouvain.be

© The Author(s), under exclusive license to Springer Nature Switzerland AG 2021

M. M. Clementino et al. (eds.), *New Perspectives in Algebra, Topology and Categories*, Coimbra Mathematical Texts 1,
https://doi.org/10.1007/978-3-030-84319-9_4

regular context, many of the well known properties of Mal'tsev varieties (see [8] for a recent survey on the subject, and the references therein).

This survey article can be seen as a first introduction to the basic categorical notions which are useful to express the exactness properties of various kinds of algebraic varieties in the sense of universal algebra. The main goal of this text is to introduce the reader to the notion of *regular category*, which is fundamental in category theory, since abelian categories, elementary toposes and varieties of universal algebras are all regular categories. Special attention will be paid to the so-called *calculus of relations*, which provides a powerful method to prove results in regular categories, possibly satisfying some additional exactness conditions. A good knowledge of the fundamentals of regular categories is useful to understand many of the recent developments in categorical algebra. The Mal'tsev axiom gives the opportunity to illustrate this method: in a regular category this axiom is equivalent to the permutability of the composition of equivalence relations, in the sense that any pair R and S of equivalence relations on a given object are such that $R \circ S = S \circ R$. Some recent results concerning the more general *Goursat categories* [10, 21] will then be explained in the last section. These aspects are useful to illustrate many of the links between exactness properties in categorical algebra, the so-called Mal'tsev conditions in universal algebra, and the validity of suitable homological lemmas [18, 19, 32].

1 Regular Categories

The notion of regular category plays an important role in the categorical understanding of algebraic structures. Regular categories capture some fundamental exactness properties shared by the categories Set of sets, Grp of groups, Ab of abelian groups, R-Mod of modules on a commutative ring R and, more generally, by any variety $\mathcal{V}$ of universal algebras. Topological models of "good" algebraic theories, such as the categories Grp(Top) of topological groups and Grp(Comp) of compact Hausdorff groups are also regular. Other examples will be considered later on in Sects. 1.3 and 3.1. The basic idea is that any arrow in a regular category can be factorized through an (essentially unique) *image*, and that these factorizations are stable under pullbacks.

Regular categories also have a prominent role in categorical logic (see [30], for instance, and the references therein). However, in this introductory course we shall only focus on the algebraic examples, with the goal of illustrating the importance of regular categories in categorical algebra.

In order to understand the notion of regular category it is useful to compare a few types of epimorphisms: this will be the subject of the following section (see [7] for further details).

1.1 Strong and Regular Epimorphisms

Definition 1.1 An arrow $f\colon A \to B$ in a category $\mathscr{C}$ is a *strong epimorphism* if, given any commutative square

$$
\begin{array}{ccc}
A & \xrightarrow{\ f\ } & B \\
{\scriptstyle g}\downarrow & \overset{t}{\nearrow} & \downarrow{\scriptstyle h} \\
C & \xrightarrow[\ m\ ]{} & D
\end{array}
$$

in $\mathscr{C}$, where $m\colon C \to D$ is a monomorphism, there exists a unique arrow $t\colon B \to C$ such that $m \circ t = h$ and $t \circ f = g$.

Remark 1.2 If the category $\mathscr{C}$ has binary products, then every strong epimorphism is an epimorphism. Indeed, if $f\colon A \to B$ is a strong epimorphism, and $u, v\colon B \to C$ are two arrows such that $u \circ f = v \circ f$, one can then consider the diagonal $(1_C, 1_C) = \Delta\colon C \to C \times C$ and the commutative square

$$
\begin{array}{ccc}
A & \xrightarrow{\ f\ } & B \\
{\scriptstyle u\circ f=v\circ f}\downarrow & & \downarrow{\scriptstyle (u,v)} \\
C & \xrightarrow[\ \Delta\ ]{} & C \times C.
\end{array}
$$

Since Δ is a monomorphism, there is a unique $t\colon B \to C$ such that $\Delta \circ t = (u, v)$ and $t \circ f = u \circ f = v \circ f$. It follows that

$$
u = p_1 \circ (u, v) = p_1 \circ \Delta \circ t = t = p_2 \circ \Delta \circ t = p_2 \circ (u, v) = v,
$$

where $p_1\colon C \times C \to C$ and $p_2\colon C \times C \to C$ are the product projections.

Lemma 1.3 *An arrow $f\colon A \to B$ is an isomorphism if and only if $f\colon A \to B$ is a monomorphism and a strong epimorphism.*

Proof If f is both a strong epimorphism and a monomorphism, one considers the commutative square

$$
\begin{array}{ccc}
A & \xrightarrow{\ f\ } & B \\
{\scriptstyle 1_A}\downarrow & \overset{t}{\nearrow} & \downarrow{\scriptstyle 1_B} \\
A & \xrightarrow[\ f\ ]{} & B.
\end{array}
$$

The unique (dotted) arrow $t\colon B \to A$ such that $f \circ t = 1_B$ and $t \circ f = 1_A$ is the inverse of f. The converse implication is obvious. $\qquad\square$

Exercise 1.4 Prove that strong epimorphisms are closed under composition, and that, if the composite $g \circ f$ of two composable arrows is a strong epimorphism, then g is a strong epimorphism. Show that if $g \circ f$ is a strong epimorphism, with g a monomorphism, then g is an isomorphism.

Definition 1.5 An arrow $f : A \to B$ is a *regular epimorphism* if it is the coequalizer of two arrows in $\mathscr{C}$.

Exercise 1.6 Prove that any regular epimorphism is an epimorphism.

Definition 1.7 A *split epimorphism* is an arrow $f : A \to B$ such that there is an arrow $i : B \to A$ with $f \circ i = 1_B$.

Observe that the axiom of choice in the category **Set** says precisely that any epimorphism is a split epimorphism. This is not the case in the categories **Grp** of groups or **Ab** of abelian groups, for instance. We are now going to prove the following chain of implications:

Proposition 1.8 *Let $\mathscr{C}$ be a category with binary products. One then has the implications*

split epimorphism $\Rightarrow$ regular epimorphism $\Rightarrow$ strong epimorphism $\Rightarrow$ epimorphism

Proof If $f : A \to B$ is split by an arrow $i : B \to A$, then f is the coequalizer of 1_A and $i \circ f$. Indeed, one sees that $f \circ (i \circ f) = f = f \circ 1_A$. Moreover, if $g : A \to X$ is such that $g \circ (i \circ f) = g \circ 1_A$, then $\phi = g \circ i$ is the only arrow with the property that $\phi \circ f = g$.

Assume then that $f : A \to B$ is a regular epimorphism. It is then the coequalizer of two arrows, say $u : T \to A$ and $v : T \to A$: consider the commutative diagram

$$
\begin{array}{ccc}
A & \xrightarrow{\ f\ } & B \\
{\scriptstyle g}\downarrow & & \downarrow{\scriptstyle h} \\
C & \xrightarrow[\ m\ ]{} & D
\end{array}
$$

with m a monomorphism. The equalities

$$ m \circ g \circ u = h \circ f \circ u = h \circ f \circ v = m \circ g \circ v $$

imply that $g \circ u = g \circ v$. The universal property of the coequalizer f implies that there is a unique $t : B \to C$ such that $t \circ f = g$. This arrow t is also such that $m \circ t = h$, so that f is a strong epimorphism.

The fact that any strong epimorphism is an epimorphism when $\mathscr{C}$ has binary products has been shown in Remark 1.2. $\square$

1.2 Quotients in Algebraic Categories

Let us then consider some examples of *quotients* in the categories of sets, of groups and of topological groups, which will be useful to explain the general construction in regular categories.

Let $f: A \to B$ be a map in Set, and

$$\mathrm{Eq}(f) = \{(x, y) \in A \times A \mid f(x) = f(y)\}$$

its *kernel pair*, i.e. the equivalence relation on A identifying the elements of A having the same image by f. This equivalence relation can be obtained by building the pullback of f along f:

$$\begin{array}{ccc}
\mathrm{Eq}(f) & \xrightarrow{\ p_2\ } & A \\
{\scriptstyle p_1}\downarrow & & \downarrow{\scriptstyle f} \\
A & \xrightarrow[\ f\ ]{} & B.
\end{array} \qquad (4.1.\mathrm{i})$$

Exercise 1.9 Show that any regular epimorphism f in a category with kernel pairs is the coequalizer of its kernel pair $(\mathrm{Eq}(f), p_1, p_2)$.

In the category Set of sets one sees that the canonical quotient $\pi: A \to A/\mathrm{Eq}(f)$ defined by $\pi(a) = \bar{a}$ is the coequalizer of p_1 and p_2. This yields a unique arrow $i: A/\mathrm{Eq}(f) \to B$ such that $i \circ \pi = f$:

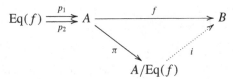

The map i is defined by $i(\bar{a}) = f(a)$ for any $\bar{a} \in A/\mathrm{Eq}(f)$. This gives a factorization $i \circ \pi$ of the arrow f, where π is a regular epimorphism (= a surjective map) and i is a monomorphism (= an injective map) in the category Set.

The same construction is possible in the category Grp of groups. Indeed, given a group homomorphism $f: G \to G'$, one can consider the kernel pair $\mathrm{Eq}(f)$ which is again obtained by the pullback (4.1.i) above, but this time computed in the category Grp. The equivalence relation $\mathrm{Eq}(f)$ is a group, as a subgroup of the product $G \times G$ of the group G with itself. The canonical quotient $\pi: G \to G/\mathrm{Eq}(f)$ is a group homomorphism, and this allows one to build the following commutative diagram in Grp

$$\mathrm{Eq}(f) \underset{p_2}{\overset{p_1}{\rightrightarrows}} G \xrightarrow{\quad f \quad} G'$$

with π going down to $G/\mathrm{Eq}(f)$ and i from $G/\mathrm{Eq}(f)$ up to G',

$$G/\mathrm{Eq}(f),$$

where π is a regular epimorphism and i is a monomorphism, exactly as in Set.

In the category Grp(Top) of topological groups, where the arrows are continuous homomorphisms, it is again possible to obtain the same kind of factorization regular epimorphism-monomorphism for any arrow. We write $(G, \cdot, \tau_G)$ for a topological group, where τ_G is the topology making both the multiplication $\cdot$ and the inversion of the group continuous. Given a continuous homomorphism $f \colon (G, \cdot, \tau_G) \to (G', \cdot, \tau_{G'})$ in Grp(Top), the kernel pair $(\mathrm{Eq}(f), \cdot, \tau_i)$ is a topological group for the topology τ_i induced by the product topology $\tau_{G \times G}$ of the topological group $(G \times G, \cdot, \tau_{G \times G})$. At the algebraic level the quotients in Grp(Top) are actually computed as in Grp, and then equipped with the quotient topology τ_q. In this way one gets the following commutative diagram

$$(Eq(f), \cdot, \tau_i) \underset{p_2}{\overset{p_1}{\rightrightarrows}} (G, \cdot, \tau_G) \xrightarrow{\quad f \quad} (G', \cdot, \tau_{G'})$$

with π going down to $(G/\mathrm{Eq}(f), \cdot, \tau_q)$ and i up to $(G', \cdot, \tau_{G'})$,

$$(G/\mathrm{Eq}(f), \cdot, \tau_q)$$

where π is the canonical quotient. It turns out that π is the coequalizer of the projections p_1 and p_2 in Grp(Top), and the induced arrow

$$i \colon (G/\mathrm{Eq}(f), \cdot, \tau_q) \to (G', \cdot, \tau_{G'})$$

is a monomorphism (since it is injective). Note that this factorization is not the one where the direct image $f(G)$ of the continuous homomorphism is equipped with the subspace topology induced by the topology of $(G', \cdot, \tau_{G'})$.

There are many other categories where the same construction as in Set, Grp and Grp(Top) is possible, for instance in the category Rng of rings, Mon of monoids, Ab of abelian groups and, more generally, in any variety $\mathcal{V}$ of universal algebras.

All these are examples of regular categories in the following sense:

Definition 1.10 [2] A finitely complete category $\mathscr{C}$ is *regular* if

- coequalizers of kernel pairs exist in $\mathscr{C}$;
- regular epimorphisms are pullback stable in $\mathscr{C}$.

1.3 Examples of Regular Categories

- The category Set is regular. We have observed that the coequalizers of kernel pairs exist in Set, and it remains to check the pullback stability of regular epimorphisms. Consider a pullback

$$
\begin{array}{ccc}
E \times_B A & \xrightarrow{\ \pi_2\ } & A \\
{\scriptstyle \pi_1}\downarrow & & \downarrow{\scriptstyle f} \\
E & \xrightarrow{\ p\ } & B
\end{array}
$$

in Set where p is a surjective map (i.e. a regular epimorphism), and let us show that π_2 is also surjective. Let a be an element in A; there exists then an $e \in E$ such that $p(e) = f(a)$. This shows that there is an $(e, a) \in E \times_B A$ such that $\pi_2(e, a) = a$, and π_2 is surjective. The same argument still works in the category Grp of groups, by taking into account the fact that regular epimorphisms therein are precisely the surjective homomorphisms, and that pullbacks are computed in Grp as in Set. For essentially the same reason the categories Rng of rings, Mon of monoids, and R-Mod of modules on a ring R are also regular categories. More generally, any variety $\mathcal{V}$ of universal algebras is a regular category, any quasivariety—such as the category $\mathsf{Ab}_{\mathrm{t.f.}}$ of torsion-free abelian groups—and also any category monadic over the category of sets, as for instance the category CHaus of compact Hausdorff spaces, and the category Frm of frames.

- The category Grp(Top) of topological groups is regular [10]. The main point here is that the canonical quotient $\pi \colon (H, \cdot, \tau_H) \to (H/\mathrm{Eq}(f), \cdot, \tau_q)$ of a topological group $(H, \cdot, \tau_H)$ by the equivalence relation $(\mathrm{Eq}(f), \cdot, \tau_i)$ which is the kernel pair of an arrow $f \colon (H, \cdot, \tau_H) \to (G, \cdot, \tau_G)$ in Grp(Top) is an *open surjective homomorphism*. To check this latter fact, let us write $K = \ker(\pi)$ for the kernel of π, and let us then show that

$$
\pi^{-1}(\pi(V)) = V \cdot K
$$

for any open $V \in \tau_H$. On the one hand if $z = v \cdot k$, where $v \in V$ and $k \in K$, one has

$$
\pi(z) = \pi(v \cdot k) = \pi(v) \cdot \pi(k) = \pi(v) \in \pi(V),
$$

so that $z \in \pi^{-1}(\pi(V))$. Conversely, if $z \in \pi^{-1}(\pi(V))$, then $\pi(z) = \pi(v_1)$, for a $v_1 \in V$, so that $v_1^{-1} \cdot z \in K$, and $z = v_1 \cdot k$, for a $k \in K$.
This implies that

$$
\pi^{-1}(\pi(V)) = (\bigcup_{k \in K} V \cdot k) \in \tau_H.
$$

Indeed, the function $m_k \colon G \to G$ defined by $m_k(x) = x \cdot k$ for any $x \in G$ (with fixed $k \in K$) is a homeomorphism, hence $V \cdot k = m_k(V) \in \tau_H$, since $V \in \tau_H$. We have then shown that $\pi(V)$ is open for any $V \in \tau_H$, and the map π is open. It follows that in Grp(Top) the regular epimorphisms are the open surjective homomorphisms. To conclude that Grp(Top) is a regular category it suffices to recall that the open surjective homomorphisms are pullback stable (a well known fact which can be easily checked). More generally, the models of any Mal'tsev theory in the category of topological spaces, i.e. any category of topological Mal'tsev algebras, is a regular category [31]. Notice that also the category Grp(Haus) of Hausdorff groups, or Grp(Comp) of compact Hausdorff groups are regular [4] (see also [13] for the categorical properties of topological semi-abelian algebras).

- As mentioned in the Introduction any abelian category [9] is a regular category, as is any elementary topos [30].
- The category Top of topological spaces, unlike Grp(Top), is not regular. The main reason is that in Top regular epimorphisms are quotient maps, and these are not pullback stable (see [3] for a counter-example, for instance).

1.4 Canonical Factorization

We are now going to show that any arrow in a regular category has a canonical factorization as a regular epimorphism followed by a monomorphism, exactly as in the examples of the categories Set, Grp and Grp(Top) recalled here above.

Theorem 1.11 *Let $\mathscr{C}$ be a regular category. Then*

1. *any arrow $f : A \to B$ in $\mathscr{C}$ has a factorization $f = m \circ q$, with q a regular epimorphism and m a monomorphism;*
2. *this factorization is unique (up to isomorphism).*

Proof 1. Let $f : A \to B$ be an arrow in $\mathscr{C}$. Consider the diagram here below where $(\mathrm{Eq}(f), f_1, f_2)$ is the kernel pair of f, q is the coequalizer of (f_1, f_2), and m the unique arrow such that $m \circ q = f$.

$$\mathrm{Eq}(f) \underset{f_2}{\overset{f_1}{\rightrightarrows}} A \overset{q}{\longrightarrow} I \qquad\qquad (4.1.\mathrm{ii})$$

with f going down-right to B and m the dotted arrow from I down to B.

We need to show that m is a monomorphism or, equivalently, that the projections $p_1 : \mathrm{Eq}(m) \to I$ and $p_2 : \mathrm{Eq}(m) \to I$ of the kernel pair of m are equal. For this, consider the diagram

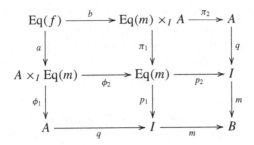

where all the squares are pullbacks. We know that the whole square is then a pullback, so that one can assume that $f_1 = \phi_1 \circ a$ and $f_2 = \pi_2 \circ b$. The arrow $\phi_2 \circ a = \pi_1 \circ b$ is then an epimorphism, as a composite of epimorphisms (we have used the pullback stability of regular epimorphisms). The fact that $\phi_1 \circ a = f_1$ and $\pi_2 \circ b = f_2$ implies that

$$p_1 \circ (\phi_2 \circ a) = q \circ \phi_1 \circ a = q \circ f_1 = q \circ f_2 = q \circ \pi_2 \circ b = p_2 \circ \pi_1 \circ b = p_2 \circ (\phi_2 \circ a).$$

Since $\phi_2 \circ a$ is an epimorphism, it follows that $p_1 = p_2$, as desired.
2. To prove the uniqueness of the factorization one can use the fact that any regular epimorphism is a strong epimorphism.

$\square$

Remark 1.12 The uniqueness of the factorization of any arrow f in Theorem 1.11 allows one to call the subobject $m : I \to B$ in diagram (4.1.ii) the (regular) *image* of f.

Proposition 1.13 *In a regular category $\mathscr{C}$ the following properties are satisfied:*

1. *regular epimorphisms coincide with strong epimorphisms;*
2. *if $g \circ f$ is a regular epimorphism, then g is a regular epimorphism;*
3. *if g and f are regular epimorphisms, then $g \circ f$ is a regular epimorphism;*
4. *if $f : X \to Y$ and $g : X' \to Y'$ are regular epimorphisms, then the induced arrow $f \times g : X \times X' \to Y \times Y'$ is also a regular epimorphism.*

Proof 1. One needs to check that any strong epimorphism $f : A \to B$ is a regular epimorphism. Consider the factorization $f = m \circ q$ of a strong epimorphism, with m a monomorphism and q a regular epimorphism (Theorem 1.11). The commutativity of the diagram

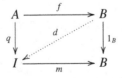

yields a unique arrow $d \colon B \to I$ such that $d \circ f = q$ and $m \circ d = 1_B$. This arrow d is the inverse of m, and f is then a regular epimorphism.

2. Follows from 1. and the properties of strong epimorphisms (Exercises 1.4).
3. Same argument as for 2.
4. If $f \colon X \to Y$ is a regular epimorphism, consider the commutative diagram

$$
\begin{array}{ccc}
X \times X' & \xrightarrow{\; f \times 1_{X'} \;} & Y \times X' \\
{\scriptstyle \pi_1} \downarrow & & \downarrow {\scriptstyle \pi_1} \\
X & \xrightarrow{\quad f \quad} & Y
\end{array}
$$

which is easily seen to be a pullback. The arrow $f \times 1_{X'}$ is then a regular epimorphism and, similarly, one checks that $1_Y \times g$ is a regular epimorphism. Since $f \times g = (1_Y \times g) \circ (f \times 1_{X'})$, this arrow is a regular epimorphism by 3.

$\square$

We are now going to give an equivalent formulation of the notion of regular category:

Theorem 1.14 *Let $\mathscr{C}$ be a finitely complete category. Then $\mathscr{C}$ is a regular category if and only if*

1. *any arrow in $\mathscr{C}$ factorizes as a regular epimorphism followed by a monomorphism;*
2. *these factorizations are* pullback stable: *if $m \circ q$ is the factorization of an arrow $p \colon E \to B$, $f \colon A \to B$ any arrow, and the squares*

$$
\begin{array}{ccccc}
E \times_B A & \xrightarrow{\; q' \;} & E' \times_B A & \xrightarrow{\; m' \;} & A \\
{\scriptstyle \pi_1} \downarrow & & \downarrow & & \downarrow {\scriptstyle f} \\
E & \xrightarrow{\quad q \quad} & E' & \xrightarrow{\quad m \quad} & B
\end{array}
$$

are pullbacks, then $m' \circ q'$ is the factorization of the pullback projection $\pi_2 \colon E \times_B A \to A$.

Proof When $\mathscr{C}$ is regular, 1. and 2. follow from Theorem 1.11.
For the converse, it is clear that 2. implies that regular epimorphisms are pullback stable. It remains to show that any kernel pair

$$
\mathrm{Eq}(f) \underset{f_2}{\overset{f_1}{\rightrightarrows}} X \tag{4.1.iii}
$$

of an arrow $f \colon X \to Y$ has a coequalizer. For this consider the regular epimorphism-monomorphism factorization $m \circ q$ of f (which exists by 1.), and observe that

(4.1.iii) is also the kernel pair of the regular epimorphism q, since m is a monomorphism. The arrow q is then the coequalizer of its kernel pair (4.1.iii) (see Exercise 1.9).   □

1.5   The Barr-Kock Theorem

The following result will be useful to prove the so-called Barr-Kock Theorem:

Lemma 1.15 *Consider a commutative diagram*

$$
\begin{array}{ccccc}
A & \xrightarrow{\ k\ } & B & \xrightarrow{\ l\ } & C \\
\Big\downarrow{\scriptstyle a} & & \Big\downarrow{\scriptstyle b} & & \Big\downarrow{\scriptstyle c} \\
A' & \xrightarrow{\ k'\ } & B' & \xrightarrow{\ l'\ } & C'
\end{array}
$$

in a regular category $\mathscr{C}$, where the left-hand square and the external rectangle are pullbacks. If k' is a regular epimorphism, then the right-hand square is a pullback.

Proof Consider the commutative diagram

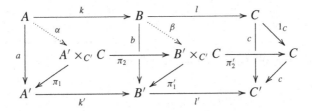

where $(B' \times_{C'} C, \pi_1', \pi_2')$ is the pullback of l' and c, and $(A' \times_{C'} C, \pi_1, \pi_2)$ is the pullback of k' and π_1', with α and β the naturally induced arrows. The fact that the external rectangle is a pullback implies that the arrow α is an isomorphism. The arrow π_2 is a regular epimorphism (because k' is one), so that $\pi_2 \circ \alpha = \beta \circ k$ is a regular epimorphism, and then β is a regular epimorphism (see Proposition 1.13). The arrow β is a monomorphism: this follows from the fact that the square

$$
\begin{array}{ccc}
A & \xrightarrow{\ k\ } & B \\
\Big\downarrow{\scriptstyle \alpha} & & \Big\downarrow{\scriptstyle \beta} \\
A' \times_{C'} C & \xrightarrow{\ \pi_2\ } & B' \times_{C'} C
\end{array}
$$

is a pullback, so that both the induced commutative squares

$$\text{Eq}(\alpha) \dashrightarrow \text{Eq}(\beta)$$

$$p_1 \Vert p_2 \qquad\qquad p_1 \Vert p_2$$

$$A \xrightarrow{\quad k \quad} B$$

are pullbacks, where the (unique) dotted arrow making them commute is then a (regular) epimorphism. The arrows $p_1 : \text{Eq}(\alpha) \to A$ and $p_2 : \text{Eq}(\alpha) \to A$ are equal (since α is a monomorphism), so that the projections $p_1 : \text{Eq}(\beta) \to B$ and $p_2 : \text{Eq}(\beta) \to B$ are also equal, and then β is a monomorphism.   □

We are now ready to prove the following interesting result, often referred to as the *Barr-Kock Theorem* [1], although it was first observed by A. Grothendieck [24] in a different context (see also [7]):

Theorem 1.16 *Let $\mathscr{C}$ be a regular category, and*

$$\text{Eq}(f) \underset{p_2}{\overset{p_1}{\rightrightarrows}} A \xrightarrow{\ f\ } X$$

$$v \downarrow \qquad u \downarrow \qquad w \downarrow$$

$$\text{Eq}(g) \underset{p_2}{\overset{p_1}{\rightrightarrows}} B \xrightarrow{\ g\ } Y$$

a commutative diagram with f a regular epimorphism. If either of the left-hand commutative squares are pullbacks, then the right-hand square is a pullback.

Proof Consider the following commutative diagram

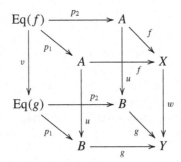

The assumptions guarantee that the left-hand face and the bottom face of the cube are pullbacks. By commutativity it follows that the rectangle

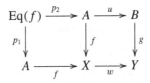

is also a pullback, as well as its left-hand square. Since f is a regular epimorphism, by Lemma 1.15 it follows that the right-hand square is a pullback. $\square$

2 Relations in Regular Categories

Definition 2.1 An internal *relation from X to Y* in a category $\mathscr{C}$ is a graph

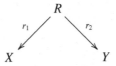

such that the pair (r_1, r_2) is jointly monomorphic. When the product $X \times Y$ exists, this is equivalent to the fact that the factorization $(r_1, r_2) : R \to X \times Y$ is a monomorphism.

As usual, we identify two (internal) relations $R \to X \times Y$ and $S \to X \times Y$ when they determine the same *subobject* of $X \times Y$, i.e. the same equivalence class of monomorphisms with codomain $X \times Y$. If $X = Y$, one says that R is a relation on X.

- A relation R on X is *reflexive* when there is an arrow $\delta : X \to R$ such that $r_1 \circ \delta = 1_X = r_2 \circ \delta$.
- R is *symmetric* if there is an arrow $\sigma : R \to R$ such that $r_1 \circ \sigma = r_2$ and $r_2 \circ \sigma = r_1$.
- Consider the pullback

$$
\begin{array}{ccc}
R \times_X R & \xrightarrow{\ p_2\ } & R \\
{\scriptstyle p_1}\downarrow & & \downarrow{\scriptstyle r_1} \\
R & \xrightarrow[\ r_2\ ]{} & X.
\end{array}
$$

The relation R is *transitive* if there is an arrow $\tau : R \times_X R \to R$ such that $r_1 \circ \tau = r_1 \circ p_1$ and $r_2 \circ \tau = r_2 \circ p_2$.

A relation R on X is an *equivalence relation* if R is reflexive, symmetric and transitive. Of course, this abstract notion of equivalence relation gives in particular the usual one when $\mathscr{C}$ is the category of sets.

When $\mathscr{C} = $ Grp, an equivalence relation $R \subset X \times X$ in Grp is an equivalence relation on the underlying set of X which is also a subgroup of the group $X \times X$. In universal algebra, an internal equivalence relation in a variety is called a *congruence*.

Lemma 2.2 *In a category with pullbacks the kernel pair* $\mathrm{Eq}(f) \underset{p_2}{\overset{p_1}{\rightrightarrows}} X$ *of an arrow* $f : X \to Y$ *is an equivalence relation on X in* $\mathscr{C}$.

Proof The arrows $p_1 : \mathrm{Eq}(f) \to X$ and $p_2 : \mathrm{Eq}(f) \to X$ are jointly monomorphic, since they are projections of a pullback. The universal property of the kernel pair $(\mathrm{Eq}(f), p_1, p_2)$ implies that there is a unique $\delta : X \to \mathrm{Eq}(f)$ such that $p_1 \circ \delta = 1_X = p_2 \circ \delta$

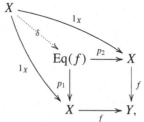

and $\mathrm{Eq}(f)$ is then reflexive. Similarly, the commutativity of the external part of the diagram

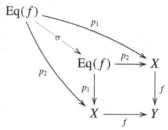

implies that there is a unique arrow $\sigma : \mathrm{Eq}(f) \to \mathrm{Eq}(f)$ such that $p_1 \circ \sigma = p_2$ and $p_2 \circ \sigma = p_1$, hence $\mathrm{Eq}(f)$ is symmetric. For the transitivity of $\mathrm{Eq}(f)$ one considers the following commutative diagram

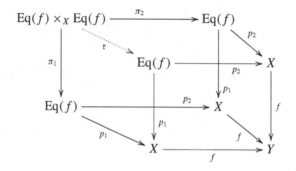

where the back face is a pullback. The universal property of the kernel pair $(\mathrm{Eq}(f), p_1, p_2)$ shows that there is a unique τ such that $p_1 \circ \tau = p_1 \circ \pi_1$ and $p_2 \circ \tau = p_2 \circ \pi_2$.   □

An important aspect of regular categories is that in these categories one can define a composition of relations, which has some nice properties.

In the category Set, if $R \to X \times Y$ is a relation from X to Y and $S \to Y \times Z$ a relation from Y to Z, one usually defines the relation $S \circ R \to X \times Z$ by setting

$$S \circ R = \{(x, z) \in X \times Z \text{ such that } \exists\, y \in Y \text{ with } x R y, y S z\}.$$

This construction is also possible in any regular category $\mathscr{C}$, thanks to the existence of regular images (Theorem 1.11). One first builds the pullback

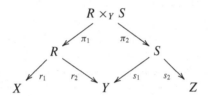

and one then factorizes the arrow $(r_1 \circ \pi_1, s_2 \circ \pi_2): R \times_Y S \to X \times Z$ as a regular epimorphism $q: R \times_Y S \to I$ followed by a monomorphism $i: I \to X \times Z$:

$$R \times_Y S \xrightarrow{\;q\;} I \xrightarrow{\;i\;} X \times Z$$

In Set, the set I consists of the element $(x, z) \in X \times Z$ such that there is a $(u, y, v) \in R \times_Y S$ with $u = x$ and $v = z$: this is precisely $S \circ R$.

This composition is actually associative:

Theorem 2.3 *Let $\mathscr{C}$ be a regular category. If $R \to A \times B$, $S \to B \times C$ and $T \to C \times D$ are relations in $\mathscr{C}$, one has the equality*

$$T \circ (S \circ R) = (T \circ S) \circ R.$$

Proof Consider the diagram obtained by building the following pullbacks:

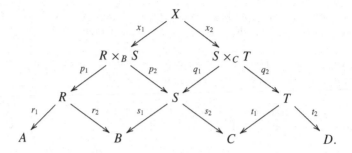

The proof consists in showing that the relations $T \circ (S \circ R)$ and $(T \circ S) \circ R$ are both given by the regular image $i: I \to A \times D$ in the factorization

as a regular epimorphism followed by a monomorphism of the arrow

$$(r_1 \circ p_1 \circ x_1, t_2 \circ q_2 \circ x_2): X \to A \times D.$$

We leave it to the reader the verification of this fact, which uses the pullback stability of regular epimorphisms in a crucial way.   □

This result allows one to define a new category starting from any regular category $\mathscr{C}$, the category $\mathsf{Rel}(\mathscr{C})$ of relations in $\mathscr{C}$. The objects are the same as the ones in $\mathscr{C}$, an arrow from X to Y is simply a relation from X to Y, and composition is the relational one defined above. For any relation R from X to Y the discrete relation (also called the equality relation) on X

$$\Delta_X : X \overset{1_X}{\underset{1_X}{\rightrightarrows}} X$$

is such that $R \circ \Delta_X = R$, and for any relation S from Z to X one has $\Delta_X \circ S = S$. It follows that the arrow Δ_X in $\mathsf{Rel}(\mathscr{C})$ is the identity on the object X for the composition in $\mathsf{Rel}(\mathscr{C})$.

There is a faithful functor $\Gamma: \mathscr{C} \to \mathsf{Rel}(\mathscr{C})$, where $\Gamma(f)$ is the *graph* of $f: X \to Y$, seen as a relation:

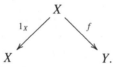

From now on we shall write 1_X for the discrete relation on X, which can also be seen as the relation $\Gamma(1_X)$.

Remark 2.4 Rel($\mathscr{C}$) is not only a category, but a (locally ordered) 2-category. Indeed, there is a natural partial ordering on its arrows, since the relations from X to Y are the subobjects of a fixed object $X \times Y$ of $\mathscr{C}$. This order is also compatible with the composition: if $R \leq S$, then $R \circ T \leq S \circ T$ whenever these composites are defined. This is the main argument to show that Rel($\mathscr{C}$) is a 2-category, which is actually locally-ordered: between any two arrows (or 1-cells) there is at most one 2-cell, and the only invertible 2-cells are the identities (see [30] for more details).

3 Calculus of Relations and Mal'tsev Categories

In this section we shall always assume that the category $\mathscr{C}$ is regular.
Given a relation $R = (R, r_1, r_2)$

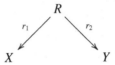

from X to Y, we write $R^\circ = (R, r_2, r_1)$ for the *opposite relation* from Y to X:

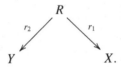

Of course $(R^\circ)^\circ = R$. It is easy to see that a relation R is symmetric if and only if $R = R^\circ$. Additionally, a relation R is transitive when $R \circ R \leq R$. Moreover, in a regular category, any relation (R, r_1, r_2) can be seen as the composite $R = r_2 \circ r_1^\circ$. By definition of the composition of relations, the relation $(X \times_Y Z, p_1, p_2)$ in a pullback

$$
\begin{array}{ccc}
X \times_Y Z & \xrightarrow{p_2} & Z \\
\downarrow{\scriptstyle p_1} & & \downarrow{\scriptstyle g} \\
X & \xrightarrow{f} & Y
\end{array}
$$

can be written as $g^\circ \circ f$. We leave the verification of the following properties to the reader:

Lemma 3.1 *In a regular category* $\mathscr{C}$:

1. *any kernel pair* $(Eq(f), f_1, f_2)$ *of an arrow* $f : X \to Y$ *can be written as* $f^\circ \circ f$;
2. $f : X \to Y$ *is a regular epimorphism if and only if* $f \circ f^\circ = 1_Y$;
3. $f : X \to Y$ *is a monomorphism if and only if* $f^\circ \circ f = 1_X$.

The relations that are "maps", i.e. of the form

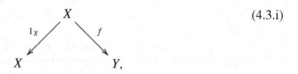

(4.3.i)

for some arrow f in $\mathscr{C}$, have the following additional property:

Lemma 3.2 *Any relation of the form (4.3.i) is difunctional:*

$$f \circ f^\circ \circ f = f.$$

Proof The relation $f \circ f^\circ \circ f = f$ is obtained as the regular image of the external graph in the following diagram,

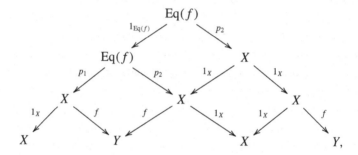

which is simply the regular image of the graph

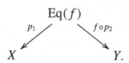

Since $p_1 : \mathrm{Eq}(f) \to X$ is a split epimorphism, thus in particular a regular epimorphism (by Proposition 1.8), we see that the relation $f \circ f^\circ \circ f$ is given by the relation $(1_X, f)$ in the commutative diagram

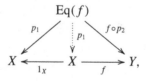

as desired. □

In the category of sets the notion of *difunctional relation* was first introduced by J. Riguet [36]. A relation R is difunctional if the fact that $(x, y) \in R$, $(z, y) \in R$ and $(z, u) \in R$ implies that $(x, u) \in R$. This property can be expressed in any regular category as follows:

Definition 3.3 A relation (R, r_1, r_2) from X to Y in a regular category is *difunctional* if

$$R \circ R^{\circ} \circ R = R.$$

The following notion was introduced by A. Carboni, J. Lambek and M.C. Pedicchio in [11], and it has been investigated in several articles in the last 30 years.

Definition 3.4 A finitely complete category $\mathscr{C}$ is called a *Mal'tsev* category if any internal reflexive relation in $\mathscr{C}$ is an equivalence relation.

The following characterization of regular Mal'tsev categories can be found in [11] (see also [34]). It is an example of a proof using the so-called calculus of relations.

Theorem 3.5 *Let $\mathscr{C}$ be a regular category. Then the following conditions are equivalent:*

1. *for any pair of equivalence relations R and S on any object X in $\mathscr{C}$,*
   $S \circ R$ is an equivalence relation;
2. *for any pair of equivalence relations R and S on any object X in $\mathscr{C}$,*
   $S \circ R = R \circ S$;
3. *for any pair of kernel pairs $\mathrm{Eq}(f)$ and $\mathrm{Eq}(g)$ on any object X in $\mathscr{C}$,*
   $\mathrm{Eq}(g) \circ \mathrm{Eq}(f) = \mathrm{Eq}(f) \circ \mathrm{Eq}(g)$;
4. *any relation U from X to Y in $\mathscr{C}$ is difunctional;*
5. *any reflexive relation R on an object X in $\mathscr{C}$ is an equivalence relation;*
6. *any reflexive relation R on an object X in $\mathscr{C}$ is symmetric;*
7. *any reflexive relation R on an object X in $\mathscr{C}$ is transitive.*

Proof 1. $\Rightarrow$ 2. By assumption the relation $S \circ R$ is an equivalence relation, thus it is symmetric:

$$(S \circ R)^{\circ} = S \circ R.$$

Since both S and R are symmetric it follows that

$$R \circ S = R^{\circ} \circ S^{\circ} = (S \circ R)^{\circ} = S \circ R.$$

2. $\Rightarrow$ 3. Obvious, since any kernel pair is an equivalence relation (Lemma 2.2).
3. $\Rightarrow$ 4. Any relation (U, u_1, u_2) can be written as $U = u_2 \circ u_1^{\circ}$. The assumption implies that the kernel pairs $\mathrm{Eq}(u_1)$ and $\mathrm{Eq}(u_2)$ of the projections commute in the sense of the composition of relations (on the object U):

$$(u_2^{\circ} \circ u_2) \circ (u_1^{\circ} \circ u_1) = (u_1^{\circ} \circ u_1) \circ (u_2^{\circ} \circ u_2).$$

By keeping in mind that the relations u_1 and u_2 are difunctional (by Lemma 3.2) it follows that

$$
\begin{aligned}
U &= u_2 \circ u_1^\circ \\
  &= (u_2 \circ u_2^\circ \circ u_2) \circ (u_1^\circ \circ u_1 \circ u_1^\circ) \\
  &= u_2 \circ (u_2^\circ \circ u_2) \circ (u_1^\circ \circ u_1) \circ u_1^\circ \\
  &= u_2 \circ (u_1^\circ \circ u_1) \circ (u_2^\circ \circ u_2) \circ u_1^\circ \\
  &= (u_2 \circ u_1^\circ) \circ (u_1 \circ u_2^\circ) \circ (u_2 \circ u_1^\circ) \\
  &= U \circ U^\circ \circ U.
\end{aligned}
$$

4. $\Rightarrow$ 5. Let (U, u_1, u_2) be a reflexive relation on an object X, so that $1_X \leq U$. By difunctionality we have:

$$
U^\circ = 1_X \circ U^\circ \circ 1_X \leq U \circ U^\circ \circ U = U,
$$

showing that U is symmetric. On the other hand:

$$
U \circ U = U \circ 1_X \circ U \leq U \circ U^\circ \circ U = U,
$$

and U is transitive.

5. $\Rightarrow$ 6. Clear.

6. $\Rightarrow$ 1. First observe that $S \circ R$ is reflexive, since both S and R are reflexive:

$$
1_X = 1_X \circ 1_X \leq S \circ R.
$$

By assumption the relation $S \circ R$ is then symmetric, so that

$$
R \circ S = R^\circ \circ S^\circ = (S \circ R)^\circ = S \circ R.
$$

The relation $S \circ R$ is transitive:

$$
S \circ R = (S \circ S) \circ (R \circ R) = S \circ (S \circ R) \circ R = S \circ (R \circ S) \circ R = S \circ R \circ S \circ R.
$$

Observe that 5. $\Rightarrow$ 7. is obvious, and let us prove that 7. $\Rightarrow$ 4. Let $U = u_2 \circ u_1^\circ$ be any relation from X to Y. The relation

$$
u_2^\circ \circ u_2 \circ u_1^\circ \circ u_1
$$

is reflexive, thus it is transitive by assumption. This gives the equality

$$
(u_2^\circ \circ u_2 \circ u_1^\circ \circ u_1) \circ (u_2^\circ \circ u_2 \circ u_1^\circ \circ u_1) = u_2^\circ \circ u_2 \circ u_1^\circ \circ u_1,
$$

yielding

$$u_2 \circ u_2^\circ \circ u_2 \circ u_1^\circ \circ u_1 \circ u_2^\circ \circ u_2 \circ u_1^\circ \circ u_1 \circ u_1^\circ = u_2 \circ u_2^\circ \circ u_2 \circ u_1^\circ \circ u_1 \circ u_1^\circ.$$

By difunctionality of u_2 and u_1° we conclude that

$$u_2 \circ u_1^\circ \circ u_1 \circ u_2^\circ \circ u_2 \circ u_1^\circ = u_2 \circ u_1^\circ,$$

and

$$U \circ U^\circ \circ U = U.$$

□

3.1   Examples of Mal'tsev Categories

The categories Grp, Ab, R-Mod, Rng and Grp(Top) are all Mal'tsev categories. By Theorem 3.5 to see this it suffices to show that any (internal) reflexive relation R on any object X in these categories is symmetric. Let us check this property for the category Grp of groups: given an element (x, y) of a reflexive relation R which is also a subgroup of $X \times X$, we know that its inverse (x^{-1}, y^{-1}) is also in R and, by reflexivity, both (x, x) and (y, y) belong to R. It follows that

$$(x, x) \cdot (x^{-1}, y^{-1}) \cdot (y, y) = (x \cdot x^{-1} \cdot y, x \cdot y^{-1} \cdot y) = (y, x) \in R$$

and Grp is a Mal'tsev category. An inspection of the proof for Grp shows that the argument is still valid if the theory of an algebraic variety has a term $p(x, y, z)$ such that $p(x, y, y) = x$ and $p(x, x, y) = y$. Varieties of algebras having such a ternary term p are called Mal'tsev varieties [37], or 2-permutable varieties, and the term p a Mal'tsev operation. This terminology is motivated by the famous Mal'tsev theorem asserting that a variety $\mathcal{V}$ of algebras has the property that each pair R and S of congruences on an algebra A in $\mathcal{V}$ permute, i.e. $R \circ S = S \circ R$ if and only if its theory has a ternary Mal'tsev operation [33].

Of course, any variety of algebras whose theory contains the operations and identities of the theory of groups is a Mal'tsev variety.

For a different example, consider the variety QGrp of quasigroups [37]: its algebraic theory has a multiplication $\cdot$, a left division $\backslash$ and a right division $/$ such that $x \backslash (x \cdot y) = y$, $(x \cdot y)/y = x$, $x \cdot (x \backslash y) = y$ and $(x/y) \cdot y = x$.
A Mal'tsev operation for the theory of quasigroups is given by the term

$$p(x, y, z) = (x/(y \backslash y)) \cdot (y \backslash z),$$

since

$$p(x, y, y) = (x/(y \backslash y)) \cdot (y \backslash y) = x,$$

and

$$p(x, x, y) = (x/(x\backslash x)) \cdot (x\backslash y) = (x \cdot (x\backslash x)/(x\backslash x)) \cdot (x\backslash y) = x \cdot (x\backslash y) = y.$$

The category **Heyt** of Heyting algebras is a Mal'tsev variety [29], with a Mal'tsev operation defined by the term

$$p(x, y, z) = ((x \to y) \to z) \wedge ((z \to y) \to x).$$

For the axioms and basic properties of Heyting algebras we refer the reader to [29], or to the Chapter *Notes on point-free topology* [35] in this volume. One observes that

$$\begin{aligned} p(x, x, y) &= ((x \to x) \to y) \wedge ((y \to x) \to x) \\ &= (1 \to y) \wedge ((y \to x) \to x) \\ &= y \wedge ((y \to x) \to x) \\ &= y \end{aligned}$$

and

$$\begin{aligned} p(x, y, y) &= ((x \to y) \to y) \wedge ((y \to y) \to x) \\ &= ((x \to y) \to y) \wedge (1 \to x) \\ &= ((x \to y) \to y) \wedge x \\ &= x. \end{aligned}$$

Other examples of regular Mal'tsev categories are: any regular additive category, therefore in particular any abelian category [9], and the dual of any elementary topos [10]. The category of **C***-algebras and the category $\mathsf{Hopf}_{K,coc}$ of cocommutative Hopf algebras over a field K are also regular Mal'tsev categories [22, 23].

On the other hand, the categories **Set** of sets and **Mon** of monoids are regular categories which are not Mal'tsev ones. Indeed, the usual order relation $\leq$ on $\mathbb{N}$ is an internal reflexive relation (both in **Set** and in **Mon**) which is not symmetric.

3.2 Regular Pushouts

An important property of regular Mal'tsev categories is expressed in terms of diagrams of the form

$$\begin{array}{ccc} C & \xrightarrow{c} & A \\ {\scriptstyle g}\big\uparrow\big\downarrow{\scriptstyle t} & & {\scriptstyle f}\big\uparrow\big\downarrow{\scriptstyle s} \\ D & \xrightarrow{d} & B \end{array} \qquad (4.3.\text{ii})$$

where $d \circ g = f \circ c$, $c \circ t = s \circ d$, $g \circ t = 1_D$, $f \circ s = 1_B$, c and d are regular epimorphisms. As observed in [15] such a square is always a pushout. The following result is due to D. Bourn (see also [10]): here we give an alternative proof using the calculus of relations as in [19]:

Proposition 3.6 [6] *A regular category $\mathscr{C}$ is a Mal'tsev category if and only if any pushout of the form (4.3.ii) has the property that the canonical morphism $(g, c): C \to D \times_B A$ to the pullback of d and f is a regular epimorphism.*

Proof The relation $(D \times_B A, p_1, p_2)$ which is the pullback of d and f can be expressed as the composite $f^\circ \circ d$. The regular image of $(g, c): C \to D \times_B A$ is $c \circ g^\circ$, so that (g, c) is a regular epimorphism if and only if $f^\circ \circ d = c \circ g^\circ$. Now, the regular image $g(\mathsf{Eq}(c))$ of $\mathsf{Eq}(c)$ along g is defined as the regular image of the arrow $g \times g \circ (p_1, p_2): \mathsf{Eq}(c) \to B \times B$, i.e. the subobject of $B \times B$ determined by the right-hand vertical arrow in the following commutative diagram

$$
\begin{array}{ccc}
\mathsf{Eq}(c) & \longrightarrow & g(\mathsf{Eq}(c)) \\
{\scriptstyle (p_1,p_2)}\downarrow & & \downarrow \\
D \times D & \underset{g \times g}{\longrightarrow} & B \times B.
\end{array}
$$

The commutativity conditions on the square (4.3.ii) imply that this relation is $\mathsf{Eq}(d)$:

$$g(\mathsf{Eq}(c)) = \mathsf{Eq}(d).$$

In a regular category this condition can be expressed by the equality

$$g \circ c^\circ \circ c \circ g^\circ = d^o \circ d.$$

Since $c \circ c^\circ = 1_A$ by Lemma 3.1 it follows that

$$
\begin{aligned}
f^\circ \circ d &= c \circ c^\circ \circ f^\circ \circ d \\
&= c \circ g^\circ \circ d^o \circ d \\
&= c \circ g^\circ \circ (g \circ c^\circ \circ c \circ g^\circ) \\
&= c \circ c^\circ \circ c \circ g^\circ \circ g \circ g^\circ \\
&= c \circ g^\circ,
\end{aligned}
$$

where the fourth equality follows from the Mal'tsev assumption:

$$g^\circ \circ g \circ c^\circ \circ c = \mathsf{Eq}(g) \circ \mathsf{Eq}(c) = \mathsf{Eq}(c) \circ \mathsf{Eq}(g) = c^\circ \circ c \circ g^\circ \circ g.$$

For the converse, by Theorem 3.5 it suffices to show that any pair of equivalence relations $\mathsf{Eq}(f)$ and $\mathsf{Eq}(g)$ which are kernel pairs of two arrows f and g permute. Note that there is no restriction in assuming that f and g are regular epimorphisms, thanks to Theorem 1.11. Consider the kernel pair $(\mathrm{Eq}(f), f_1, f_2)$ of $f: X \to Y$ and the kernel pair $(\mathrm{Eq}(g), g_1, g_2)$ of $g: X \to Z$. We then consider the regular image of $\mathrm{Eq}(f)$ along g inducing the following commutative diagram

$$
\begin{array}{ccc}
\mathrm{Eq}(f) & \xrightarrow{\;\gamma\;} & g(\mathrm{Eq}(f)) \\
{\scriptstyle f_1}\big\updownarrow{\scriptstyle f_2} & & {\scriptstyle r_1}\big\updownarrow{\scriptstyle r_2} \\
X & \xrightarrow[\;g\;]{} & Z,
\end{array}
\qquad (4.3.\mathrm{iii})
$$

and observe that the assumption implies that $f_2 \circ \gamma^\circ = g^\circ \circ r_2$ and $\gamma \circ f_1^\circ = r_1^\circ \circ g$. We then have the following identities:

$$
\begin{aligned}
\mathrm{Eq}(f) \circ \mathrm{Eq}(g) &= f_2 \circ f_1^\circ \circ g^\circ \circ g \\
&= f_2 \circ \gamma^\circ \circ r_1^\circ \circ g \\
&= g^\circ \circ r_2 \circ r_1^\circ \circ g \\
&= g^\circ \circ r_2 \circ \gamma \circ f_1^\circ \\
&= g^\circ \circ g \circ f_2 \circ f_1^\circ. \\
&= \mathrm{Eq}(g) \circ \mathrm{Eq}(f).
\end{aligned}
$$

$\square$

4 Goursat Categories

In universal algebra a weaker property than the Mal'tsev axiom is the so-called 3-permutability of congruences. Given any two congruences R and S on an algebra A in a variety $\mathcal{V}$, the following equality holds:

$$
R \circ S \circ R = S \circ R \circ S.
$$

Definition 4.1 [10, 11] A regular category $\mathscr{C}$ is a *Goursat category* if

$$
R \circ S \circ R = S \circ R \circ S
$$

for any pair of equivalence relations R and S on any object X in $\mathscr{C}$.

Any regular Mal'tsev category $\mathscr{C}$ is a Goursat category: indeed, given any two equivalence relations R and S on an object X in $\mathscr{C}$, one has:

$$
R \circ (S \circ R) = R \circ (R \circ S) = R \circ S = R \circ (S \circ S) = (S \circ R) \circ S.
$$

An example of a Goursat category which is not a Mal'tsev one will be given at the end of this section, where we shall prove that implication algebras form a Goursat variety.

Among regular categories, Goursat categories are characterized by the property that equivalence relations are stable under regular images along regular epimorphisms [10]. Here below we give a direct proof which uses the calculus of relations:

Proposition 4.2 *For a regular category $\mathscr{C}$ the following conditions are equivalent:*

1. *$\mathscr{C}$ is a Goursat category;*
2. *for any regular epimorphism $f : X \rightarrow Y$ and any equivalence relation R on X the regular image $f(R)$ of R along f is an equivalence relation.*

Proof 1. $\Rightarrow$ 2. When (R, r_1, r_2) is an equivalence relation it is always true that the regular image $f(R) = f \circ R \circ f^\circ$ along a regular epimorphism $f : X \rightarrow Y$ is both reflexive and symmetric. Let us then prove that $f(R)$ is also transitive: one has the equalities

$$
\begin{aligned}
f(R) \circ f(R) &= f \circ R \circ f^\circ \circ f \circ R \circ f^\circ \\
&= f \circ (f^\circ \circ f) \circ R \circ (f^\circ \circ f) \circ f^\circ \\
&= f \circ R \circ f^\circ \\
&= f(R)
\end{aligned}
$$

where the second equality follows from the Goursat assumption, and the third one from Lemma 3.2.

2. $\Rightarrow$ 1. Conversely, consider two equivalence relations (R, r_1, r_2) and (S, s_1, s_2) on a same object X in $\mathscr{C}$, and observe that the arrow $r_2 : R \rightarrow X$ is a split epimorphism, thus in particular a regular epimorphism. Then:

$$
\begin{aligned}
R \circ S \circ R &= (r_2 \circ r_1^\circ) \circ (s_2 \circ s_1^\circ) \circ (r_2 \circ r_1^\circ) \\
&= (r_2 \circ r_1^\circ) \circ (s_2 \circ s_1^\circ) \circ (r_2 \circ r_1^\circ)^\circ \\
&= r_2 \circ (r_1^\circ \circ s_2 \circ s_1^\circ \circ r_1) \circ r_2^\circ \\
&= r_2(r_1^\circ \circ s_2 \circ s_1^\circ \circ r_1) \\
&= r_2(r_1^{-1}(S)).
\end{aligned}
$$

Recall that the inverse image $r_1^{-1}(S)$ of the equivalence relation S along r_1 is obtained by taking the pullback

$$
\begin{array}{ccc}
r_1^{-1}(S) & \longrightarrow & S \\
\downarrow & & \downarrow {\scriptstyle (s_1, s_2)} \\
R \times R & \xrightarrow[r_1 \times r_1]{} & X \times X,
\end{array}
$$

and $r_1^{-1}(S)$ is always an equivalence relation. By taking into account this observation and the assumption 2., one deduces that the relation $r_2(r_1^{-1}(S)) = R \circ S \circ R$ is transitive. It follows that

$$S \circ R \circ S \leq R \circ S \circ R \circ S \circ R$$
$$\leq (R \circ S \circ R) \circ (R \circ S \circ R)$$
$$\leq R \circ S \circ R$$

and, symmetrically, $R \circ S \circ R \leq S \circ R \circ S$, hence $S \circ R \circ S = R \circ S \circ R$.

□

Exercise 4.3 Show that the regular image of an equivalence relation in Set is not necessarily transitive.

4.1 Goursat Pushouts

In a Goursat category there is a class of pushouts that has a similar role to the one of regular pushouts in a regular Mal'tsev category:

Definition 4.4 Consider a commutative diagram (4.3.ii), and the induced arrow $\hat{c}$ making the following diagram commute:

$$
\begin{array}{ccc}
\mathrm{Eq}(g) & \overset{\hat{c}}{\dashrightarrow} & \mathrm{Eq}(f) \\
{\scriptstyle p_1}\big\Vert{\scriptstyle p_2} & & {\scriptstyle p_1}\big\Vert{\scriptstyle p_2} \\
C & \underset{c}{\longrightarrow} & A
\end{array}
$$

Then the square (4.3.ii) is called a *Goursat pushout* [18] when the arrow $\hat{c}$ is a regular epimorphism.

The following result was proved in [18]. Here we give a different proof of one of the two implications, based on the calculus of relations:

Proposition 4.5 [18] *For a regular category $\mathscr{C}$ the following conditions are equivalent:*

1. *$\mathscr{C}$ is a Goursat category;*
2. *any square (4.3.ii) is a Goursat pushout.*

Proof 1. $\Rightarrow$ 2. If $\mathscr{C}$ is a Goursat category then

$$c(\mathsf{Eq}(g)) = c \circ g^\circ \circ g \circ c^\circ$$
$$= c \circ (c^\circ \circ c) \circ (g^\circ \circ g) \circ (c^\circ \circ c) \circ c^\circ$$
$$= c \circ (g^\circ \circ g) \circ (c^\circ \circ c) \circ (g^\circ \circ g) \circ c^\circ$$
$$= c \circ g^\circ \circ d^\circ \circ d \circ g \circ c^\circ$$
$$= c \circ c^\circ \circ f^\circ \circ f \circ c \circ c^\circ$$
$$= f^\circ \circ f$$
$$= \mathsf{Eq}(f)$$

where the third equality follows from the Goursat assumption, the fourth one from $g(\mathsf{Eq}(c)) = \mathsf{Eq}(d)$, and the sixth one from the fact that c is a regular epimorphism (Lemma 3.1).

2. $\Rightarrow$ 1. Conversely, given a commutative diagram

$$
\begin{array}{ccc}
R & \xrightarrow{\overline{f}} & f(R) = T \\
{\scriptstyle r_1}\big\updownarrow\big\updownarrow{\scriptstyle r_2} & & {\scriptstyle t_1}\big\updownarrow\big\updownarrow{\scriptstyle t_2} \\
X & \xrightarrow{f} & Y
\end{array}
$$

where (R, r_1, r_2) is an equivalence relation, f is a regular epimorphism and (T, t_1, t_2) is the regular image of R along f. We are to show that the relation $f(R) = T$ is an equivalence relation (by Proposition 4.2). Since the regular image of a reflexive and symmetric relation is always reflexive and symmetric, it suffices to show that T is transitive. This follows from the computation:

$$T \circ T = T \circ T^\circ$$
$$= t_2 \circ t_1^\circ \circ t_1 \circ t_2^\circ$$
$$= t_2 \circ (\overline{f} \circ r_1^\circ \circ r_1 \circ \overline{f}^\circ) \circ t_2^\circ$$
$$= f \circ r_2 \circ r_1^\circ \circ r_1 \circ r_2^\circ \circ f^\circ$$
$$= f \circ R \circ R^\circ \circ f^\circ$$
$$= f \circ R \circ f^\circ$$
$$= T.$$

Remark that the assumption that any square of the form (4.3.ii) is a Goursat pushout has been used in the third equality, where it has been applied to the diagram

$$
\begin{array}{ccc}
R & \xrightarrow{\overline{f}} & T \\
{\scriptstyle r_1}\big\updownarrow & & {\scriptstyle t_1}\big\updownarrow \\
X & \xrightarrow{f} & Y.
\end{array}
$$

$\square$

To conclude this short introduction to Goursat categories we give a characterization of those varieties of universal algebras which are 3-permutable by using the notion of Goursat pushout. This proof, originally discovered in [26], has a categorical version which has first been given in [18].

When $\mathscr{V}$ is a variety of universal algebras, we shall denote by $X = F(1)$ the free algebra on the one-element set.

Theorem 4.6 *For a variety $\mathscr{V}$ of universal algebras the following conditions are equivalent:*

1. *$\mathscr{V}$ is 3-permutable: for any pair R, S of congruences on any algebra A in $\mathscr{V}$ one has the equality*
$$R \circ S \circ R = S \circ R \circ S;$$

2. *the theory of $\mathscr{V}$ contains two quaternary operations p and q satisfying the identities*
$$p(x, y, y, z) = x, \quad q(x, y, y, z) = z, \quad p(x, x, y, y) = q(x, x, y, y).$$

Proof 1. $\Rightarrow$ 2. Consider the commutative diagram

$$
\begin{array}{ccc}
X + X + X + X & \xrightarrow{\;1+\nabla_2+1\;} & X + X + X \\[2pt]
\Big\updownarrow{\scriptstyle \nabla_2+\nabla_2}\;\;\Big\updownarrow{\scriptstyle i_2+i_1} & & \Big\updownarrow{\scriptstyle \nabla_3}\;\;\Big\updownarrow{\scriptstyle i_2} \\[2pt]
X + X & \xrightarrow[\;\nabla_2\;]{} & X
\end{array}
$$

where ∇_k is the codiagonal from the k-indexed copower of X to X (for $k \in \{2, 3\}$). The vertical arrows $\nabla_2 + \nabla_2$ and ∇_3 are split epimorphisms, whereas the horizontal arrows are regular epimorphisms, so that the diagram is a Goursat pushout by Proposition 4.5. It follows that the unique morphism

$$\overline{1 + \nabla_2 + 1} \colon \mathsf{Eq}(\nabla_2 + \nabla_2) \to \mathsf{Eq}(\nabla_3)$$

in $\mathscr{V}$ making the diagram

$$
\begin{array}{ccc}
\mathsf{Eq}(\nabla_2 + \nabla_2) & \xrightarrow{\;\overline{1+\nabla_2+1}\;} & \mathsf{Eq}(\nabla_3) \\[2pt]
\Big\downarrow{\scriptstyle p_1}\;\Big\downarrow{\scriptstyle p_2} & & \Big\downarrow{\scriptstyle p_1}\;\Big\downarrow{\scriptstyle p_2} \\[2pt]
X + X + X + X & \xrightarrow[\;1+\nabla_2+1\;]{} & X + X + X
\end{array}
\qquad (4.4.\mathrm{i})
$$

commute is a regular epimorphism (here p_1 and p_2 are the kernel pair projections), thus it is surjective. Observe that the terms $p_1(x, y, z) = x$ and $p_3(x, y, z) = z$ are identified by ∇_3, so that $(p_1, p_3) \in \text{Eq}(\nabla_3)$. The surjectivity of $\overline{1 + \nabla_2 + 1}$ then implies that there are terms $(p, q) \in \text{Eq}(\nabla_2 + \nabla_2)$ such that $\overline{1 + \nabla_2 + 1}(p, q) = (p_1, p_3)$. This latter property means exactly that

$$p(x, y, y, z) = x, \quad q(x, y, y, z) = z,$$

while the fact that $(p, q) \in \text{Eq}(\nabla_2 + \nabla_2)$ gives the identity

$$p(x, x, y, y) = q(x, x, y, y).$$

2. $\Rightarrow$ 1. For the converse implication, take R and S two congruences on an algebra A in $\mathcal{V}$, and let us show that $R \circ S \circ R \leq S \circ R \circ S$. For $(a, b) \in R \circ S \circ R$, let x and y be such that $(a, x) \in R$, $(x, y) \in S$ and $(y, b) \in R$. Then the fact that $(a, a), (x, a), (y, b), (b, b)$ are in R implies that both $(p(a, x, y, b), p(a, a, b, b))$ and $(q(a, x, y, b), q(a, a, b, b))$ are in R. Since $p(a, a, b, b) = q(a, a, b, b)$ we deduce that $(p(a, x, y, b), q(a, x, y, b)) \in R$. On the other hand, the elements $(a, a), (x, x), (x, y), (b, b)$ are all in S so that $(p(a, x, x, b), p(a, x, y, b)) \in S$, $(q(a, x, x, b), q(a, x, y, b)) \in S$, hence $(a, p(a, x, y, b))$ and $(b, q(a, x, y, b))$ are both in S. We then observe that

$$(a, p(a, x, y, b)) \in S$$
$$(p(a, x, y, b), q(a, x, y, b)) \in R$$
$$(q(a, x, y, b), b) \in S$$

we conclude that (a, b) belongs to $S \circ R \circ S$. It then follows that $R \circ S \circ R = S \circ R \circ S$, as desired. $\square$

Remark 4.7 Note that one can give a proof of the Mal'tsev theorem characterizing 2-permutable varieties by using some categorical arguments similar to the ones in Theorem 4.6. This was first observed in [12] and, more recently, in [8].

Remark 4.8 A wide generalization of Theorem 4.6 was obtained by P.-A. Jacqmin and D. Rodelo in [27], where a categorical approach to n-permutability was developed. Thanks to their approach the authors have been able to characterize the property of n-permutability in terms of some specific stability properties of regular epimorphisms, which extend the one considered in [20] to study Goursat categories.

4.2   Implication Algebras

A typical example of 3-permutable variety, thus of a Goursat category, is provided by
the variety ImplAlg of *implication algebras* [1]. The algebraic theory of the variety
ImplAlg has a binary operation such that

(A) $(xy)x = x$,
(B) $(xy)y = (yx)x$,
(C) $x(yz) = y(xz)$.

As explained in [25], to see that ImplAlg is 3-permutable, one first checks that the
term xx is a constant: indeed, the identities

$$
\begin{aligned}
xx &= [(xy)x]x & \text{(by (A))}\\
&= [x(xy)](xy) & \text{(by (B))}\\
&= x[[x(xy)]y] & \text{(by (C))}\\
&= x[[((xy)x)(xy)]y] & \text{(by (A))}\\
&= x[(xy)y] & \text{(by (A))}\\
&= (xy)(xy) & \text{(by (C))}
\end{aligned}
$$

imply that

$$
xx = [x(yy)][x(yy)] = [y(xy)][y(xy] = yy,
$$

and one denotes such an equationally defined constant by 1. This notation is justified
by the fact that

$$
1y = (yy)y = y.
$$

One then verifies that the terms $p(x, y, z, u) = (zy)x$ and $q(x, y, z, u) = (yz)u$ are
such that

$$
\begin{aligned}
p(x, y, y, z) &= (yy)x = 1x = x,\\
q(x, y, y, z) &= (yy)z = 1z = z,
\end{aligned}
$$

and

$$
p(x, x, z, z) = (zx)x = (xz)z = q(x, x, z, z).
$$

4.3   Diagram Lemmas and Goursat Categories

We conclude these notes by mentioning a connection between the validity of
some suitable *diagram lemmas* and the permutability conditions on a regular category considered above. The classical 3×3-Lemma in abelian categories [14] has
been extended to several non-additive contexts by various authors (see [5, 28], for

instance). An original extension to a non-pointed context was first established by D. Bourn in the context of regular Mal'tsev categories [6]. The main point in order to formulate the 3×3-Lemma in a category which does not have a 0-object is to replace the classical notion of short exact sequence with the notion of exact fork: a diagram of the form

$$R \underset{r_2}{\overset{r_1}{\rightrightarrows}} X \overset{f}{\longrightarrow} Y$$

is an *exact fork* if and only if (R, r_1, r_2) is the kernel pair of f, and f is the coequalizer of r_1 and r_2. With this notion at hand the appropriate way of expressing the 3×3-Lemma is then the following, which is called the *denormalized* 3×3-*Lemma*: given any commutative diagram

(4.4.ii)

in $\mathscr{C}$ such that

- $y_i \circ a_j = b_j \circ z_i, y \circ b_i = c_i \circ z, b \circ y_i = k_i \circ a, x \circ b = c \circ y$ (for $i, j \in \{1, 2\}$),
- the three columns and the middle row are exact forks,

then the upper row is an exact fork if and only if the lower row is an exact fork.
S. Lack observed in [32] that this denormalized 3×3-Lemma holds not only in regular Mal'tsev categories (as observed by D. Bourn [6]) but also in Goursat categories. Later on it turned out that the validity of the denormalized 3×3-Lemma actually characterizes Goursat categories among regular ones:

Theorem 4.9 [18, 32] *For a regular category $\mathscr{C}$ the following conditions are equivalent:*

1. *$\mathscr{C}$ is a Goursat category;*
2. *if the lower row in a diagram (4.4.ii) is an exact fork then the upper row is an exact fork;*
3. *if the upper row in a diagram (4.4.ii) is an exact fork then the lower row is an exact fork;*
4. *the denormalized 3×3-Lemma holds in $\mathscr{C}$: the lower row is an exact fork if and only if the upper row is an exact fork.*

We would like to point out that both the calculus of relations and the notion of Goursat pushout play a central role in the proof of this result. Note that a unification of both the classical 3 × 3-Lemma and of the denormalized one in the context of star-regular categories is also possible [17]. Further results linking the Goursat property to natural conditions appearing in universal algebra—in relationship to congruence modularity—have been investigated in [21] (see also the references therein). Finally, let us mention that also Mal'tsev categories can be characterized via a suitable diagrammatic condition that is stronger than the denormalized 3 × 3-Lemma, called the *Cuboid Lemma* [19].

Acknowledgements A part of the material presented in this survey article is based on [7, 8, 16]. The author is grateful to Tomas Everaert for an important suggestion concerning Theorem 1.16. Many thanks also to Maria Manuel Clementino, Diana Rodelo, Idriss Tchoffo Nguefeu, David Broodryk and the anonymous referee for carefully proofreading a first version of the article and suggesting some useful changes and corrections.

References

1. Abbott, J.C.: Algebras of implication and semi-lattices. Séminaire Dubreil. Algèbre et théorie des nombres **20**(2), 1–8 (1966–1967)
2. Barr, M., Grillet, P.A., van Osdol, D.H.: Exact Categories and Categories of Sheaves. Springer Lecture Notes in Mathematics, vol. 236. Springer, Heidelberg (1971)
3. Borceux, F.: Handbook of Categorical Algebra. 2. Categories and Structures. Encyclopedia of Mathematics and Its Applications, vol. 51. Cambridge University Press, Cambridge (1994)
4. Borceux, F., Clementino, M.M.: Topological semi-abelian algebras. Adv. Math. **190**, 425–453 (2005)
5. Bourn, D.: 3 × 3 lemma and protomodularity. J. Algebra **236**, 778–795 (2001)
6. Bourn, D.: The denormalized 3 × 3-lemma. J. Pure Appl. Algebra **177**(2), 113–129 (2003)
7. Bourn, D., Gran, M.: Regular, protomodular and abelian categories. In: Categorical Foundations - Special Topics in Order, Topology, Algebra and Sheaf Theory. Encyclopedia of Mathematics and Its Applications, vol. 97, pp. 165–211. Cambridge University Press (2004)
8. Bourn, D., Gran, M., Jacqmin, P.-A.: On the naturalness of Mal'tsev categories. In: Casadio, C., Scott, P. (eds.) Joachim Lambek: The Interplay of Mathematics, Logic, and Linguistics, Outstanding Contributions to Logic, vol. 20, pp. 59–104. Springer (2021)
9. Buchsbaum, D.: Exact categories and duality. Trans. Am. Math. Soc. **80**, 1–34 (1955)
10. Carboni, A., Kelly, G.M., Pedicchio, M.C.: Some remarks on Mal'tsev and Goursat categories. Appl. Categ. Struct. **4**, 385–421 (1993)
11. Carboni, A., Lambek, J., Pedicchio, M.C.: Diagram chasing in Mal'cev categories. J. Pure Appl. Algebra **69**, 271–284 (1990)
12. Carboni, A., Pedicchio, M.C.: A new proof of the Mal'cev theorem. Categorical studies in Italy (Perugia, 1977). Rend. Circ. Mat. Palermo **2**(Suppl. No. 64), 13–16 (2000)
13. Clementino, M.M.: An invitation to topological semi-abelian algebras. In: Clementino, M.M., Facchini, A., Gran, M. (eds.) New Perspectives in Algebra, Topology and Categories, Coimbra Mathematical Texts 1, pp. 27–66. Springer Nature and University of Coimbra (2021)
14. Freyd, P.J.: Abelian Categories. An Introduction to the Theory of Functors. Harper's Series in Modern Mathematics, New York (1964)
15. Gran, M.: Central extensions and internal groupoids in Maltsev categories. J. Pure Appl. Algebra **155**, 139–166 (2001)
16. Gran, M.: Notes on regular, exact and additive categories. Notes for a mini-course given at the Summer School on Category Theory and Algebraic Topology, Ecole Polytechnique Fédérale de Lausanne (2014)

17. Gran, M., Janelidze, Z., Rodelo, D.: 3 × 3-lemma for star-exact sequences. Homology Homotopy Appl. **14**(2), 1–22 (2012)
18. Gran, M., Rodelo, D.: A new characterisation of Goursat categories. Appl. Categ. Struct. **20**, 229–238 (2012)
19. Gran, M., Rodelo, D.: The cuboid lemma and Mal'tsev categories. Appl. Categ. Struct. **22**, 805–816 (2014)
20. Gran, M., Rodelo, D.: Beck-Chevalley condition and Goursat categories. J. Pure Appl. Algebra **221**, 2445–2457 (2017)
21. Gran, M., Rodelo D., Tchoffo Nguefeu, I.: Variations of the Shifting Lemma and Goursat categories. Algebra Univers. **80**(2) (2019)
22. Gran, M., Rosický, J.: Semi-abelian monadic categories. Theory Appl. Categ. **13**(6), 106–113 (2004)
23. Gran, M., Sterck, F., Vercruysse, J.: A semi-abelian extension of a theorem by Takeuchi. J. Pure Appl. Algebra **223**, 4171–4190 (2019)
24. Grothendieck, A.: Technique de construction en géométrie analytique. IV. Formalisme général des foncteurs représentables., Sém. Henri Cartan **13**(1), 1–28 (1962)
25. Gumm, H.P., Ursini, A.: Ideals in universal algebra. Algebra Univers. **19**, 45–54 (1984)
26. Hagemann, J., Mitschke, A.: On n-permutable congruences. Algebra Univers. **3**, 8–12 (1973)
27. Jacqmin, P.-A., Rodelo, D.: Stability properties characterising n-permutable categories. Theory Appl. Categ. **32**, 1563–1587 (2017)
28. Janelidze, Z.: The pointed subobject functor, 3 × 3 lemmas and subtractivity of spans. Theory Appl. Categ. **23**, 221–242 (2010)
29. Johnstone, P.T.: Stone Spaces. Cambridge Studies in Advanced Mathematics, vol. 3. Cambridge University Press, Cambridge (1982)
30. Johnstone, P.T.: Sketches of an Elephant: A Topos Theory Compendium. Oxford Logic Guides, vol. 43. Oxford University Press, Oxford (2002)
31. Johnstone, P.T., Pedicchio, M.C.: Remarks on continuous Mal'cev algebras. Rend. Ist. Matem. Univ. Trieste **25**, 277–287 (1995)
32. Lack, S.: The 3-by-3 lemma for regular Goursat categories. Homology Homotopy Appl. **6**(1), 1–3 (2004)
33. Mal'tsev, A.I.: On the general theory of algebraic systems. Matematicheskii Sbornik N.S. **35**(77), 3–20 (1954)
34. Meisen, J.: Relations in categories. Thesis, McGill University (1972)
35. Picado, J., Pultr, A.: Notes on point-free topology. In: Clementino, M.M., Facchini, A., Gran, M. (eds.) New Perspectives in Algebra, Topology and Categories, Coimbra Mathematical Texts 1, pp. 173–223. Springer Nature and University of Coimbra (2021)
36. Riguet, J.: Relations binaires, fermetures, correspondances de Galois. Bull. de la Société Mathématique de France **76**, 114–155 (1948)
37. Smith, J.D.H.: Mal'cev Varieties. Springer Lecture Notes in Mathematics, vol. 554. Springer, Heidelberg (1976)

Chapter 5
Categorical Commutator Theory

Sandra Mantovani and Andrea Montoli

Abstract In these notes, we introduce the reader to the categorical commutator theory (of subobjects), following the *formal approach* given by Mantovani and Metere in 2010. Such an approach is developed along the lines provided by Higgins, based on the notion of *commutator word*, introduced by the author in the context of varieties of Ω-groups (groups equipped with additional algebraic operations of signature Ω). An internal interpretation of the commutator words is described, providing an intrinsic notion of Higgins commutator, which reveals to have good properties in the context of ideal determined categories. Furthermore, we will illustrate some applications of commutator theory in categorical algebra, such as a useful way to test the normality of subobjects on one side, and the construction of the abelianization functor on the other.

Keywords Commutator theory · Semi-abelian category · Normal subobject

Math. Subj. Classification: 08A30 · 18A20 · 18E13

Introduction

The theory of commutators [9] can be considered as an extension of the classical commutator theory for groups to more general varieties of algebras. A description of commutator of congruences in Mal'tsev varieties was developed by Smith [20],

S. Mantovani (✉) · A. Montoli
Dipartimento di Matematica "Federigo Enriques", Università degli Studi di Milano, Via Saldini 50, 20133 Milan, Italy
e-mail: sandra.mantovani@unimi.it

A. Montoli
e-mail: andrea.montoli@unimi.it

© The Author(s), under exclusive license to Springer Nature Switzerland AG 2021
M. M. Clementino et al. (eds.), *New Perspectives in Algebra, Topology and Categories*, Coimbra Mathematical Texts 1,
https://doi.org/10.1007/978-3-030-84319-9_5

and then extended to a categorical context by Pedicchio [19], while a first categorical notion of commutator of subobjects was given by Huq in [13].

In these notes, following [18], we first recall the Higgins commutator, based on the notion of *commutator word*, introduced by Higgins [12] in the context of varieties of Ω-groups (groups equipped with additional algebraic operations of signature Ω). We will describe also the internal interpretation of these commutator words given in [12] by means of the so-called *formal commutator*, which allows us to provide an intrinsic notion of Higgins commutator $[H, K]$ of two subobjects H, K of A in any regular and unital category with finite colimits (see Definition 3.2). Such a commutator $[H, K]$ is not in general a normal subobject of A (see Example 2.6), but if we move into the context of ideal determined categories [15], we easily see that such a commutator is always normal in the join $H \vee K$ of H and K in A.

In Sect. 4, we revisit also Huq commutator [13], showing that in a unital and normal category [16], Huq commutator is nothing but the normal closure of Higgins commutator. The two commutators are different in general, even in the category of groups, if H and K are not normal in A, as Example 1.2 shows. But they coincide when one of the two subobjects is the whole A.

The case with $H = A$ is special also for another reason. In the category of groups, the commutator $[A, K]$ can be used to test whether the subgroup K of A is normal in A. Actually K is normal in A if, and only if, $[A, K]$ is a subgroup of K. A natural question is to ask if the internal formulation of this connection is still valid in a categorical setting. In Proposition 3.7 we recall from [18] that, in an ideal determined and unital category $\mathscr{C}$, any normal subobject K of A contains the commutator $[A, K]$. In order to get the converse, we need to move into the world of semi-abelian categories [14], where the full characterization of normality via commutators holds.

Furthermore, in Sect. 5, we recall from [2] the categorical notions of commutative and abelian object and how they are related to the previous notion of commutator. Referring to [6], we show in Theorem 5.8 that, in the realm of pointed Mal'tsev categories, the two notions coincide. We conclude by describing how to obtain the abelianization functor (left adjoint to the inclusion of the subcategory of abelian objects), by means of the cokernel of the commutator $[X, X]$ over X, in any pointed normal Mal'tsev category.

1 Commutators of Groups

We begin by recalling the notion of commutator of two elements in a group and of two subgroups. All the material of this section can be found in any textbook about group theory. Let G be a group, and let $g, h \in G$. The commutator of g and h is the element

$$[g, h] = ghg^{-1}h^{-1}.$$

If H and K are subgroups of G, the commutator of H and K is the subgroup of G generated by all the elements of the form $[h, k]$ with $h \in H$ and $k \in K$:

$$[H, K] = \langle [h, k] \mid h \in H, k \in K \rangle.$$

In particular, the subgroup $[G, G]$ is called the derived subgroup of G. It is clear that G is an abelian group if and only if $[G, G] = 0$, where 0 denotes the trivial group. More generally, given $h, k \in G$, one has $[h, k] = 1$ if and only if h and k permute: $hk = kh$.

Remark 1.1 Given two subgroups H and K of a group G, if we denote by $H \vee K$ the smallest subgroup of G containing both H and K (namely the supremum of H and K in the lattice of subgroups of G), then we have that the commutator $[H, K]$ is a normal subgroup of $H \vee K$.

Proof If $hkh^{-1}k^{-1}$ is a generator of $[H, K]$ and $\bar{h} \in H$, then

$$\bar{h}(hkh^{-1}k^{-1})\bar{h}^{-1} = \bar{h}hkh^{-1}\bar{h}^{-1}\bar{h}k^{-1}\bar{h}^{-1}$$

$$= \bar{h}hkh^{-1}\bar{h}^{-1}k^{-1}k\bar{h}k^{-1}\bar{h}^{-1} = ((\bar{h}h)k(\bar{h}h)^{-1}k^{-1})(k\bar{h}k^{-1}\bar{h}^{-1}) \in [H, K],$$

since both $(\bar{h}h)k(\bar{h}h)^{-1}k^{-1}$ and $k\bar{h}k^{-1}\bar{h}^{-1}$ belong to $[H, K]$. Similarly, if $\bar{k} \in K$, then $\bar{k}(hkh^{-1}k^{-1})\bar{k}^{-1} \in [H, K]$, and this is enough to conclude. □

However, if H and K are not normal subgroups of G, $[H, K]$ is not normal in general. The following example is borrowed from Alan Cigoli's Ph.D. thesis [7]:

Example 1.2 Let G be the alternating group A_5 and let H and K be the following subgroups of G:

$$H = \langle (12)(34) \rangle, \quad K = \langle (12)(45) \rangle.$$

These subgroups are not normal in A_5 (actually A_5 is a simple group, i.e. it has no non-trivial subgroups). Let us put $h = (12)(34)$ and $k = (12)(45)$. Then $h = h^{-1}$ and $k = k^{-1}$ and so

$$[h, k] = hkhk = (12)(34)(12)(45)(12)(34)(12)(45)$$

$$= (34)(45)(34)(45) = (354)(354) = (345).$$

So, $[H, K] = \langle (345) \rangle$ is not normal in A_5.

The situation improves when H and K are normal subgroups of G. Indeed, the following property holds:

Proposition 1.3 *If H and K are normal subgroups of G, then $[H, K]$ is normal in G as well.*

Proof If $h \in H, k \in K$ and $g \in G$, then $ghg^{-1} = \bar{h} \in H$ and $gkg^{-1} = \bar{k} \in K$. Hence

$$g(hkh^{-1}k^{-1})g^{-1} = (ghg^{-1})(gkg^{-1})(gh^{-1}g^{-1})(gk^{-1}g^{-1}) = \bar{h}\bar{k}\bar{h}^{-1}\bar{k}^{-1} \in [H, K].$$

□

We will come back later, in a more general framework, to this last property, which will be called *normality of the Higgins commutator*. For the moment, we observe that this property allows one to get two important chains of normal subgroups of G, namely the *derived series*:

$$G \triangleright [G, G] \triangleright [[G, G], [G, G]] \triangleright [[[G, G], [G, G]], [[G, G], [G, G]]] \triangleright \ldots$$

and the *lower central series*:

$$G \triangleright [G, G] \triangleright [[G, G], G] \triangleright [[[G, G], G], G] \triangleright \ldots$$

The derived series allows one to define solvable groups: a group is solvable if its derived series reaches the trivial group after a finite number of steps. Similarly, the lower central series allows one to define nilpotent groups: a group is nilpotent if its lower central series reaches the trivial group after a finite number of steps.

Before moving to a more general context than the one of groups, we list some important properties of the commutator of subgroups. Their proofs are left to the reader.

Proposition 1.4 *1. Given a subgroup K of G, K is normal in G if and only if $[K, G] \subseteq K$.*
2. If H, K, H', K' are subgroups of G, with $H \subseteq H'$ and $K \subseteq K'$, then $[H, K] \subseteq [H', K']$.
3. $f : G \to G'$ is a surjective group homomorphism, and H and K are subgroups of G, then $f([H, K]) = [f(H), f(K)]$.

2 The Case of Ω-groups

In order to extend, in a unified way, the notions of ideal and commutator to a wide range of algebraic structures, Higgins introduced in [12] the notion of Ω-*group*. An Ω-group G is a group $(G, +, -, 0)$ (written in additive notation, although it is not necessarily abelian) equipped with a set Ω of additional operations, of finite arity $n \geq 1$, such that, for all $\omega \in \Omega$:

$$\omega(0, 0, \ldots 0) = 0.$$

For any fixed Ω, we get a variety in the sense of universal algebra. We will denote by Ω-Grp the category whose objects are the Ω-groups (for the fixed Ω) and whose morphisms are the group homomorphisms that preserve any $\omega \in \Omega$. Every such category is pointed, with the initial and terminal object given by $0 = \{0\}$.

Let X be a set of indeterminates and let $\underline{x}$ denote a finite sequence $(x_1, x_2, ..., x_n)$ of elements of X. A *word* in X is an expression obtained by formally applying the operations $\omega \in \Omega, +, -$ to elements of X and to 0 a finite number of times. As it usually happens for varieties of universal algebras, considering words is the first step in order to build the free Ω-group on the set X (some identifications, according to the equations of the corresponding variety, will be needed). Now we can consider two special families of words:

Definition 2.1 Let $f(\underline{x}, \underline{y})$ be a word in two disjoint sets of indeterminates X and Y. We shall say that

1. $f(\underline{x}, \underline{y})$ is an *ideal word* of X w.r.t. Y if $f(\underline{x}, \underline{y})$ satisfies the equation $f(\underline{0}, \underline{y}) = 0$, where $f(\underline{0}, \underline{y})$ is the word obtained from $f(\underline{x}, \underline{y})$ by replacing every x_i in $\underline{x}$ by 0.
2. $f(\underline{x}, \underline{y})$ is a *commutator word* in X and Y if $f(\underline{0}, \underline{y}) = f(\underline{x}, \underline{0}) = 0$.

We will denote by X^Y the set of ideal words of X w.r.t. Y, and by $[X, Y]$ the set of commutator words in X and Y. It is clear that $[X, Y] = X^Y \cap Y^X$.

Example 2.2 In the variety Grp of groups (and, more generally, in every variety of Ω-groups), the word $y + x - y$ is an ideal word of $\{x\}$ w.r.t. $\{y\}$, and $x + y - x - y$ is a commutator word in $\{x\}$ and $\{y\}$. In the variety Rng of (non-unitary) rings, the word xy is both an ideal and a commutator word in $\{x\}$ and $\{y\}$.

The reason for the name *ideal word* is that, given an Ω-group G and an Ω-subgroup H of G, H is an *ideal* (i.e. the kernel of a morphism) if and only if, for any ideal word $f(\underline{h}, \underline{g})$ in H and G, one has that $f(\underline{h}, \underline{g}) \in H$ (where, now, we do not see $f(\underline{h}, \underline{g})$ as a formal combination of elements, but we compose the symbols using the operations in G). Following the same spirit, given two Ω-subgroups H and K of G, we will denote by $[H, K]$ the set of the realizations in G of all the commutator words $f(\underline{h}, \underline{k})$ in H and K. In the variety of groups, it is not difficult to see that

$$[H, K] = \langle [h, k] \mid h \in H, k \in K \rangle,$$

while, in the variety CRng of commutative rings

$$[H, K] = HK.$$

In general, we have the following:

Lemma 2.3 [12, Lemma 4.1] $[H, K]$ *is an Ω-subgroup of G. Moreover, it is an ideal in the join $H \vee K$ of H and K in G.*

In the particular case of *distributive* Ω-groups, namely those Ω-groups in which every $\omega \in \Omega$ is distributive w.r.t. $+$:

$$\forall a_1, \dots a_n, b \quad \omega(a_1, \dots a_i + b, \dots a_n) = \omega(a_1, \dots a_i, \dots a_n) + \omega(a_1, \dots b, \dots a_n),$$

there is an easy description of ideals:

Theorem 2.4 [12, Theorem 4a] *If G is a distributive Ω-group and H is an Ω-subgroup of G, H is an ideal of G if and only if it is a normal subgroup of G and, for all $g_1, \ldots g_n \in G$, $h \in H$ and $\omega \in \Omega$, $\omega(g_1, \ldots g_{i-1}, h, g_{i+1}, \ldots g_n) \in H$.*

If, moreover, we suppose that the group operation $+$ is commutative, then there is also an easy description of the commutator of two subobjects:

Theorem 2.5 [12, Theorem 4c] *If G is a distributive Ω-group, with $+$ commutative, and H and K are Ω-subgroups of G, then $[H, K]$ consists of all polynomials in elements of H and K each term of which contains both a factor from H and a factor from K.*

In the previous section, we observed that the commutator subgroup $[H, K]$, of two normal subgroups H and K of a group G, is always normal in G. It is a natural question whether the same property holds for Ω-groups. Unfortunately, the answer is negative, even for distributive Ω-groups, as the following example shows:

Example 2.6 Consider the variety of abelian groups endowed with an additional binary operation $*$ which is distributive w.r.t. the group operation $+$. Consider the free abelian group on three elements $A = \mathbb{Z}_x + \mathbb{Z}_y + \mathbb{Z}_t$ with the operation $*$ defined in the following way on the generators:

$*$	x	y	t
x	x	0	y
y	0	0	x
t	y	x	t

Then the free abelian subgroup $K = \mathbb{Z}_x + \mathbb{Z}_y$ is an ideal of A, since any product which involves elements of K still belongs to K, but the commutator $[K, K] = \mathbb{Z}_x$ is not an ideal of A, because, for instance, $x * t$ does not belong to $\mathbb{Z}_x$.

3 The Categorical Higgins Commutator

In this section we extend the commutator defined by Higgins for Ω-groups to a categorical context. For this purpose, we will consider a pointed category $\mathscr{C}$ with finite limits and finite colimits. In this context, for any pair of objects H and K of $\mathscr{C}$, we get canonical inclusions

$$H \xrightarrow{\langle 1, 0 \rangle} H \times K \xleftarrow{\langle 0, 1 \rangle} K$$

determined by the universal property of the product. Similarly, we have canonical morphisms

$$H \xleftarrow{(1,0)} H + K \xrightarrow{(0,1)} K$$

determined by the universal property of the coproduct. Combining them, we get a canonical morphism

$$\Sigma \colon H + K \to H \times K.$$

If $\mathscr{C}$ is the category Grp of groups, $H + K$ is the free product of H and K, while $H \times K$ is the usual direct product. Then Σ is defined as follows:

$$\Sigma(h_1, k_1, h_2, k_2, \ldots h_n, k_n) = (h_1 h_2 \ldots h_n, k_1 k_2 \ldots k_n),$$

where $h_i \in H, k_i \in K$ and the chain $h_1, k_1, h_2, k_2, \ldots h_n, k_n$ is a word in the alphabet $H \cup K$, which represents an element of $H + K$. In Grp such a morphism Σ is always surjective. In Grp, for a morphism being surjective is equivalent to being a regular epimorphism (i.e. the coequalizer of a pair of morphisms) and to being a normal epimorphism (i.e. the cokernel of a morphism). Moreover, a morphism of groups f is surjective if and only if it is an extremal epimorphism: this means that, if $f = mg$, with m a monomorphism, then m is an isomorphism. Let us give a name to those categories for which the canonical morphism Σ is always an extremal epimorphism:

Definition 3.1 [3] Let $\mathscr{C}$ be a pointed category with finite limits and finite colimits. $\mathscr{C}$ is *unital* if, for every pair of objects H and K, the canonical morphism Σ is an extremal epimorphism.

Actually, the definition of a unital category can be given even in absence of finite coproducts. Indeed, it suffices to ask that, for every pair of objects H and K, the canonical morphisms $\langle 1, 0 \rangle \colon H \to H \times K$ and $\langle 0, 1 \rangle \colon K \to H \times K$ are jointly extremal epimorphic: if they factor through a common monomorphism m, then m is an isomorphism.

Let us now consider the kernel of $\Sigma \colon H + K \to H \times K$. We will denote it by $H \diamond K$ (it was introduced in [5] under the name *cosmash product* of H and K). It is not difficult to show that, in Grp, $H \diamond K$ is given by commutator words in H and K. To see this, one can consider what is the image under Σ of words of the form h_1, k_1, h_2, k_2: they are sent to the neutral element of $H \times K$ if and only if $h_1 h_2 = 1$ and $k_1 k_2 = 1$, i.e. $h_2 = h_1^{-1}$ and $k_2 = k_1^{-1}$. Actually the same fact holds in every category of Ω-groups. For this reason, we can call $H \diamond K$ the *formal commutator* of H and K.

Let us now assume that the category $\mathscr{C}$ is not only unital, but also *regular* [1]. This means that every morphism f in $\mathscr{C}$ admits a pullback-stable factorization $f = me$, where e is a regular epimorphism and m is a monomorphism (for further information about regular categories, the reader is addressed to the chapter *An introduction to regular categories* of this book). In this setting, consider two subobjects $h \colon H \rightarrowtail A$ and $k \colon K \rightarrowtail A$ of the same object A. These two arrows induce a morphism $(h, k) \colon H + K \to A$, which we call the *realization map*. The name comes from the fact that, in the case of varieties of universal algebra, this arrow turns a formal word belonging to $H + K$ into the element in A obtained by realizing such word w.r.t. the operations in A. Consider then the following commutative square:

$$
\begin{array}{ccc}
H \diamond K & \xrightarrow{\ e\ } & [H, K] \\
{\scriptstyle \ker(\Sigma)} \downarrow & & \downarrow {\scriptstyle m} \\
H + K & \xrightarrow[(h,k)]{} & A,
\end{array}
$$

where (e, m) is the (regular epi, mono) factorization of the restriction of (h, k) to $H \diamond K$, i.e. of the composite $(h, k)\ker(\Sigma)$.

Definition 3.2 The object $[H, K]$ defines the categorical Higgins commutator of H and K in A.

Once the categorical version of the Higgins commutator is defined, our goal becomes to check which of the properties the classical commutator in Grp has, still hold in this setting. First of all, we ask ourselves whether the commutator $[H, K]$ of two subobjects in A is a normal subobject of the join $H \vee K$ (by "normal subobject" we simply mean a kernel of a morphism. This notion should not be confused with the one of subobject normal to an internal equivalence relation in the sense of Bourn (see [2])). In order to answer this question, we first need to recall how the join can be constructed in our categorical setting. It is obtained as the monomorphic part of the (regular epi, mono) factorization of $(h, k)\colon H + K \to A$ given by:

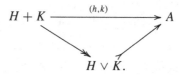

Clearly H is a subobject of $H \vee K$, because we have the following commutative diagram:

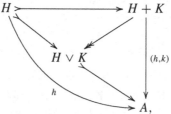

and the same holds for K. Moreover, if Z is a subobject of A having both H and K as subobjects, as in the following diagram:

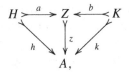

then we get the commutative diagram

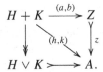

By the uniqueness, up to isomorphisms, of the (regular epi, mono) factorization, we obtain that (a, b) factors through $H \vee K$, which is then a subobject of Z:

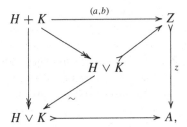

and this tells us that $H \vee K$ is the smallest subobject of A containing both H and K. In order to show that the commutator $[H, K]$ is normal in $H \vee K$ (i.e. it is the kernel of a morphism with domain $H \vee K$), we need a further assumption on our category $\mathscr{C}$, namely that it is *ideal determined* [15]:

Definition 3.3 A pointed regular category $\mathscr{C}$ with finite colimits is *ideal determined* if the following two conditions hold:

1. $\mathscr{C}$ is *normal* in the sense of [16], which means that every regular epimorphism in $\mathscr{C}$ is a normal epimorphism (i.e. the cokernel of a morphism);
2. the regular image of a normal monomorphism is a normal monomorphism. This means that, given a commutative square

$$
\begin{array}{ccc}
A & \longrightarrow & B \\
\downarrow & & \downarrow \\
C & \longrightarrow & D,
\end{array}
$$

where the horizontal arrows are regular epimorphisms and the vertical ones are monomorphisms, if the left vertical arrow is normal, then the right one is normal as well.

It is immediate to see that, in an ideal determined category, the commutator $[H, K]$ is a normal subobject of the join $H \vee K$; it suffices to consider the following commutative square:

$$
\begin{array}{ccc}
H \diamond K & \xrightarrow{\ e\ } & [H, K] \\
{\scriptstyle \ker(\Sigma)}\Big\downarrow & & \Big\downarrow \\
H + K & \longrightarrow\!\!\!\!\!\longrightarrow & H \vee K.
\end{array}
$$

Let us now explore more in detail the second condition defining ideal determined categories. Let $h\colon H \to X$ be a normal subobject, and let $f\colon X \to Y$ be a regular epimorphism in an ideal determined category $\mathscr{C}$. If we consider the (regular epi, mono) factorization of fh:

$$
\begin{array}{ccc}
H & \xrightarrow{\ f_1\ } & K \\
{\scriptstyle h}\Big\downarrow & & \Big\downarrow{\scriptstyle k} \\
X & \xrightarrow[\ f\ ]{} & Y,
\end{array}
$$

we know, from Condition 2, that k is a normal subobject. So it will be the kernel of its cokernel. The cokernel of k can be built in the following way: let $q\colon X \to Q$ be the cokernel of h, and consider the pushout of q along f:

$$
\begin{array}{ccc}
H & \xrightarrow{\ f_1\ } & K \\
{\scriptstyle h}\Big\downarrow & & \Big\downarrow{\scriptstyle k} \\
X & \xrightarrow{\ f\ } & Y \\
{\scriptstyle q}\Big\downarrow & & \Big\downarrow{\scriptstyle q'} \\
Q & \xrightarrow[\ f'\ ]{} & Q'.
\end{array}
$$

Then q' is the cokernel of k. Indeed:

$$
q'kf_1 = f'qh = 0 = 0f_1,
$$

from which we get $q'k = 0$, since f_1 is an epimorphism. Moreover, if $t\colon Y \to T$ is such that $tk = 0$, then

$$
0 = tkf_1 = tfh.
$$

Since q is the cokernel of h, there exists a unique morphism $t'\colon Q \to T$ such that $tf = t'q$. Finally, by the universal property of the pushout, we get a unique $s\colon Q' \to T$ such that $sq' = t$ (and $sf' = t'$). Hence q' is the cokernel of k.

Thanks to this observation, we can give an alternative characterization of ideal determined categories [17, Proposition 3.1]:

Proposition 3.4 *Let $\mathscr{C}$ be a normal category with finite colimits. $\mathscr{C}$ is ideal determined if and only if, for any commutative diagram*

$$
\begin{array}{ccc}
H & \xrightarrow{\;f_1\;} & K \\
\big\downarrow{\scriptstyle h} & & \big\downarrow{\scriptstyle k} \\
X & \xrightarrow{\;f\;} & Y \\
\big\downarrow{\scriptstyle q} & & \big\downarrow{\scriptstyle q'} \\
Q & \xrightarrow{\;f'\;} & Q'
\end{array}
\tag{5.3.i}
$$

such that the lower square is a pushout of regular epimorphisms, $h = \ker(q)$ and $k = \ker(q')$, one has that f_1 is a regular epimorphism, too.

Proof Suppose $\mathscr{C}$ is ideal determined and the diagram (5.3.i) is given. Consider the (regular epi, mono) factorization of fh:

$$
\begin{array}{ccc}
H & \xrightarrow{\;g\;} & K' \\
\big\downarrow{\scriptstyle h} & & \big\downarrow{\scriptstyle k'} \\
X & \xrightarrow{\;f\;} & Y.
\end{array}
$$

From what we observed before, we obtain that k' is the kernel of its cokernel, which is necessarily q'. Then k and k' are isomorphic, and so f_1 is a regular epimorphism.

Conversely, let $h \colon H \to X$ be the kernel of its cokernel $q \colon X \to Q$, and let $f \colon X \to Y$ be a regular epimorphism. Considering the pushout of f along q:

$$
\begin{array}{ccc}
X & \xrightarrow{\;f\;} & Y \\
\big\downarrow{\scriptstyle q} & & \big\downarrow{\scriptstyle q'} \\
Q & \xrightarrow{\;f'\;} & Q'
\end{array}
$$

and completing the diagram with $k = \ker(q')$ and with the morphism f_1 induced by the universal property of the kernel (since $q'fh = f'qh = 0$), we obtain a diagram like (5.3.i). By assumption, f_1 is a regular epimorphism. Hence the regular image of the normal monomorphism h is the normal monomorphism k, and the category is ideal determined.   □

Another property of the commutator we already observed for groups is the following: a subgroup K of A is normal if and only if the commutator $[A, K]$ is a subgroup of K. In order to see if this property still holds in our general context, we first need to recall some facts concerning the construction of cokernels. Given a subobject $k\colon K \to A$, one way to build its cokernel in $\mathscr{C}$ is by means of the pushout

$$
\begin{array}{ccc}
K & \xrightarrow{\ k\ } & A \\
\downarrow & & \downarrow{\scriptstyle q} \\
0 & \longrightarrow & Q.
\end{array}
$$

Another one (see [18]) is via the following, alternative pushout:

$$
\begin{array}{ccc}
A + K & \xrightarrow{\ (1,k)\ } & A \\
{\scriptstyle (1,0)}\downarrow & & \downarrow{\scriptstyle q} \\
A & \xrightarrow{\ p\ } & Q.
\end{array}
$$

Indeed, from the equality $p(1,0) = q(1,k)$ we get, precomposing with the first coproduct inclusion $\iota_A\colon A \to A + K$:

$$
p = p(1,0)\iota_A = q(1,k)\iota_A = q.
$$

Moreover, precomposing with the second coproduct inclusion $\iota_K\colon K \to A + K$, we obtain

$$
0 = p(1,0)\iota_K = q(1,k)\iota_K = qk = pk,
$$

and, for every morphism $t\colon A \to T$ such that $tk = 0$, one has $t(1,0) = t(1,k)$ and so, by the universal property of the pushout, there is a unique $s\colon Q \to T$ such that the following diagram commutes:

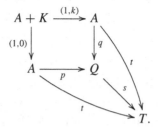

Hence $p = q$ is the cokernel of k.

The latter construction of the cokernel as a pushout gives an easy description of the normal closure of a subobject $k\colon K \to A$ (i.e. of the smallest normal subobject $\bar{k}\colon \overline{K} \to A$ containing K): it is the kernel of the cokernel of k. It is clear that a subobject is normal if and only if it coincides with its normal closure.

Let us now denote by $A \flat K$ the kernel of the canonical morphism $(1, 0)\colon A + K \to A$, where $k\colon K \to A$ is a subobject of A. Considering the following diagram

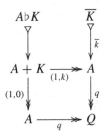

we see that, in an ideal determined category $\mathscr{C}$, the normal closure $\overline{K}$ of K is the regular image of the kernel $A \flat K$ of the canonical morphism $(1, 0)\colon A + K \to A$. Such object $A \flat K$ represents the "formal conjugator" of A over K: indeed, in the category Grp of groups, $A \flat K$ is the subgroup of the free product $A + K$ formed by the ideal words in K and A, and it is generated by words of the form (a, k, a^{-1}) with $a \in A$ and $k \in K$. The fact that $A \flat K$ is the subalgebra of the coproduct formed by the ideal words actually holds in every category of Ω-groups (see [12]). It is easy to see that the following fact, already observed for Ω-groups, holds also in our categorical context:

Remark 3.5 Given two subobjects $h\colon H \to A$ and $k\colon K \to A$ of the same object A, one has

$$H \diamond K = (H \flat K) \cap (K \flat H).$$

Moreover, the normal closure $\overline{K}$ of K is obtained from $A \flat K$ via the realization morphism, which, in Grp, sends the word (a, k, a^{-1}) to the element $aka^{-1} \in A$. Using this fact, one can prove the following result (whose proof is omitted, and can be found in [18, Proposition 5.10]):

Proposition 3.6 *Given a subobject* $k\colon K \to A$ *in a unital, ideal determined category* $\mathscr{C}$, *one has that* $[A, K] = [A, \overline{K}]$.

Now we are ready to prove the following

Proposition 3.7 [18, Proposition 6.1] *Given a unital, ideal determined category* $\mathscr{C}$, *if K is a normal subobject of A, then $[A, K]$ is a subobject of K.*

Proof Since $\mathscr{C}$ is ideal determined, we already know that $[A, K]$ is a normal subobject of $A \vee K = A$. Consider then the following commutative diagram:

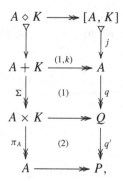

where the squares (1) and (2) are pushouts. Then the rectangle (1) + (2) is a pushout as well. Moreover, $\pi_A \Sigma = (1, 0) \colon A + K \to A$, and we denote by p the composite $q'q$. Thanks to the previous observation, we have that p is the cokernel of $k \colon K \to A$. Since K is normal in A, $K = \overline{K}$, and so k is the kernel of p. To conclude that $[A, K]$ is a subobject of K, it suffices to observe that $pj = 0$, because already $qj = 0$. □

The converse implication does not hold, in general, in a unital, ideal determined category. In order to recover it, we need to add a third condition, usually called *Hofmann axiom*, to the two conditions defining ideal determined categories. Hofmann axioms says that, given a commutative square

$$
\begin{array}{ccc}
X & \xrightarrow{f'} & X' \\
\downarrow x & & \downarrow x' \\
Y & \xrightarrow{f} & Y,
\end{array}
$$

where the horizontal arrows are regular epimorphisms, the vertical ones are monomorphisms and x' is normal, if $k = \ker(f)$ factors through x, then x is normal as well.

Definition 3.8 [14] An ideal determined category $\mathscr{C}$ which satisfies the Hofmann axiom is called a *semi-abelian* category.

Among the many examples of semi-abelian categories there are the category Grp of groups, as well as the category Rng of (not necessarily unitary) rings and every category of Ω-groups. Further examples are the dual of the category of pointed sets and every abelian category. Actually the following characterization of abelian categories holds:

Remark 3.9 [14] A category $\mathscr{C}$ is abelian if and only if both $\mathscr{C}$ and its dual $\mathscr{C}^{\text{op}}$ are semi-abelian.

The previous remark explains the name "semi-abelian". Semi-abelian categories have many good properties. One of them is that they are always exact in the sense of Barr [1]: a category is Barr-exact if it is regular and, moreover, every internal equivalence relation is effective (which means that it is the kernel pair of a morphism). Another interesting property of semi-abelian categories is that they are Mal'tsev categories: we will explain this notion in the next section. Semi-abelian categories can be characterized as follows:

Proposition 3.10 *A pointed, Barr-exact category $\mathscr{C}$ with finite coproducts is semi-abelian if and only if the short five lemma holds in it: given a commutative diagram*

$$
\begin{array}{ccccccccc}
0 & \longrightarrow & A & \longrightarrow & B & \longrightarrow & C & \longrightarrow & 0 \\
 & & \alpha \downarrow & & \beta \downarrow & & \gamma \downarrow & & \\
0 & \longrightarrow & A' & \longrightarrow & B' & \longrightarrow & C' & \longrightarrow & 0
\end{array}
$$

whose rows are short exact sequences, if α and γ are isomorphisms, then β also is.

In the setting of semi-abelian categories we can state the converse of Proposition 3.7. We are not going to give a proof, which can be found in [18, Theorem 6.3].

Theorem 3.11 *In a semi-abelian category $\mathscr{C}$, a subobject K of an object A is normal if and only if $[A, K]$ is a subobject of K.*

Remark 3.12 This way to test the normality of a subobject K of A via the commutator $[A, K]$ actually provides a characterization of semi-abelian categories among finitely cocomplete homological categories, as proved in [11].

4 The Huq Commutator

The aim of this section is to introduce another notion of commutator in a categorical context, and to compare it with the Higgins commutator we studied in the previous section. This alternative notion of commutator was first considered by Huq in [13] and further studied in the context of unital categories in [4]. Before going to our general categorical context, let us start with an observation in the case of groups:

Proposition 4.1 *Given two subgroups H and K of the same group A, one has that $[H, K] = 0$ if and only if there exists a (necessarily unique) morphism $\varphi : H \times K \to A$ such that the following diagram commutes:*

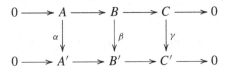

$$
\begin{array}{ccccc}
H & \xrightarrow{\langle 1,0 \rangle} & H \times K & \xleftarrow{\langle 0,1 \rangle} & K \\
 & \searrow & \downarrow \varphi & \swarrow & \\
 & & A. & &
\end{array}
$$

(5.4.i)

Proof Since every element $(h, k) \in H \times K$ can be decomposed as $(h, k) = (h, 1) \cdot (1, k)$, a morphism φ making the diagram above commute must satisfy the following equality:

$$\varphi(h, k) = \varphi((h, 1) \cdot (1, k)) = \varphi(h, 1) \cdot \varphi(1, k) = \varphi\langle 1, 0\rangle(h) \cdot \varphi\langle 0, 1\rangle(k) = hk.$$

This shows that such a morphism, when it exists, is uniquely determined. Let us now prove that φ is a morphism if and only if $[H, K] = 0$. Given $h, h' \in H$ and $k, k' \in K$, one has

$$\varphi(h, h') \cdot \varphi(k, k') = hh'kk', \quad \text{while} \quad \varphi((h, h') \cdot (k, k')) = hkh'k'$$

and it is clear that the two expressions are equal for all $h, h' \in H$ and $k, k' \in K$ if and only if $[H, K] = 0$.   □

With this property of groups in mind, we can give the following categorical definition:

Definition 4.2 [4] Let $\mathscr{C}$ be a unital category. Two subobjects $h \colon H \to A$ and $k \colon K \to A$ of the same object A are said to *cooperate* if there exists a morphism $\varphi \colon H \times K \to A$ such that diagram (5.4.i) commutes.

The morphism φ as above, when it exists, is called the *cooperator* of h and k. It is always unique; indeed, if two morphisms φ and φ' make diagram (5.4.i) commute, then

$$\varphi \Sigma = \varphi' \Sigma,$$

since Σ is induced by $\langle 1, 0\rangle$ and $\langle 0, 1\rangle$. But, in a unital category, Σ is an extremal epimorphism, and this implies $\varphi = \varphi'$.

Let us see when two subobjects cooperate in the category Rng of rings. Using the same argument we explained for groups, one can conclude that, if a cooperator φ between subrings H and K of A exists, then it must be defined by $\varphi(h, k) = h + k$ for all $h \in H, k \in K$. But then

$$hk = \varphi(h, 0)\varphi(0, k) = \varphi(0, 0) = 0,$$

and, in the same way, $kh = 0$. It is not difficult to check that the converse is also true. So, H and K cooperate if and only if $hk = kh = 0$ for all $h \in H, k \in K$.

We observe that, in order to Definition 4.2 make sense, there is no need that the morphisms h and k are monomorphisms. So, the definition can be extended to the one of cooperating morphisms with the same codomain, and no other restriction:

Definition 4.3 Let $\mathscr{C}$ be a unital category. Two morphisms $f \colon H \to A$ and $g \colon K \to A$ with the same codomain *cooperate* if there exists a morphism $\varphi \colon H \times K \to A$ such that the diagram

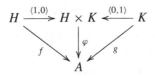

commutes.

Actually, if the category $\mathscr{C}$ is normal, this last definition is not really more general than Definition 4.2. In fact, the following Proposition holds:

Proposition 4.4 *Let $\mathscr{C}$ be a normal and unital category. Two morphisms $f: H \to A$ and $g: K \to A$ cooperate if and only if their regular images cooperate as subobjects of A.*

Proof Consider the following commutative diagram:

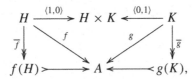

and suppose that $f(H)$ and $g(K)$ cooperate as subobjects of A. Then there is a cooperator $\varphi: f(H) \times g(K) \to A$. It is immediate to check that, composing it with the morphism $\overline{f} \times \overline{g}: H \times K \to f(H) \times g(K)$, one gets a cooperator for f and g (see the diagram below).

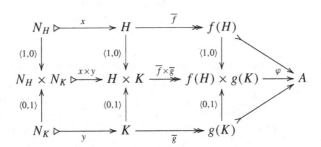

Conversely, suppose that there is a cooperator $\psi: H \times K \to A$. The morphism $\overline{f} \times \overline{g}$ is a regular epimorphism, hence it is the cokernel of its kernel (because the category is normal). It is immediate to check that its kernel is $x \times y: N_H \times N_K \to H \times K$. Since

$$\psi(x \times y)\langle 1, 0 \rangle = 0 = \psi(x \times y)\langle 0, 1 \rangle$$

and the category is unital, one has $\psi(x \times y) = 0$. Then, by the universal property of the cokernel, one gets a unique morphism $\varphi: f(H) \times g(K) \to A$ such that $\varphi(\overline{f} \times \overline{g}) = \psi$. It is easy to check that such a morphism is a cooperator for $f(H)$ and $g(K)$ as subobjects of A. $\square$

Now we have everything we need to define the Huq commutator:

Definition 4.5 Let $h\colon H \to A$ and $k\colon K \to A$ be two subobjects of the same object A in a normal and unital category $\mathscr{C}$. The Huq commutator of h and k, denoted by $[H, K]_{\mathrm{Huq}}$, is the smallest normal subobject $n\colon N \to A$ of A such that, denoting by $q\colon A \to A/N$ the cokernel of n, the morphisms qh and qk cooperate.

First of all, we should show that such smallest normal subobject always exists. In order to do that, consider the following pushout:

$$
\begin{array}{ccc}
H + K & \xrightarrow{\;(h,k)\;} & A \\
{\scriptstyle \Sigma}\downarrow & {\scriptstyle (qh,qk)} & \downarrow{\scriptstyle q} \\
H \times K & \xrightarrow{\;\varphi\;} & Q.
\end{array}
$$

We want to show that $[H, K]_{\mathrm{Huq}}$ is the kernel of q. Let us complete the previous diagram as follows:

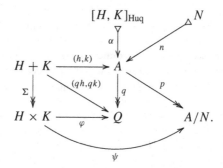

First of all, we observe that φ is the cooperator of qh and qk, because $\Sigma = (\langle 1, 0\rangle, \langle 0, 1\rangle)$, and the commutativity of the pushout square says precisely that $\varphi(\langle 1, 0\rangle, \langle 0, 1\rangle) = (qh, qk)$. Moreover, if $n\colon N \to A$ is a normal subobject of A such that ph and pk cooperate (where p is a cokernel of n), one has a cooperator $\psi\colon H \times K \to A/N$, which is a morphism such that $\psi\Sigma = p(h, k)$. The universal property of the pushout gives then a unique morphism $\gamma\colon Q \to A/N$ such that $\gamma q = p$ (and $\gamma\varphi = \psi$). From the universal property of n as a kernel of p, we get a unique $\beta\colon [H, K]_{\mathrm{Huq}} \to N$ such that $n\beta = \alpha$, and such a β is necessarily a monomorphism.

Proposition 4.6 [18] *Let $h\colon H \to A$ and $k\colon K \to A$ be two subobjects of the same object A in a normal and unital category $\mathscr{C}$. The Huq commutator $[H, K]_{\mathrm{Huq}}$ is the normal closure of the Higgins commutator $[H, K]$ of h and k.*

Proof It suffices to consider the following commutative diagram:

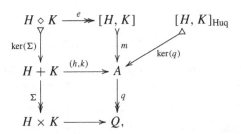

where the lower square is a pushout, and to observe that q is the cokernel of m.   □

So, if $[H, K]$ is normal in A (in particular, when $A = H \vee K$), then $[H, K] = [H, K]_{\text{Huq}}$. However, as we observed in the case of groups, it can be $[H, K] \neq [H, K]_{\text{Huq}}$ if H and K are not normal in A. In our general categorical context, it can happen that $[H, K] \neq [H, K]_{\text{Huq}}$ even if H and K are normal in A. We will say that the category $\mathscr{C}$ satisfies the condition of *normality of the Higgins commutator* if the Higgins commutator $[H, K]$ of two normal subobjects of an object A is normal in A. The category **Grp** has this property. We refer to [8] for more examples and counterexamples of semi-abelian categories with respect to this property.

5 Abelian Objects

The aim of this section is to introduce the notions of commutative and abelian object in a categorical context, and to compare them. We start talking about commutative objects.

Definition 5.1 An object X in a unital category $\mathscr{C}$ is *commutative* if the identity morphism 1_X cooperates with itself.

Thanks to the observations at the end of the previous section, we can conclude that, in a normal and unital category, an object X is commutative if and only if the Huq commutator $[X, X]_{\text{Huq}}$ is the zero object. Moreover, since $X \vee X = X$ and X is clearly normal in itself, we have that the Higgins commutator $[X, X]$ coincides with $[X, X]_{\text{Huq}}$, and so X is commutative if and only if $[X, X] = 0$. Another characterization of commutative objects, in terms of internal algebraic structures, is possible. In order to describe it, we first need to recall the following definition:

Definition 5.2 An *internal unitary magma* in a category $\mathscr{C}$ with finite limits is a triple (X, m, e), where X is an object of $\mathscr{C}$, and $m \colon X \times X \to X$, $e \colon 1 \to X$ are morphisms in $\mathscr{C}$ (by 1 we denote the terminal object of $\mathscr{C}$) such that e "behaves like a unit for the internal operation m", namely the following diagram commutes:

$$1 \times X \xrightarrow{e \times 1_X} X \times X \xleftarrow{1_X \times e} X \times 1$$

$$\sim \| \qquad \downarrow m \qquad \| \sim$$

$$X = X = X.$$

Proposition 5.3 *An internal unitary magma structure on an object X in a unital category $\mathscr{C}$, when it exists, is unique. It exists if and only if the object X is commutative.*

Proof Since a unital category is pointed, the morphism $e : 1 = 0 \to X$ is uniquely determined and, moreover, we have the following commutative triangles:

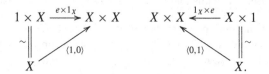

Hence the multiplication m of an internal unitary magma (X, m, e) must make the following diagram commute:

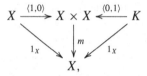

and so m must be a cooperator for the pair $(1_X, 1_X)$. Then it is unique, when it exists. Moreover, it exists if and only if 1_X cooperates with itself, i.e. if and only if X is commutative. □

Actually we can say more:

Proposition 5.4 *Every internal unitary magma structure (X, m, e) in a unital category $\mathscr{C}$ is an internal commutative monoid structure, i.e. the internal multiplication m is associative and commutative.*

Proof In order to prove the associativity of m, we need to show that the following square commutes:

$$X \times X \times X \xrightarrow{1_X \times m} X \times X$$

$$m \times 1_X \downarrow \qquad \qquad \downarrow m$$

$$X \times X \xrightarrow{\quad m \quad} X,$$

or, in other terms, that $m(m \times 1_X) = m(1_X \times m)$. To do that, we will show that $m(m \times 1_X)$ and $m(1_X \times m)$ are cooperators for the same pair of morphisms. Consider the following diagram:

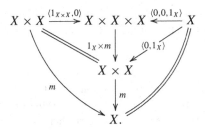

The lower left-hand side triangle clearly commutes, the lower right-hand side one also does, since m is a cooperator for the pair $(1_X, 1_X)$, as observed in the proof of the previous proposition. For the same reason, the upper right-hand side triangle commutes as well. To check whether the remaining triangle commutes, we precompose with the jointly epimorphic pair of morphisms $\langle 1, 0 \rangle, \langle 0, 1 \rangle \colon X \to X \times X$. We have

$$(1_X \times m)\langle 1_{X \times X}, 0 \rangle \langle 1, 0 \rangle = (1_X \times m)\langle 1, 0, 0 \rangle = \langle 1, 0 \rangle$$

and

$$(1_X \times m)\langle 1_{X \times X}, 0 \rangle \langle 0, 1 \rangle = (1_X \times m)\langle 0, 1, 0 \rangle = \langle 0, 1 \rangle,$$

again using the fact that m is a cooperator for the pair $(1_X, 1_X)$. Hence the whole diagram commutes, and this tells us that $m(1_X \times m)$ is a cooperator for the pair $(m, 1_X)$. In a similar way, one can check that $m(m \times 1_X)$ is a cooperator for the same pair, and so these two morphisms coincide.

In order to show that m is commutative, we have to check that $m = m \circ tw$, where $tw = \langle \pi_2, \pi_1 \rangle \colon X \times X \to X \times X$ is the "twisting" isomorphism (in set-theoretic terms, it sends a pair (x, y) to the pair (y, x)). We have the following commutative diagram:

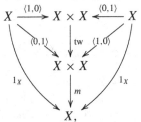

which tells us that $m \circ tw$ is a cooperator for the pair $(1_X, 1_X)$, as well as m, hence these two morphisms coincide.   □

Once we know that every commutative object in a unital category has a (unique) structure of internal commutative monoid (and conversely), a natural question arises: to understand when these internal commutative monoids are internal abelian groups. Internal abelian groups deserve a specific name:

Definition 5.5 An *abelian object* in a unital category $\mathscr{C}$ is an object X equipped with a (necessarily unique) structure of internal abelian group.

The terminology is justified by the fact that, in the category Grp of groups, the abelian objects are precisely the abelian groups. So, in Grp, commutative objects and abelian objects coincide. Unfortunately this is not true in every unital category: there can be commutative objects that are not abelian. In order to get the equivalence between the two notions, we need to impose further assumptions on our category. Before doing it, we recall some terminology.

Given two sets X and Y, a relation R from X to Y is *difunctional* if the following condition holds:

$$\forall\, x, x' \in X, y, y' \in Y \quad xRy', x'Ry', x'Ry \implies xRy.$$

This notion is important because, for example, it allows an easy characterization of equivalence relations among reflexive ones. Indeed, a reflexive relation on a set X is an equivalence relation if and only if it is difunctional. We can actually talk about relations internally to every category with finite limits (the reader may again refer to the chapter *An introduction to regular categories* of this volume for a full treatment of relations in regular categories). Indeed, an internal relation between two objects X and Y in a finitely complete category $\mathscr{C}$ is nothing but a subobject of $X \times Y$, which can be represented by a monomorphism $R \rightarrowtail X \times Y$. All the classical properties of relations can be easily expressed categorically. For instance, an internal relation R on an object X is reflexive if the diagonal morphism $\langle 1, 1 \rangle \colon X \to X \times X$ factors through R. An internal relation $R \rightarrowtail X \times Y$ is difunctional if, considering the commutative diagram

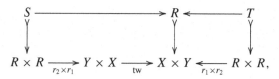

where $r_1 \colon R \to X$ and $r_2 \colon R \to Y$ are the composites of the monomorphism $R \rightarrowtail X \times Y$ with the product projections on X and Y, respectively, tw is the twisting isomorphism, and both squares are pullbacks, one has that the canonical inclusion $S \cap T \rightarrowtail S$ is an isomorphism. However, for our purposes, this internal description of difunctionality is not so important. Indeed, the notion of internal relation, as well as the main properties of relations (like reflexivity or difunctionality), can be expressed only by means of finite limits. Hence, if these properties of relations hold in the category Set of sets, then they hold in every category with finite limits (we do not enter the details of this fact; the interested reader can find a self-contained explanation of this in Chapter 0 of [2]).

Now we have everything we need to give the following

Definition 5.6 [6] A finitely complete category $\mathscr{C}$ is a *Mal'tsev category* if every internal relation in $\mathscr{C}$ is difunctional.

In fact, one can define equivalently Mal'tsev categories as those finitely complete categories in which every internal reflexive relation is an equivalence relation (see [2]). The first property of Mal'tsev categories we are interested in is the following:

Proposition 5.7 *Every pointed Mal'tsev category is unital.*

Proof Let $\mathscr{C}$ be a pointed Mal'tsev category. For any pair of objects X, Y in $\mathscr{C}$, we have to show that the morphisms

$$X \xrightarrow{\langle 1,0 \rangle} X \times Y \xleftarrow{\langle 0,1 \rangle} Y$$

are jointly extremal epimorphic. So, suppose they both factorize through a common monomorphism $m: R \rightarrowtail X \times Y$.

Such an m gives rise to an internal relation R in $\mathscr{C}$, which is then difunctional. The fact that $\langle 1, 0 \rangle$ and $\langle 0, 1 \rangle$ factor through R can be expressed in set-theoretic terms saying that for all $x \in X$ and $y \in Y$ one has $x R 0$ and $0 R y$. By difunctionality we get

$$x R 0, \; 0 R 0, \; 0 R y \;\; \Longrightarrow \;\; x R y.$$

Hence R coincides, up to an isomorphism, with the total relation $X \times Y$. This means that m is an isomorphism, proving that $\langle 1, 0 \rangle$ and $\langle 0, 1 \rangle$ are jointly extremal epimorphic. $\qquad\square$

Every semi-abelian category is a Mal'tsev category (see, for example, [2]). So, Grp, Rng, as well as every category of Ω-groups in the sense of [12] are Mal'tsev categories.

Theorem 5.8 [6] *Every commutative object in a pointed Mal'tsev category is abelian.*

Proof Let X be a commutative object in a pointed Mal'tsev category $\mathscr{C}$, and let (X, m, e) be its unique internal commutative monoid structure. We have to show that this structure is the one of an internal abelian group, i.e. there exists a morphism $i: X \to X$ making the following diagram commute:

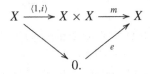

$$X \xrightarrow{\langle 1,i \rangle} X \times X \xrightarrow{m} X$$

0.

We use again the elementwise approach. First of all, we show that the multiplication m satisfies the two following conditions:

(i) $m(x, y) = m(x, z) \implies y = z$;
(ii) $m(y, x) = m(z, x) \implies y = z$.

In order to do that, consider the relation R from $X \times X$ to X defined by

$$(y, z)Rx \iff m(x, y) = m(x, z).$$

Suppose that $m(x, y) = m(x, z)$. Denoting by 0 the neutral element of m, we have that:

– $(y, z)Rx$, since $m(x, y) = m(x, z)$;
– $(0, 0)Rx$, since $m(x, 0) = m(x, 0) = x$;
– $(0, 0)R0$, since $m(0, 0) = m(0, 0) = 0$.

By difunctionality we get that $(y, z)R0$, which means that $y = m(0, y) = m(0, z) = z$. This proves (i); the proof of (ii) is analogous.

Let us now define another relation S on X by putting xSy if and only if there exists z such that $m(z, y) = x$. Such a z is unique because of (i) above. We have xSx, with $z = 0$, since $m(0, x) = x$. Moreover, $xS0$, with $z = x$, since $m(x, 0) = x$, and finally $0S0$. By difunctionality, we conclude that $0Sx$, i.e. there exists a (unique) z such that $m(z, x) = 0$. This element z allows us to define the morphism i we are looking for. $\square$

We conclude by observing that, in a pointed, normal Mal'tsev category $\mathscr{C}$, the full subcategory $\mathrm{Ab}(\mathscr{C})$ of abelian objects is reflective. The reflection is performed by the so-called *abelianization functor*

$$\mathrm{ab} \colon \mathscr{C} \to \mathrm{Ab}(\mathscr{C}).$$

It is obtained as follows: given an object X of $\mathscr{C}$, $\mathrm{ab}(X) = \frac{X}{[X,X]_{\mathrm{Huq}}}$ as in the following pushout:

$$
\begin{array}{ccc}
X + X & \xrightarrow{(1,1)} & X \\
{\scriptstyle \Sigma}\downarrow & & \downarrow{\scriptstyle q} \\
X \times X & \longrightarrow & \frac{X}{[X,X]_{\mathrm{Huq}}}.
\end{array}
$$

Indeed, we already observed that $[X, X]_{\mathrm{Huq}}$ is the smallest normal subobject of X such that the pair $(1_X, 1_X)$, composed with the projection q, commutes in $\frac{X}{[X,X]_{\mathrm{Huq}}}$. Hence, being $\mathscr{C}$ Mal'tsev, the object $\frac{X}{[X,X]_{\mathrm{Huq}}}$ is abelian. The functoriality of this construction is obvious. Let us check that it has the universal property of the reflection. Given a morphism $f \colon X \to A$, where A is an abelian object, we have that f cooperates with itself in A. Indeed, if m is the internal multiplication of A, $\psi = m \circ (f \times f)$ is a cooperator for the pair (f, f), because the following diagram commutes:

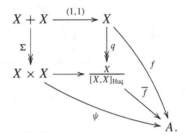

Then, by the universal property of the pushout, we get a unique $\overline{f} \colon [X, X]_{\mathrm{Huq}} \to A$ making the following diagram commute:

$$
\begin{array}{ccc}
X + X & \xrightarrow{(1,1)} & X \\
\Sigma \downarrow & & \downarrow q \\
X \times X & \longrightarrow & \dfrac{X}{[X,X]_{\mathrm{Huq}}}
\end{array}
$$

Whence the universality of the abelianization.

Acknowledgements The authors are grateful to Maria Manuel Clementino, Marino Gran and the anonymous referee for carefully proofreading a first version of the article and suggesting some useful changes and corrections.

References

1. Barr, M.: Exact categories. In: Lecture Notes in Mathematics, vol. 236, pp. 1–120. Springer, Heidelberg (1971)
2. Borceux, F., Bourn, D.: Mal'cev, protomodular, homological and semi-abelian categories. Kluwer Academic Publishers, Mathematics and Its Applications (2004)
3. Bourn, D.: Mal'cev categories and fibration of pointed objects. Appl. Categ. Struct. **4**, 307–327 (1996)
4. Bourn, D.: Intrinsic centrality and associated classifying properties. J. Algebra **256**, 126–145 (2002)
5. Carboni, A., Janelidze, G.: Smash product of pointed objects in lextensive categories. J. Pure Appl. Algebra **183**, 27–43 (2003)
6. Carboni, A., Pedicchio, M.C., Pirovano, N.: Internal graphs and internal groupoids in Mal'cev categories. In: CMS Conference Proceedings, Category Theory, vol. 1991, no. 13, pp. 97–109 (1992)
7. Cigoli, A.S.: Centrality via internal actions and action accessibility via centralizers. Ph.D. thesis, Università degli Studi di Milano (2010)

8. Cigoli, A.S., Gray, J.R.A., Van der Linden, T.: On the normality of Higgins commutators. J. Pure Appl. Algebra **219**, 897–912 (2015)
9. Freese, R., McKenzie, R.: Commutator Theory for Congruence Modular Varieties. London Mathematical Social Lecture Notes Series, vol. 125. Cambridge University Press, Cambridge (1987)
10. Gumm, H.P., Ursini, A.: Ideals in universal algebras. Algebra Universalis **19**, 45–54 (1984)
11. Hartl, M., Loiseau, B.: On actions and strict actions in homological categories. Theory Appl. Categ. **27**(15), 347–392 (2012)
12. Higgins, P.J.: Groups with multiple operators. Proc. Lond. Math. Soc. **6**(3), 366–416 (1956)
13. Huq, S.A.: Commutator, nilpotency and solvability in categories. Q. J. Math. **19**(2), 363–389 (1968)
14. Janelidze, G., Márki, L., Tholen, W.: Semi-abelian categories. J. Pure Appl. Algebra **168**(2–3), 367–386 (2002)
15. Janelidze, G., Márki, L., Tholen, W., Ursini, A.: Ideal determined categories. Cahiers Top. Geom. Diff. Cat. LI **2**, 115–125 (2010)
16. Janelidze, Z.: The pointed subobject functor, 3 × 3 lemmas, and subtractivity of spans. Theory Appl. Categ. **23**(11), 221–242 (2010)
17. Mantovani, S.: The Ursini commutator as normalized Smith-Pedicchio commutator. Theory Appl. Categ. **27**, 174–188 (2012)
18. Mantovani, S., Metere, G.: Normalities and commutators. J. Algebra **324**, 2568–2588 (2010)
19. Pedicchio, M.C.: A categorical approach to commutator theory. J. Algebra **177**, 647–657 (1995)
20. Smith, J.D.H.: Mal'cev varieties. In: Lecture Notes in Mathematics, vol. 554. Springer, Heidelberg (1976)

Chapter 6
Notes on Point-Free Topology

Jorge Picado and Aleš Pultr

Abstract Point-free topology is the study of the category of locales and localic maps and its dual category of frames and frame homomorphisms. These notes cover the topics presented by the first author in his course on *Frames and Locales* at the *Summer School in Algebra and Topology*. We give an overview of the basic ideas and motivation for point-free topology, explaining the similarities and dissimilarities with the classical setting and stressing some of the new features.

Keywords Point-free topology · Category of frames · Category of locales · Heyting algebra · Sober space

Math. Subj. Classification 06D22 · 18F70 · 54-02

Introduction

Topology can be cultivated as the natural geometry of places ("spots") and their interrelations. One does not have to think of them as sets of points: they can be thought as entities in their own right (similarly as lines in classical geometry are not sets of incident points). Such (point-free) approach to general geometry appeared in topology already in the late thirties and forties, started to be systematically cultivated in the last decades of previous century, and flourishes since. It has turned out that by forgetting about points one does not lose really important information (unless, of course, when needing a topology as technical means for solving a problem in a concrete—typically otherwise structured—set, which is another matter). On the contrary, one gains important new insights into the general structure of space and obtains results that are in some respects better than the classical counterparts, or such

J. Picado (✉)
CMUC, Department of Mathematics, University of Coimbra, 3001-501 Coimbra, Portugal
e-mail: picado@mat.uc.pt

A. Pultr
Department of Applied Mathematics and ITI, MFF, Charles University, Prague, Czech Republic
e-mail: pultr@kam.mff.cuni.cz

© The Author(s), under exclusive license to Springer Nature Switzerland AG 2021
M. M. Clementino et al. (eds.), *New Perspectives in Algebra, Topology and Categories*, Coimbra Mathematical Texts 1,
https://doi.org/10.1007/978-3-030-84319-9_6

that have no classical counterparts at all. Also, the point-free approach opens door to applications in theoretical computer science.

In this text we would like to introduce the reader to point-free thinking, to illustrate the reasoning (for this, we present some proofs hoping to persuade the reader that the techniques are in fact fairly friendly) and present some results to illustrate the merits of the approach. In Prologue we expound point-free topology as a natural synthetic general geometry and briefly outline the history. Then we discuss the necessary order-theoretic background, the relation to classical spaces, and some categorical aspects. Next we introduce the reader to some concrete facts about point-free spaces (locales), and finally we present a few facts showing their advantages and merits.

1 Prologue

1.1 General Topology

General (set-theoretic) topology is a generalized geometry. Note, however, that it fundamentally differs from the classical geometry: it lacks a synthetic variant preceding the current analytic one. Let us explain.

In the classical (synthetic Euclidean) geometry we work with entities like points, lines, or planes, entities of their own right. One studies their interrelations, in particular the incidence, which should not be confused with the set-theoretic $\in$. A point can be incident (or non-incident) with a line. A line p, however, is **not** identified with the set of all the points incident with p, and similarly in stereometry, the relation between lines and planes is not a set-theoretic inclusion.

Only much later, in the *analytic* version, one starts with a beforehand given set of elements (say, pairs of numbers); points, lines, etc., are defined as specific types of subsets (the fact that points are represented as one-element subsets is more or less accidental and not important for what we want to emphasize).

General topology, like many modern structures, comes right away in the analytic form: one starts with a set, and the structure is given by assigning specific roles to some of the subsets.

1.2 A Synthetic Generalized Geometry

The following is not an account of what historically happened. Let us just pretend designing a *synthetic generalized geometry* from scratch. We will do it modelling the intuition of the behavior of "pieces of space" (we will call them *spots*) as we think about them in the "space around us".

1.2.1 Order First, the system L of such spots is naturally ordered:

a spot can be an extension of another one

(we do not have in mind a set-theoretic inclusion—a spot is not a set of elements—just think of a spot being larger than another one).

1.2.2 Glueing Spots Together Next, given a system A of spots we can think of conglutinating (merging, pasting) them together and obtaining a new one. It will be the smallest spot larger than all the $a \in A$ and hence, formally, this amounts to assuming that the ordered L is a complete lattice, and that the systems A combine to their suprema (joins) $\bigvee A$.

1.2.3 Meeting Conglutinated Spots It is natural to assume that

a spot b meets the result of pasting the $a \in A$ together, the conglutination $\bigvee A$, only if it meets some of the constituents $a \in A$.

Formally,

$$(\bigvee A) \wedge b \neq 0 \quad \text{only if} \quad a \wedge b \neq 0 \text{ for some } a \in A. \tag{meet}$$

It is easy to see that this means precisely that our complete lattice admits pseudo-complements. Thus, our (we hope admittedly natural) assumptions lead us to the conclusion that

a general synthetic geometry can be viewed as a complete pseudocomplemented lattice.

1.2.4 One More Assumption Finally, let us agree that at least some naturally defined sublattices of L are synthetic geometries as well. Confining ourselves to the up-sets $\uparrow u = \{x \mid u \leq x\} \subseteq L$ we obtain the condition (meet) strengthened to

$$\forall u \in L \; ((\bigvee A) \wedge b \nleq u \quad \text{only if} \quad a \wedge b \nleq u \text{ for some } a \in A) \tag{MEET}$$

and this is easily seen to be equivalent to the *frame distribution law*

$$\forall A \subseteq L, \; \forall b \in L, \; (\bigvee A) \wedge b = \bigvee \{a \wedge b \mid a \in A\}. \tag{frm}$$

Complete lattices satisfying (frm) are called *frames*; taking into account that this is equivalent with the Heyting structure (see 2.6 below) and that it is a more expedient condition than the mere existence of pseudocomplements, we can view frames as representatives of fairly general synthetic geometries. And there will be a strong corroboration of this view in the next subsections.

1.3   A More Realistic Account of the Events

The development of topology did not follow the reckoning outlined in the subsection 1.2. General topology has been created as an *analytic* theory, *preceding a synthetic one*; but the synthetic successor turns out to be precisely what we have obtained above. Moreover it leads to a suitable definition of mappings between generalized geometries, which we have in 1.2 not even started to contemplate.

Modern topology originated in the pioneering Hausdorff's article [23] published in 1914. The intuition behind the introduced general concept of a space is based on the natural distinction between a set surrounding a point as opposed to a set just including it: think of a ship in a lake surrounded by water from all the sides as opposed to a landed one touching a pier. The structure of a space, carried by a set, is constituted by determining the *neighborhoods* $U \ni x$ as special subsets of X containing x, satisfying very natural assumptions.

Soon (already in the twenties) an equivalent alternative approach based on the notion of an *open set* (in the original concept setting: a set that is a *neighborhood of all its points*; and on the other hand, if we start with the notion of an open set we can define a neighborhood of x as a $U \ni x$ such that there is an open V with $x \in V \subseteq U$). Comparing it with the neighborhood intuition it might look somewhat less transparent to start with (trading an obvious intuition for technical advantages which are indisputable), but nowadays we know better: in fact it is the intuition of a synthetic version of topology as outlined above: an open set, a set without boundary, is a good model of a spot, and the system of open sets constitutes a complete sublattice $\Omega(X)$ of the powerset of X closed under arbitrary unions and finite meets, and hence satisfying the frame distribution law. Thus we have here an example (in fact a typical one) of a frame, a general geometry presented in subsection 1.2.

1.3.1   A Few Historical Notes Although one did not necessarily have in mind developing synthetic topology, the ideas of harnessing lattice theory in topology (via the lattices of open resp. closed sets) appeared already in the late thirties and in the forties. The Stone duality [53] replacing (very special) topological spaces and continuous maps by Boolean algebras and homomorphisms is deservedly a most cited example. But one should not forget the outstanding Wallman's article [55] where the lattice technique allowed for an ingenious compactification (and not only that: the author is consequent in the lattice technique even to the point of using a specific point-free separation axiom).

In the fifties the attempts to develop a variant of topology without points became more and more frequent. It is not our intention to present here a detailed history. The reader can find a short account in the introduction to our monograph [44], but we can particularly recommend the excellent Johnstone's "The point of pointless topology" [29] and "Elements of the history of locale theory" [33]. Here let us just state that the basic concepts started to settle in late fifties [14, 16, 42] and continued in works of Banaschewski, who had been regarding to topology from a lattice-theoretic point of view since 1953 [3], Dowker, Isbell, Johnstone, Joyal and Tierney, that the first stage of the theory culminated in the monograph [28], and that the theory flourishes.

1.4 Frame Homomorphisms

We have spoken of spaces, either classical or point-free, as objects. The most important concept of classical topology, however, is continuity. What is the counterpart of continuous maps in the point-free context? The characteristic property of continuous maps is that they preserve openness by preimage. Thus, with a continuous map $f: X \to Y$ there is associated a mapping $\Omega(f): \Omega(Y) \to \Omega(X)$ sending each U to $\Omega(f)(U) = f^{-1}[U]$. Since preimage preserves unions and intersections and since *all* unions of open sets are open and *finite* intersections of open sets are open as well, we see that $\Omega(f)$ preserves arbitrary joins and finite meets. This has been adopted for the definition of a *frame homomorphism*: it is a mapping $h: L \to M$ between frames preserving all joins and finite meets.[1]

1.4.1 Note The general geometry (leading, ultimately, to frames) was based on the idea of spots that can merge and the relation of intersecting. Thus, while the former really has to do with joins, the latter involves only the question whether $x \wedge y = 0$ or $x \wedge y \neq 0$. Thus we can naturally ask whether the appropriate maps between geometries should not be those that

(1) preserve all joins, and
(2) satisfy the implication $x \wedge y = 0 \Rightarrow h(x) \wedge h(y) = 0$.

It turns out that in a large (and important) class of frames such maps are frame homomorphisms anyway (see [13]). But in full generality these conditions are weaker.

2 Background

2.1 Posets

When dealing with posets we will use the standard notation. If necessary we use different symbols for different orders, but if there is no danger of confusion we write simply $\leq$ (like in saying that "$f: (X, \leq) \to (Y, \leq)$ is monotone if $x \leq y$ implies

---

[1] In fact this property is characteristic for the representation of continuous maps: if the spaces are *sober*, which is a very general condition (see 3.4 below), each frame homomorphism $h: \Omega(Y) \to \Omega(X)$ is an $\Omega(f)$ for some continuous f [49].

that $f(x) \leq f(y)$" even if there is in fact on Y a relation different from that on X—similarly like we do not hesitate to use the same arrow symbol "$\rightarrow$" when indicating morphisms in two distinct categories).

We write

$$\downarrow A \text{ for } \{x \mid \exists\, a \in A, x \leq a\} \quad \text{and} \quad \uparrow A \text{ for } \{x \mid \exists\, a \in A, x \geq a\}$$

and abbreviate $\downarrow\{a\} = \downarrow a, \uparrow\{a\} = \uparrow a$. An element a is an *upper* (resp. *lower*) *bound* of $A \subseteq (X, \leq)$ if $A \subseteq \downarrow a$ (resp. $A \subseteq \uparrow a$) and the least upper (resp. largest lower) bound, if it exists is called the *supremum* or *join* (resp. *infimum* or *meet*) of A and denoted by $\bigvee A$ resp. $\bigwedge A$. We also use symbols $a \vee b$ for $\bigvee\{a, b\}$, $a_1 \vee \cdots \vee a_n$, $\bigvee_{i \in I} a_i$, and, similarly, with $\bigwedge$ or $\wedge$, in the obvious sense.

The least resp. largest element of $(X, \leq)$, that is, $\bigvee \emptyset$ resp. $\bigwedge \emptyset$, if it exists will be denoted by 0 resp. 1.

The poset obtained by reversing the order, that is $(X, \leq')$ with $x \leq' y \equiv y \leq x$ is called the dual of $(X, \leq)$ and denoted as $(X, \leq)^{\mathrm{op}}$. We may also write $\leq^{\mathrm{op}}$ for thus defined $\leq'$.

2.1.1 **Lattices** A poset in which all the subsets have infima and suprema are called *complete lattices*. If all finite sets have suprema and infima we speak of *bounded lattices*, in case of *non-void* finite sets simply of *lattices*.

2.2 Adjunctions

We say that monotone maps

$$f : (X, \leq) \rightarrow (Y, \leq) \quad \text{and} \quad g : (Y, \leq) \rightarrow (X, \leq)$$

are *(Galois) adjoint, f to the left and g to the right*, and write $f \dashv g$, if

$$f(x) \leq y \iff x \leq g(y). \tag{adj1}$$

It is standard and (very easy) to see that this is equivalent with assuming that

$$fg \leq \mathrm{id}_Y \quad \text{and} \quad \mathrm{id}_X \leq gf. \tag{adj2}$$

Note that from these inequalities it readily follows that

$$fgf = f \quad \text{and} \quad gfg = g. \tag{adj3}$$

2.2.1 **Facts** (a) *Generally, left adjoints preserve all existing suprema and right adjoints preserve all existing infima.*
(b) *On the other hand, if X, Y are complete lattices then each $f : X \to Y$ preserving all suprema is a left adjoint (i.e. it has an adjoint on the right), and each $g : Y \to X$ preserving all infima is a right adjoint.*

Proof (a) Let f be a left adjoint and $s = \bigvee A$ in X. Then obviously $f(s)$ is an upper bound of $f[A]$. Now let b be a general upper bound of $f[A]$, that is, let $f(a) \le b$ for all $a \in A$. Then for all $a \in A$, $a \le g(b)$, hence $g(b)$ is an upper bound of A, hence $s \le g(b)$, and finally $f(s) \le b$.
(b) Let f preserve all joins. Set $g(y) = \bigvee\{z \mid f(z) \le y\}$. If $f(x) \le y$ then trivially $x \le g(y)$. On the other hand if $x \le g(y) = \bigvee\{z \mid f(z) \le y\}$ then $f(x) \le f(\bigvee\{z \mid f(z) \le y\}) = \bigvee\{f(z) \mid f(z) \le y\} \le y$. □

2.3 For Category Minded Readers: Posets as Special Categories

A category is said to be *thin* if for any two objects A, B there is at most one morphism $A \to B$. A poset $(X, \le)$ (more generally, a preordered set) is a thin category in which a morphism $x \to y$ is the statement that $x \le y$, if it holds true: reflexivity of $\le$ provides the identity morphisms and transitivity provides the composition of morphisms.

Note that, on the other hand

every thin category is a preordered partially ordered class,

(which differs from a partially ordered set by possibly being carried by a proper class, and by $a \le b$ and $b \le a$ for isomorphic distinct objects a, b).

Monotone maps $f : (X, \le) \to (Y, \le)$ are in this perspective precisely the functors between such categories.

Now the adjunction from 2.2 is a special case of the adjunction of functors $L : \mathscr{A} \to \mathscr{B}$ and $R : \mathscr{B} \to \mathscr{A}$ in categories. Recall (adj1) and compare

$$f(a) \le b \text{ iff } a \le g(b) \quad \text{with} \quad \mathscr{B}(L(A), B) \cong \mathscr{A}(A, R(B));$$

further compare (adj2) with the adjunction units

$$\lambda : LR \to \mathrm{Id}_{\mathscr{B}} \text{ and } \rho : \mathrm{Id}_{\mathscr{A}} \to RL,$$

and (adj3) with

$$\left(L \xrightarrow{L\rho} LRL \xrightarrow{\lambda_L} L\right) = \mathrm{id}_L \quad \text{and} \quad \left(R \xrightarrow{\rho_R} RLR \xrightarrow{R\lambda} R\right) = \mathrm{id}_R$$

(in the general case the latter has to be assumed, in the thin case it comes for free and becomes the former).

Further realize that the (partial order) upper or lower bounds are precisely the categorical upper or lower bounds in the thin case, and infima resp. suprema coincide with products and coproducts:

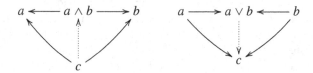

Thus, the first fact in 2.2.1 is a special case of preservation of colimits and limits by left and right adjoint functors. Since equalizers (resp. coequalizers) exist trivially in thin categories, the *existence of limits* (resp. *colimits*) means *existence of products* (resp. *coproducts*), that is, infima (resp. suprema). Hence, complete lattices are the complete (and cocomplete) such categories.

The correspondence of the second fact in 2.2.1 with the theorems on the existence of adjoint functors is not quite so straightforward. One has to keep in mind that in general categories we do not have a proper counterpart to the concept of a complete lattice: in complete and cocomplete categories we consider small (set) diagrams in large (class carried) categories (and large diagrams would not make sense) while in complete lattices we admit (and need) "diagrams of the same size". But there are theorems that present adjoint functors under reasonable circumstances.

To finish this short excursion to categorial reasoning: we have seen posets and Galois adjunction as a special case of adjunction in categories. It is, however, sometimes also profitable to look at the situation the other way, namely as primarily a phenomenon in posets extended to categories where instead of one arrow between nodes one has labelled ones, with structured labelling.

2.4 Some Special Posets

A *meet* resp. *join semilattice* has $a \wedge b$ resp. $a \vee b$ for any a, b (and consequently all non-empty finite meets resp. joins); if it is obvious from the context whether the meets or joins are meant, one speaks simply of a *semilattice*.

We have already introduced lattices, bounded lattices and complete lattices in 2.1.1. Further, a lattice is *distributive* if we have

$$(a \vee b) \wedge c = (a \wedge c) \vee (b \wedge c) \quad \text{which is equivalent to} \quad (a \wedge b) \vee c = (a \vee c) \wedge (b \vee c)$$

(the equivalence may be slightly surprising; it is important to realize that it amounts to the fact that the dual of a distributive lattice is also distributive).

2.5 Pseudocomplements, Supplements and Complements

We might consider a more general situation, but for our purposes everything in the remaining part of this section will happen in bounded lattices L. A *pseudocomplement* (resp. *supplement*) of an element $a \in L$ is an element b such that

$$a \wedge x = 0 \text{ iff } x \leq b \quad (\text{resp. } a \vee x = 1 \text{ iff } x \geq b).$$

None of them has to exist, but if it does it is obviously uniquely determined. If it exists we usually denote it by a^* resp. $a^\#$.

A *complement* of a is an element b such that $a \wedge b = 0$ and $a \vee b = 1$. It does not have to exist and in general it is not even uniquely determined. But in a distributive lattice there is at most one and if it exists it is simultaneously a pseudocomplement and a supplement. One then speaks of a *complemented* element and the complement is usually denoted a^* (if there is no danger of confusion, otherwise another symbol, e.g. a^c, is used).

2.6 Heyting Algebras

A bounded lattice L is called a *Heyting algebra* if there is a binary operation $x \to y$ (the *Heyting operation*) such that for all a, b, c in L,

$$a \wedge b \leq c \text{ iff } a \leq b \to c. \tag{Hey}$$

Recall 2.2 and realize that (Hey) says precisely that for every b

the mapping $b \to (-) = (x \mapsto b \to x): L \to L$ *is a right Galois adjoint of* $(-) \wedge b = (x \mapsto x \wedge b): L \to L$

and hence

(H1) *the operation* $\to$, *if it exists, is uniquely determined; thus, being Heyting is in fact a condition on the meet in* L,

(H2) *in a Heyting algebra one has* $(\bigvee A) \wedge b = \bigvee_{a \in A} (a \wedge b)$ *for any* $A \subseteq L$ *such that* $\bigvee A$ *exists, and* $b \to \bigwedge A = \bigwedge_{a \in A} (b \to a)$ *for any* $A \subseteq L$ *such that* $\bigwedge A$ *exists,*

(H3) *and if* L *is complete then the distributivity rule*

$$(\bigvee A) \wedge b = \bigvee_{a \in A} (a \wedge b)$$

is a necessary and sufficient condition for the existence of a Heyting operation on L.

2.6.1 **Notes** (a) Unlike the plain distributivity $(a \vee b) \wedge c = (a \wedge c) \vee (b \wedge c)$ the distributivity from (H3) is not carried over to the dual. See 3.2.2 below.
(b) It follows immediately from (Hey) that $a \leq b \rightarrow c$ iff $b \leq a \rightarrow c$. This is a contravariant Galois adjunction that yields moreover the rule

(H4) $(\bigvee A) \rightarrow b = \bigwedge_{a \in A} (a \rightarrow b)$.

(c) Dually one defines a coHeyting algebra as a bounded lattice with a binary operation (*coHeyting operation*) $c \smallsetminus b$ (called the *difference*) such that

$$a \vee b \geq c \quad \text{iff} \quad a \geq c \smallsetminus b.$$

The importance of this concept will be apparent in 5.4 below.

2.6.2 **A Few Heyting Rules** In the sequel we will often need to compute with the Heyting operation. Here are some formulas that are immediate consequences of (Hey).

(1) $a \leq b \rightarrow a$ (since $a \wedge b \leq a$).
(2) $1 \rightarrow a = a$ (since $x \leq a$ iff $x \wedge 1 \leq a$, that is, $x \leq 1 \rightarrow a$).
(3) $a \rightarrow b = 1$ iff $a \leq b$ (since $1 \leq a \rightarrow b$ iff $1 \wedge a \leq b$).
(4) $a \wedge (a \rightarrow b) \leq b$ (since $a \rightarrow b \leq a \rightarrow b$, the well known "*modus ponens*" rule).
(5) $a \wedge (a \rightarrow b) = a \wedge b$ ($\leq$ by (4) and $\geq$ by (1)).

And here are three further useful ones, perhaps slightly less trivial, but still very simple.

(6) $a \rightarrow (b \rightarrow c) = (a \wedge b) \rightarrow c = b \rightarrow (a \rightarrow c)$
(we have $x \leq a \rightarrow (b \rightarrow c)$ iff $x \wedge a \leq b \rightarrow c$ iff $x \wedge a \wedge b \rightarrow c$ iff $x \leq (a \wedge b) \rightarrow c$).
(7) $a \rightarrow b = a \rightarrow c$ iff $a \wedge b = a \wedge c$
($\Rightarrow$: By (5) and (4), $a \wedge b = a \wedge (a \rightarrow b) = a \wedge (a \rightarrow c) = a \wedge c$. $\Leftarrow$: By (3) and (H2), $a \rightarrow b = (a \rightarrow a) \wedge (a \rightarrow b) = a \rightarrow (a \wedge b) = a \rightarrow (a \wedge c) = a \rightarrow c$).
(8) $x = (x \vee a) \wedge (a \rightarrow x)$
(by (4) and (1), $(x \vee a) \wedge (a \rightarrow x) = (a \wedge (a \rightarrow x)) \vee (x \wedge (a \rightarrow x)) \leq x$; on the other hand, by (1), $x \leq (x \vee a) \wedge (a \rightarrow x)$).

2.6.3 **Pseudocomplement Rules** In a Heyting algebra one obviously has a pseudocomplement, namely $a^* = a \rightarrow 0$, with the following obvious properties:

(1) $a \leq b \Rightarrow b^* \leq a^*$.
(2) $a \leq a^{**}$ and $a^{***} = a^*$.
(3) $(\bigvee A)^* = \bigwedge_{a \in A} a^*$ for any $A \subseteq L$ such that $\bigvee A$ exists (*De Morgan law*). Caution: the dual law for $\bigwedge A$ does not hold in general.

(Note that in $\Omega(X)$, U^* is the interior of $X \smallsetminus U$.)
Dually, in a coHeyting algebra one has the supplement $a^{\#} = 1 \smallsetminus a$.
On the other hand we easily prove

2.6.4 **Observation** *Let b have a complement b^c in a distributive lattice L. Then*

$$a \wedge b \leq c \text{ iff } a \leq b^c \vee c, \quad \text{and} \quad a \vee b \geq c \text{ iff } a \geq c \wedge b^c.$$

Thus, in a Heyting algebra we have for any complemented element b, $b \rightarrow c = b^* \vee c$ and in a coHeyting algebra we have for any complemented element b, $c \smallsetminus b = c \wedge b^\#$.

Note All the assumptions are essential, though. In particular the formulas $b \rightarrow c = b^* \vee c$ resp. $c \smallsetminus b = c \wedge b^\#$ hold for complemented elements only; the Heyting resp. coHeyting operation cannot be thus reduced to pseudocomplementing resp. supplementing in no other case.

2.7 Boolean Algebras

A *Boolean algebra* is a distributive lattice in which every element is complemented. From 2.6.4 we immediately obtain

Corollary *A Boolean algebra is both a Heyting and a coHeyting algebra.*

3 Frames and Spaces

3.1 The Category of Frames

A *frame* (resp. *coframe*) is a complete lattice L satisfying the distributivity law

$$(\bigvee A) \wedge b = \bigvee \{a \wedge b \mid a \in A\} \tag{frm}$$
$$(\text{resp. } (\bigwedge A) \vee b = \bigwedge \{a \vee b \mid a \in A\}) \tag{cofrm}$$

for all $A \subseteq L$ and $b \in L$. A *frame homomorphism* $h : L \rightarrow M$ between two frames is a mapping preserving all joins and all finite meets. The resulting category will be denoted by

$$\mathsf{Frm}.$$

Similarly we have *coframe homomorphisms* between coframes preserving all meets and finite joins.

3.2 Spaces and Frames. The Functor Ω

A typical frame is the lattice $\Omega(X)$ of all open subsets of a topological space. Furthermore, if $f: X \to Y$ is a continuous mapping we have a frame homomorphism $\Omega(f): \Omega(Y) \to \Omega(X)$ defined by $\Omega(f)(U) = f^{-1}[U]$. Thus we obtain a *contravariant* functor

$$\Omega: \mathsf{Top} \to \mathsf{Frm},$$

a basic link between classical spaces and what will turn out to be the generalized ones (see already in 3.5.2 below, then in 3.9, and again and again).

3.2.1 A Warning Let us agree that our spaces will be, from now on, always T_0: the frames $\Omega(X)$ will be central in our approach to spaces and it will make no sense to discuss classical points that cannot be distinguished by open sets. In particular we will, without further particular mentioning, use the following

Observation *Let $f, g: X \to Y$ be distinct continuous maps, and let Y be T_0. Then $\Omega(f) \neq \Omega(g)$.*

(Indeed, if $f(x) \neq g(x)$ consider a U such that, say, $g(x) \notin U \ni f(x)$. Then $x \in \Omega(f)(U) \smallsetminus \Omega(g)(U)$.)

3.2.2 Notes (a) Unlike plain (finite) distributivity, the frame distributivity typically does not carry over to the dual, that is, a frame is seldom simultaneously a coframe. Take e.g. any T_1-space X with a non-isolated point x and an open set $W \ni x$. Set $V = W \smallsetminus \{x\}$ and $\mathcal{U} = \{U \in \Omega(X) \mid x \in U\}$. Then $\bigwedge \mathcal{U} = \mathrm{int} \bigcap \mathcal{U} = \emptyset$ and hence $(\bigwedge \mathcal{U}) \cup V = V$ while $\bigwedge \{U \cup V \mid U \in \mathcal{U}\} = W \neq V$.
(b) As the example shows, coframes will seldom come as models of (generalized) spaces. They will play, however, a fundamental role in the study of the structure of generalized subspaces.

3.3 The Heyting Structure

Recall 2.6 (and 2.2). The distributivity rule (frm) makes a frame a Heyting algebra and computing with the Heyting operation will be extensively used; similarly we will use computing with the difference in coframes. But we have to keep in mind that the category Frm is not that of (complete) Heyting algebras: frame homomorphisms generally do not respect the Heyting operation.

3.4 Prime Elements and Sobriety

Recall that an element $p < 1$ in a distributive lattice is *prime* if

$$a \wedge b = p \text{ implies that either } a = p \text{ or } b = p$$

(compare with primeness of numbers); equivalently, $a \wedge b \leq p$ only if $a \leq p$ or $b \leq p$ (readily deduced replacing $a \wedge b \leq p$ by $(a \vee p) \wedge (b \vee p) = (a \wedge b) \vee p = p$).

Typical prime elements in $\Omega(X)$ are the open sets $X \smallsetminus \overline{\{x\}}$. A T_0-space is *sober* [44] if there are no other primes in $\Omega(X)$.

3.4.1 **Notes** (a) Sobriety is a very common property of topological spaces. For instance every Hausdorff space is sober:

Suppose a prime $P \in \Omega(X)$ lacks two distinct points x and y. Separate them by disjoint $U \ni x$ and $V \ni y$ and consider the intersection $P = (P \cup U) \cap P \cup V)$ where P contains none of $P \cup U$, $P \cup V$.

Or, Scott spaces are mostly sober. On the other hand, sobriety is incomparable with the axiom T_1.

(b) Because of the relation with the Hausdorff axiom, sobriety is sometimes viewed as one of the so-called separation axioms. But as it was rightly pointed out by Marcel Erné, it is, rather, a completion condition akin to the completion in metric or more generally uniform spaces. As we will see in the following proposition, it amounts to the assumption that filters that have the natural property of a neighborhood system have "a point in the center" that is, are really neighborhood systems.

Recall that a filter F in a distributive lattice is *prime* if $a \vee b \in F$ implies that $a \in F$ or $b \in F$, and *completely prime* if $\bigvee_{i \in I} a_i \in F$ implies that some $a_i \in F$ for any system $\{a_i \mid i \in I\}$. A typical completely prime filter in $\Omega(X)$ is the system $\mathcal{U}(x) = \{U \mid x \in U\}$ of all neighborhoods of a point x.

3.4.2 **Proposition** *A $(T_0\text{-})$space is sober iff each completely prime filter in $\Omega(X)$ is $\mathcal{U}(x)$ for some $x \in X$.*

Proof $\Rightarrow$: Let X be sober and let $\mathcal{F} \subseteq \Omega(X)$ be a completely prime filter. Set $V_0 = \bigcup \{V \in \Omega(X) \mid V \notin \mathcal{F}\}$. By completeness, $V_0 \notin \mathcal{F}$, hence it is the largest such open set, and

$$U \in \mathcal{F} \quad \text{iff} \quad U \not\subseteq V_0. \tag{$*$}$$

Now if $U_1 \cap U_2 \subseteq V_0$ then both of the U_i cannot be in $\mathcal{F}$ (since it is a filter); thus, for some i, $U_i \subseteq V_0$ and V_0 is prime, and by sobriety $V_0 = X \smallsetminus \overline{\{x_0\}}$ for some x_0. Thus by $(*)$, $U \in \mathcal{F}$ iff $U \not\subseteq X \smallsetminus \overline{\{x_0\}}$ which holds precisely when $x_0 \in U$, that is, $U \in \mathcal{U}(x_0)$.

$\Leftarrow$: Let V_0 be prime. Set $\mathcal{F} = \{U \in \Omega(X) \mid U \not\subseteq V_0\}$. Obviously $\mathcal{F}$ is a complete prime filter, hence $\mathcal{U}(x_0)$ for some x_0, so that $U \not\subseteq V_0$ iff $x_0 \in U$, that is, $U \subseteq V_0$ iff $U \subseteq X \smallsetminus \overline{\{x_0\}}$ and we conclude that $V_0 = X \smallsetminus \overline{\{x_0\}}$. $\square$

3.5   Theorem

Let Y be sober and let $h \colon \Omega(Y) \to \Omega(X)$ be a frame homomorphism. Then there is precisely one continuous map $f \colon X \to Y$ such that $h = \Omega(f)$.

Proof Obviously a preimage $h^{-1}[\mathcal{F}]$ of a completely prime filter is a completely prime filter. Take an $x \in X$ and consider $h^{-1}[\mathcal{U}(x)]$. By sobriety and 3.4.2 it is $\mathcal{U}(y)$ for some $y \in Y$. Choose such y and denote it by $f(x)$. Thus, $h^{-1}[\mathcal{U}(x)] = \mathcal{U}(f(x))$, that is,

$$h(U) \ni x \text{ iff } U \ni f(x), \text{ that is, iff } x \in f^{-1}[U],$$

hence thus defined f is continuous and $h = \Omega(f)$. Uniqueness of f follows immediately from T_0 property: if $f(x) \neq g(x)$ choose a U such that, say, $f(x) \in U \not\ni g(x)$ showing that $\Omega(f)(U) \neq \Omega(g)(U)$. $\square$

Thus the restriction $\Omega \colon \mathsf{Sob} \to \mathsf{Frm}$ of Ω is a full embedding.

3.5.1   Corollary *A sober space X can be homeomorphically reconstructed from the frame $\Omega(X)$ as the set*

$$\{h \colon \Omega(X) \to \mathbf{2} = \{0, 1\} \mid h \text{ is a frame homomorphism}\}$$

endowed with the topology consisting of the $\widetilde{U} = \{h \mid h(U) = 1\}$ with $U \in \Omega(X)$. (Indeed, consider the one-point space $P = \{*\}$. Then $\Omega(P) = \{\emptyset, P\} \cong \{0, 1\}$ and we can consider the elements $x \in X$ represented by the continuous maps f_x with $f_x(*) = x$. Those are then by 3.5 in a one-to-one correspondence with the $h_x = \Omega(f_x)$, and $h_x(U) = 1$ iff $x = f_x(*) \in U$.)

3.5.2   Locales—So Far Formally Consider, so far just formally, the dual of the category of frames. It is called the category of locales and it will be studied later in a more transparent and useful concrete form. For the purposes of this section, however, it will be simply

$$\mathsf{Loc} = \mathsf{Frm}^{\mathrm{op}}$$

with frame homomorphisms understood in opposite direction for morphisms. Then we have a covariant functor

$$\Omega \colon \mathsf{Top} \to \mathsf{Loc}$$

and Theorem 3.5 can be interpreted as that this functor embeds sober spaces, a substantial part of the category of spaces, into Loc as a full subcategory. This justifies viewing frame theory, the "point-free topology", as an extension of (at least a substantial part) of the classical one. This point of view will be corroborated and confirmed in the sequel; in this section we only wish to demonstrate the basic linkage between the two.

3.6 Points and Spectra

The role of the sobriety in 3.5 and 3.5.1 was in the one-to-one correspondences, not in detecting (classical) points in the lattice $\Omega(X)$: any point x in any space X is represented by the map $(* \mapsto x)\colon P \to X$. This leads to the definition of a *point* in a frame L as a map of locales $\Omega(P) \to L$, that is, a frame homomorphism $L \to \Omega(P) = \mathbf{2} = \{0, 1\}$ (cf. Clementino [15]).

The following representations of points will come handy.

3.6.1 **Proposition** (1) *Points h in L are in a one-to-one correspondence with the completely prime filters F in L given by $h \mapsto F_h = \{x \mid h(x) = 1\}$ and $F \mapsto h_F$ with $h_F(x) = 1$ iff $x \in F$.*
(2) *Points h in L are in a one-to-one correspondence with the prime elements p of L given by $h \mapsto p_h = \bigvee\{x \mid h(x) = 0\}$ and $p \mapsto h_p$ with $h_p(x) = 1$ iff $x \not\leq p$.*

Proof It is a matter of straightforward checking.   □

3.6.2 **Spectra** In the following construction we will represent points as completely prime filters (briefly, *cp-filters*). This has technical advantages but it is also fairly intuitive: think of points represented by their systems of neighborhoods. The morphisms in Loc will be (so far) represented as frame homomorphisms, one has only to be careful with the interchanged domain and codomain.

The spectrum of a frame L is the topological space

$$\Sigma(L) = \Big(\{F \mid F \text{ cp-filter in } L\}, \{\Sigma_a \mid a \in L\} \Big)$$

where $\Sigma_a = \{F \mid a \in F\}$. Note that

$$\Sigma_0 = \emptyset, \quad \Sigma_1 = \Sigma(L), \quad \Sigma_{a \wedge b} = \Sigma_a \cap \Sigma_b \text{ and } \Sigma_{\bigvee a_i} = \bigcup \Sigma_{a_i} \tag{Σ1}$$

so that $\{\Sigma_a \mid a \in L\}$ is really a topology. For each frame homomorphism $h\colon M \to L$ ($L \to M$ in Loc) set $\Sigma(h)(F) = h^{-1}[F]$. We have

$$(\Sigma(h))^{-1}[\Sigma_a] = \{F \mid a \in h^{-1}[F]\} = \{F \mid h(a) \in F\} = \Sigma_{h(a)}; \qquad (\Sigma 2)$$

hence, each $\Sigma(h)$ is continuous and we have obtained a functor

$$\Sigma: \mathsf{Loc} \to \mathsf{Top}.$$

Observation *Each $\Sigma(L)$ is a sober space.*

(If $F \nsubseteq G$, with $a \in F, a \notin G$ then $G \notin \Sigma_a \ni F$. Thus $\Sigma(L)$ is T_0. Let Σ_a be a prime in $\Omega\Sigma(L)$. Set $p = \bigvee\{b \in L \mid \Sigma_b \subseteq \Sigma_a\}$. In particular, $\Sigma_p = \Sigma_a$. If $x \wedge y \leq p$ then $\Sigma_x \cap \Sigma_y \subseteq \Sigma_p = \Sigma_a$ and hence, say, $\Sigma_x \subseteq \Sigma_p$ so that $x \leq p$. Thus, p is a prime in L. Now note that $F \in \overline{\{G\}}$ iff $F \subseteq G$. Consider the $F = F_{h_p}$ from 3.6.1. We have $G \notin \overline{\{F\}}$ iff $G \nsubseteq F$ iff $h_p(c) = 0$ for some $c \in G$ iff $c \leq p$ for some $c \in G$ iff $G \in \Sigma_p = \Sigma_a$. Hence, $\Sigma_a = \Sigma(L) \smallsetminus \overline{\{F\}}$.)

Theorem Σ *is a right adjoint to Ω.*

Proof Consider the mappings

$$\sigma_L: L \to \Omega\Sigma(L) \quad (\Omega\Sigma(L) \to L \text{ in } \mathsf{Loc}) \quad \text{and} \quad \rho_X: X \to \Sigma\Omega(X)$$

given by $\sigma_L(a) = \Sigma_a$ and $\rho_X(x) = \mathcal{U}(x) = \{U \mid x \in U\}$. We have already seen in $(\Sigma 1)$ that σ_L is a homomorphism, and since

$$\rho_X^{-1}[\Sigma_U] = \{x \mid \mathcal{U}(x) \in \Sigma_U\} = \{x \mid \mathcal{U}(x) \ni U\} = \{x \mid x \in U\} = U, \qquad (\Sigma 3)$$

ρ_X is continuous.

Next, they constitute natural transformations $\sigma: \Omega\Sigma \to \mathrm{Id}$ (viewed as in Loc) and $\rho: \mathrm{Id} \to \Sigma\Omega$: indeed,

$$\Omega\Sigma(h)(\sigma_L(a)) = (\Sigma(h))^{-1}[\Sigma_a] = \Sigma_{h(a)} = \sigma_M(h(a))$$

(recall $(\Sigma 2)$), and $\Sigma\Omega(f)(\rho_X(x)) = \Omega(f)^{-1}[\mathcal{U}(x)] = \{U \mid f^{-1}[U] \in \mathcal{U}(x)\} = \{U \mid x \in f^{-1}[U]\} = \{U \mid f(x) \in U\} = \rho_Y(f(x))$.

Finally, we have to check that the compositions

$$\Sigma(L) \xrightarrow{\ \rho_{\Sigma(L)}\ } \Sigma\Omega\Sigma(L) \xrightarrow{\ \Sigma\sigma_L\ } \Sigma(L) \qquad (\Sigma 4)$$

and

$$\Omega(X) \xrightarrow{\ \sigma_{\Omega(X)}\ } \Omega\Sigma\Omega(X) \xrightarrow{\ \Omega(\rho_X)\ } \Omega(X) \qquad (\Sigma 5)$$

result in identities. We have

$$\Sigma\sigma_L(\rho_{\Sigma(L)}(F)) = \sigma_L^{-1}[\mathcal{U}(F)] = \sigma_L^{-1}[\{\Sigma_a \mid F \in \Sigma_a\}]$$
$$= \{x \mid \Sigma_x \ni F\} = \{x \mid x \in F\} = F$$

and

$$\Omega(\rho_X)(\sigma_{\Omega(X)}(U)) = \rho_X^{-1}[\Sigma_U] = \{x \mid \mathcal{U}(x) \in \Sigma_U\}$$
$$= \{x \mid U \in \mathcal{U}(x)\} = \{x \mid x \in U\} = U.$$

□

3.7 Spatial Frames

A frame L is said to be *spatial* if it is isomorphic to $\Omega(X)$ for some space X. The adjointness counit σ of the spectrum offers an expedient criterion of spatiality. We have that

   a frame L is spatial iff σ_L is one-to-one

(which is the same as saying that it is an isomorphism). Indeed, by the definition of the space $\Sigma(L)$, σ_L is always onto, hence if the condition holds we have, trivially, $L \cong \Omega\Sigma(L)$. On the other hand in the identity in (Σ4) above, $\sigma_{\Omega(X)}$ is a coretract and hence an isomorphism. Now if there is an isomorphism $\phi : L \to \Omega(X)$ we obtain from the transformation commutativity an isomorphism

$$\sigma_L = (\Omega\Sigma(\phi))^{-1} \cdot \sigma_{\Omega(X)} \cdot \phi.$$

Note that in view of 3.6.1 this condition can be reformulated as saying that for any two $a, b \in L$ with $a \not\leq b$ there is a prime p such that $b \leq p$ and $a \not\leq p$, and hence

$$\text{for every } a \in L, \; a = \bigwedge\{p \mid p \text{ prime}, \; a \leq p\}. \qquad \text{(spatial)}$$

3.8 Sober Reflection

The unit ρ constitutes a reflection of **Top** to **Sob**. We have that

   a space X is sober iff ρ_X is a homeomorphism.

Indeed, ρ_X is invertible by 3.5 and by (Σ3), $\rho_X[U] = \rho_X[\rho_X^{-1}(\Sigma_U)] = \Sigma_U$, so that an invertible ρ_X is an open map. The converse follows from Observation 3.6.2.

3.9   Classical and Generalized (Point-Free) Spaces

Now we are ready for a rough outline of the relation of the point-free and classical
spaces.

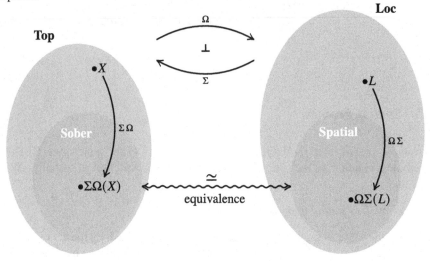

"sobrification" of a space   *"spatialization" of a locale*

On the one hand the point-free theory extends the classical one (or, at least a
substantial part of it; precisely, the subcategory of sober spaces). On the other hand,
the scope of the generalized is much larger; we will present two simple examples
shortly. The reader may of course ask whether such an extension is useful. It turns
out that it is, as will be hopefully apparent from the following sections.

3.9.1   Two Easy Examples First, consider a complete Boolean algebra B without
atoms (e.g., the lattice of all *regular open sets*[2] of any Euclidean space). There are
no primes at all: indeed, let p be one. Since it is not an atom, there is an x with
$p < x < 1$. We have $x^* \wedge x = 0 \leq p$, hence $x^* = 0$ and $x = 1$, a contradiction.

Next, take such a B again and consider $L = \{(x, y) \in B \times B \mid x \leq y\}$. Since all
the $(0, x)$ are in L we obtain by the same reasoning as above that a prime in L has
to be of the form $(q, 1)$ and hence there are no primes to separate distinct $(0, a)$ and
$(0, b)$.

The latter example seems to be very similar to the former, but it is in fact much
more interesting. While Boolean frames are something like a generalization of discrete spaces (albeit constituting a much more colorful class), the frames L of this
example are geometrically rather peculiar (do not forget that a subframe is, due to
the contravariance, geometrically more like a quotient space, not like a subspace):

---

[2] That is, the opens $U = \text{int } \overline{U}$, "open sets without lesions", the open sets one thinks about first.

interpreted as spaces they are behaving like Hausdorff ones, while on the other hand they are not even subfit (see 7.6.2 below) which is a property weaker than T_1!

4 Categorical Remarks

4.1 Semilattices and a Free Functor

Under a *semilattice* we here understand a *meet semilattice with 0 and 1*, and *semilattice homomorphisms* preserve $\wedge$, 0 and 1.

4.1.1 Biproduct Note that in the category of semilattices (similarly like in abelian groups), the cartesian product with the injections and projections as in the following diagram

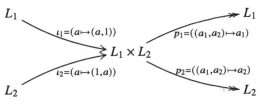

constitutes a biproduct (note that $p_1\iota_1 \wedge p_2\iota_2 = \mathrm{id}$ and check the coproduct and product properties).

4.1.2 A Free Construction For a semilattice L consider the down-set lattice

$$\mathfrak{D}(L) = \{U \subseteq L \mid \emptyset \neq U = \downarrow U\}$$

ordered by inclusion. Further, define $\lambda = \lambda_L \colon L \to \mathfrak{D}(L)$ by setting $\lambda(a) = \downarrow a$. Obviously,

$\mathfrak{D}(L)$ is a frame and λ is a semilattice homomorphism

(since we take only the non-empty down-sets, the zero of $\mathfrak{D}(L)$ is $\{0\}$; all the other joins are the unions).

Proposition *Let M be a frame and let $h \colon L \to M$ be a semilattice homomorphism. Then there is precisely one frame homomorphism $\widetilde{h} \colon \mathfrak{D}(L) \to M$ such that the diagram*

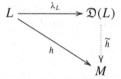

commutes. This $\widetilde{h}$ is given by the formula $\widetilde{h}(U) = \bigvee\{h(a) \mid a \in U\}$.

Proof Since for a down-set U, $U = \bigcup\{\downarrow a \mid a \in U\} = \bigcup\{\lambda(a) \mid a \in U\}$ we have for a (possible) frame homomorphism $\widetilde{h}$ commuting as desired $\widetilde{h}(U) = \bigvee\{\widetilde{h}\lambda(a) \mid a \in U\} = \bigvee\{h(a) \mid a \in U\}$; hence the formula and the unicity. Obviously, this formula gives a mapping $\widetilde{h}\colon \mathfrak{D}(L) \to M$ that preserves all joins, and $\widetilde{h}(L) = 1$. Finally, we have

$$\widetilde{h}(U) \wedge \widetilde{h}(V) = \bigvee\{h(a) \mid a \in U\} \wedge \bigvee\{h(b) \mid b \in V\}$$
$$= \bigvee\{h(a) \wedge h(b) \mid a \in U, b \in V\} = \bigvee\{h(a \wedge b) \mid a \in U, b \in V\}$$
$$\leq \bigvee\{h(c) \mid c \in U \cap V\} = \widetilde{h}(U \cap V) \leq \widetilde{h}(U) \wedge \widetilde{h}(V),$$

so $\widetilde{h}$ is indeed a frame homomorphism.   □

4.2   Free Objects in **Frm**

For a set X define
$$F(X) = \{A \subseteq X \mid A \text{ finite}\}$$

ordered by $\leq \;=\; \supseteq$ so that we have the meet $A \wedge B = A \cup B$. Denote by β_X the mapping
$$\beta_X = (x \mapsto \{x\})\colon X \to F(X).$$

Then we have for each meet-semilattice S with 1 and each mapping $f\colon X \to S$ precisely one meet-semilattice homomorphism $\overline{f}\colon F(X) \to S$ such that the diagram

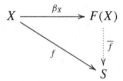

commutes and $\overline{f}(\emptyset) = 1$, namely the homomorphism defined by $\overline{f}(A) = \bigwedge_{x \in A} f(x)$.

The free frame over a set can be now obtained combining F and $\mathfrak{D}$, that is, as $\mathfrak{D}F(X)$. This provides a functor $\mathfrak{D}F\colon$ Set $\to$ Frm right adjoint to the forgetful functor Frm $\to$ Set.

4.3 Algebraic Aspects of **Frm**

The category **Frm** is clearly *equationally presentable* i.e. its objects are described by a (proper class of) operations, namely

- *0-ary:* $0, 1 \colon L^0 \to L$,
- *binary:* $L^2 \to L$, $(a, b) \mapsto a \wedge b$,
- *κ-ary (any cardinal κ):* $L^\kappa \to L$, $(a_i)_\kappa \mapsto \bigvee_\kappa a_i$,

and equations

- $(L, \wedge, 1)$ is an idempotent commutative monoid,
- with a zero element satisfying the absorption law $a \wedge 0 = 0 = 0 \wedge a$,
- $\bigvee_0 a_i = 0$, $a_j \wedge \bigvee_\kappa a_i = a_j$, $a \wedge \bigvee_\kappa a_i = \bigvee_\kappa (a \wedge a_i)$.

Then, by general results of category theory (see Manes [40], Chapter 1, or Johnstone [28]), it follows that

4.3.1 Proposition Frm *has all (small) limits (i.e., it is a complete category) and they are constructed exactly as in the category* **Set** *of sets (i.e., the forgetful functor* **Frm** $\to$ **Set** *preserves them).*

Combining this with the fact that it has free objects over **Set**, **Frm** is a *monadic category over* **Set** [40]. This means that, in particular,

4.3.2 Proposition (1) **Frm** *has all (small) colimits (i.e., it is a cocomplete category).*
(2) *The monomorphisms in* **Frm** *are exactly the injective homomorphisms.*
(3) *Epimorphisms in* **Frm** *need not be surjective but the regular epimorphisms are precisely the surjective homomorphisms.*
(4) *Every morphism in* **Frm** *can be factored (uniquely up to isomorphism) as a regular epimorphism followed by a monomorphism.*
(5) *Quotients are described by congruences.*

4.3.3 A Consequence: Presentations by Generators and Relations The fact that one has free frames and quotient frames implies, in particular, that, like in traditional categories of algebras, we may present frames by generators and relations: take the quotient of the free frame on the given set of generators modulo the congruence generated by the pairs (u, v) for the given relations $u = v$.

For example, the point-free space of reals is introduced as the *frame of reals* $\mathfrak{L}(\mathbb{R})$ [6, 34] generated by all ordered pairs (p, q) $(p, q \in \mathbb{Q})$, subject to the relations

(R1) $(p, q) \wedge (r, s) = (p \vee r, q \wedge s)$,
(R2) $(p, q) \vee (r, s) = (p, s)$ whenever $p \le r < q \le s$,
(R3) $(p, q) = \bigvee\{(r, s) \mid p < r < s < q\}$,

(R4) $\bigvee_{p,q\in\mathbb{Q}}(p,q) = 1$.

This provides a development of the "theory of function rings $C(X)$" in frames and locales [6] and the treatment of more general point-free real functions (see e.g. [22]).

As another example, the *product* of two locales L and M (see 4.5 below) is the frame generated by all pairs $a \otimes b$, $a \in L$, $b \in M$, subject to the relations

(P1) $1 \otimes 1 = 1$,

(P2) $a \otimes 0 = 0 \otimes b = 0$,

(P3) $(a \otimes b) \wedge (a' \otimes b') = (a \wedge a') \otimes (b \wedge b')$,

(P4) $\bigvee_{i\in I}(a_i \otimes b) = (\bigvee_{i\in I} a_i) \otimes b$, $\bigvee_{i\in I}(a \otimes b_i) = a \otimes (\bigvee_{i\in I} b_i)$.

For further examples, see e.g. [21, 36].

4.4 Taking Quotients

In general, extending a binary relation to a congruence (and subsequent factorizing) in an algebra can be a hard task. In frames, however, it is surprisingly easy.

4.4.1 Saturation Let $R \subseteq L \times L$ be an arbitrary relation on a frame L. An element $s \in L$ is said to be *R-saturated* (briefly, *saturated*) if

$$aRb \quad \Rightarrow \quad a \to s = b \to s.$$

The set of all R-saturated elements of L is a frame: since $b \to (-)$ is a right adjoint, it is closed under meets, and by 2.6.2(6) we have for any x and aRb, $a \to (x \to s) = x \to (a \to s) = x \to (b \to s) = b \to (x \to s)$, hence it is also closed under the Heyting operation and therefore it is a complete Heyting algebra, hence a frame[3] (with the same meets and the same Heyting operation as in L but not necessarily the same joins). It will be denoted by L/R.

We will show that L/R is the quotient of L by the congruence generated by R, and more.

4.4.2 The Associated Nucleus For any $a \in L$ set

$$\nu(a) = \nu_R(a) = \bigwedge\{s \in L/R \mid a \le s\}.$$

We have

---

[3] This proof shows indeed more: L/R is a *sublocale* of L, see 5.4 below.

Proposition (1) *For every $a \in L$ and $s \in L/R$, $a \to s = v(a) \to s$.*

(2) *v is a nucleus, that is, it is monotone, $a \leq v(a)$, $vv(a) = v(a)$ and $v(a \wedge b) = v(a) \wedge v(b)$.*

Proof (1) For any x we have trivially $x \leq a \to s$ iff $a \leq x \to s$ and since this last is in L/R, this is the same as $v(a) \leq x \to s$, and this is equivalent with $x \leq v(a) \to s$. (2) The first three formulas are trivial, and also trivially $v(a \wedge b) \leq v(a) \wedge v(b)$. Now since $a \wedge b \leq v(a \wedge b)$, we have, by (1), $a \leq b \to v(a \wedge b) = v(b) \to v(a \wedge b)$, hence $v(a) \leq v(b) \to v(a \wedge b)$ and finally $v(a) \wedge v(b) \leq v(a \wedge b)$. $\qquad\square$

4.4.3 Proposition *v understood as a mapping $L \to L/R$ is an onto frame homomorphism, we have for aRb, $v(a) = v(b)$, and if a frame homomorphism $h: L \to M$ is such that $h(a) = h(b)$ for all aRb then there is an $\overline{h}: L/R \to M$ such that $\overline{h} \cdot v = h$.*

Furthermore, $\overline{h}(s) = h(s)$ for all $s \in L/R$.

Proof v preserves finite meets by 4.4.2(2). The joins $\bigsqcup$ in L/R are given by $\bigsqcup s_i = v(\bigvee s_i)$ (if $t \in L/R$ and $t \geq s_i$ for all i then $t \geq \bigvee s_i$ and $t = v(t) \geq v(\bigvee s_i)$) and hence $v(\bigvee a_i) \leq v(\bigvee v(a_i)) = \bigsqcup v(a_i) \leq v(\bigvee a_i)$. Hence v is a frame homomorphism.

Next, if aRb then $1 = a \to v(a) = b \to v(a)$ and hence $b \leq v(a)$, and $v(b) \leq v(a)$; equality by symmetry.

Finally, let $h: L \to M$ be such that $h(a) = h(b)$ for aRb. Set $\sigma(a) = \bigvee\{x \mid h(x) \leq h(a)\}$. Then obviously

$$a \leq \sigma(a), \quad \text{and} \quad h\sigma(a) \leq h(a) \text{ and hence } h\sigma = h. \qquad (*)$$

Hence we have $x \leq \sigma(a)$ iff $h(x) \leq h(a)$ ('$\Rightarrow$' by $(*)$ and '$\Leftarrow$' by the definition of σ) so that for any uRv we have for any $x, x \leq u \to \sigma(a)$ iff $x \wedge u \leq \sigma(a)$ iff $h(x \wedge v) = h(x \wedge u) \leq h(a)$ iff $x \wedge v \leq \sigma(a)$ iff $x \leq v \to \sigma(a)$. Thus, $\sigma(a)$ is saturated, hence $a \leq v(a) \leq \sigma(a)$ and we have

$$h(a) \leq hv(a) \leq h\sigma(a) = h(a)$$

so that $h(a) = hv(a)$ and the statement follows. $\qquad\square$

4.4.4 Proposition *Let there be a join-basis $C \subseteq L$ such that for all $c \in C$ and aRb we have $(a \wedge c)R(b \wedge c)$. Then s is R-saturated iff for all aRb, $a \leq s$ iff $b \leq s$.*

Proof If the statement holds we have for every $c \in C$ and aRb, $c \wedge a \leq s$ iff $c \wedge b \leq s$, that is, $c \leq a \to s$ iff $c \leq b \to s$, and $a \to s = b \to s$. On the other hand, if s is saturated then in particular $a \to s = 1$ iff $b \to s = 1$. $\qquad\square$

4.5   Product in **Loc** *(Coproduct in* **Frm***) Concretely*

We will present a construction of the coproduct in the category of frames [28, 45]. It will be done for just two factors; the idea of the general case is precisely the same, only one has to use a more complicated notation which makes the presentation less transparent. The reader may do the general construction as a simple exercise taking instead of the $\iota_i \colon L_i \to L_1 \times L_2$ below in the role of the coproduct in the category of semilattices the general coproduct in that category, namely

$$\iota_i \colon L_i \to \coprod_{j \in J} L_j = \{(a_j)_{j \in J} \in \prod_{j \in J} L_j \mid \text{for all but finitely many } j \in J, a_j = 1\}$$

where $\iota_i(a) = (x_j)_{j \in J}$ with $x_i = a$ and $x_j = 1$ otherwise.

On the frame $\mathfrak{D}(L_1 \times L_2)$ define a relation R by setting

$$R = \left\{ \left(\bigcup_{i \in I} \downarrow(a_i, b), \downarrow(\bigvee_{i \in I} a_i, b) \right) \mid a_i \in L_1, b \in L_2 \right\}$$
$$\cup \left\{ \left(\bigcup_{i \in I} \downarrow(a, b_i), \downarrow(a, \bigvee_{i \in I} b_i) \right) \mid a \in L_1, b_i \in L_2 \right\}.$$

Note that

- the void index set is not excluded, hence we have

$$\{(0, 0)\} R \downarrow (0, b) \quad \text{and} \quad \{(0, 0)\} R \downarrow (a, 0)$$

  for all $a \in L_1$ and $b \in L_2$;
- it is easy to check that the R-saturated $U \in \mathfrak{D}(L_1 \times L_2)$ are precisely those that

$$\text{for any } (a_i, b) \in U, i \in I, \text{ also } (\bigvee_{i \in I} a_i, b) \in U,$$

$$\text{and, for any } (a, b_i) \in U, i \in I, \text{ also } (a, \bigvee_{i \in I} b_i) \in U$$

  (the relation satisfies the conditions of 4.4.4, hence we can use the simplified saturation formula).

Theorem *The maps*

$$\overline{\iota}_i \colon \nu_R \cdot \lambda_{L_1 \times L_2} \cdot \iota_i \colon L_i \to L_1 \oplus L_2 = \mathfrak{D}(L_1 \times L_2)/R \quad (i = 1, 2)$$

are frame homomorphisms and constitute a coproduct in **Frm**.

Proof Let $h_i \colon L_i \to M$ be frame homomorphisms. Consider first the semilattice homomorphism $h' \colon L_1 \times L_2 \to M$ obtained for the h_i understood as semilattice homomorphisms (recall 4.1) and, using 4.1.2, lift it to a frame homomorphism $g = \widetilde{h}' : \mathfrak{D}(L_1 \times L_2) \to M$. Consider the following diagram.

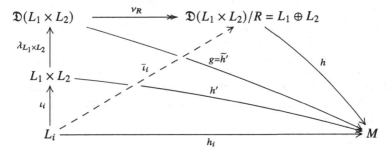

The $\lambda\iota_i$ are semilattice homomorphisms that generally do not need to preserve the joins in L_i. The nucleus homomorphism ν_R, however, obviously provides the necessary equalities, and since it preserves meets, we obtain frame homomorphisms $\nu_R \lambda\iota_i$. Using the formula for $\widetilde{h}'$ from 4.1.2 (and taking into account that obviously $h'(a, b) = h_1(a) \wedge h_2(b)$) we easily check that it respects the relation R and hence we have, by 4.4.3, a frame homomorphism h such that $h\nu_R = g$ and hence $h\nu_R\lambda\iota_i = h_i$. Finally, the

$$\downarrow(a, b) = \downarrow(a, 1) \cap \downarrow(1, b) = \lambda\iota_1(a) \cap \lambda\iota_2(b)$$

obviously generate $\mathfrak{D}(L_1 \times L_2)$ by joins, and ν_R is onto, so that h is uniquely determined by $h\bar{\iota}_1$ and $h\bar{\iota}_2$. $\qquad\square$

5 Loc as a Concrete Category. Localic Maps and Sublocales

5.1 Localic maps

Since frame homomorphisms $h \colon M \to L$ preserve all joins they have uniquely defined right adjoints $f = h_* \colon L \to M$. We will use them for a concrete representation of the category Loc of locales [43, 44]. Thus, from now on we will speak of the meet preserving maps $f \colon L \to M$ between frames with left adjoints f^* that are frame homomorphisms as of *localic maps*. The category Loc will be that with frames as objects (in this context we often—although not always—speak of them as of *locales*) and localic maps as morphisms.

5.2 Proposition

A meet preserving map $f\colon L \to M$ is a localic map iff

(a) $f(x) = 1$ *only if* $x = 1$, *and*
(b) *for all* $y \in M$ *and* $x \in L$, $f(f^*(y) \to x) = y \to f(x)$ (*this identity is often referred to as the Frobenius identity*).

Proof Set $h = f^*$. The point is in determining when h preserves finite meets. First, we have to have $h(1) = 1$; this makes in the adjunction the condition $1 = h(1) \leq x$ iff $1 \leq f(x)$, hence (a).

We have

$$h(x) \wedge h(y) \leq z \quad \text{iff} \quad h(x) \leq h(y) \to z \quad \text{iff} \quad x \leq f(h(y) \to z), \text{ and}$$
$$h(x \wedge y) \leq z \quad \text{iff} \quad x \wedge y \leq f(z) \quad \text{iff} \quad x \leq y \to f(z).$$

If $h(x) \wedge h(y) = h(x \wedge y)$ the first inequalities coincide and we have for all x, $x \leq f(h(y) \to z)$ iff $x \leq y \to f(z)$, hence $f(h(y) \to z) = y \to f(z)$. On the other hand, if $f(h(y) \to z) = y \to f(z)$ we have for all z, $h(x) \wedge h(y) \leq z$ iff $h(x \wedge y) \leq z$, hence $h(x) \wedge h(y) = h(x \wedge y)$.   □

5.2.1 Examples (1) For each continuous map $f\colon X \to Y$, the localic map right adjoint to $\Omega(f)$ is given by

$$\Omega(f)_*(U) = Y \smallsetminus \overline{f[X \smallsetminus U]}.$$

(2) Recall 3.6. A point of a locale L is a localic map $p\colon \mathbf{2} \to L$. Then $p(1) = 1$ and $p(0) = a \neq 1$ is a prime in L: $x \wedge y \leq a = p(0)$ iff $p^*(x) \wedge p^*(y) \leq 0$ hence $p^*(x) = 0$ or $p^*(y) = 0$, that is, $x \leq p(0) = a$ or $y \leq p(0) = a$.

5.3 Aside: Spectrum in Thus Represented Category of Locales

Recall 3.6.2. Let us represent points as primes. We have

Observation *Let* $f\colon L \to M$ *be a localic map. Then for every prime p in L, $f(p)$ is prime in M.*

(*Indeed, since $p \neq 1$, $f(p) \neq 1$, and $a \wedge b \leq f(p)$ iff $f^*(a) \wedge f^*(b) \leq p$ iff $f^*(a) \leq p$ or $f^*(b) \leq p$ iff $a \leq f(p)$ or $b \leq f(p)$.*)

Set

$$\Sigma(L) = \Big(\{p \mid p \in L, \, p \text{ prime}\}, \{\Sigma_a \mid a \in L\}\Big)$$

where $\Sigma_a = \{p \mid a \not\leq p\}$ and take (using Observation) for $\Sigma(f)\colon \Sigma(L) \to \Sigma(M)$
simply the restriction of f. By the adjunction we have

$$(\Sigma(f))^{-1}[\Sigma_a] = \{p \mid a \not\leq f(p)\} = \{p \mid f^*(a) \not\leq p\} = \Sigma_{f^*(a)}$$

and we easily see that $\Sigma_0 = \emptyset$, $\Sigma_1 = \Sigma(L)$, $\Sigma_{a \wedge b} = \Sigma_a \cap \Sigma_b$ and $\Sigma_{\bigvee a_i} = \bigcup \Sigma_{a_i}$.
Thus, the $\Sigma(L)$ are topological spaces and the $\Sigma(f)$ are continuous maps, and we
have a functor $\mathsf{Loc} \to \mathsf{Top}$ (this time without any formal reversals). If we now define
$\sigma_L(a) = \Sigma_a$ and $\rho_X\colon X \to \Sigma\Omega(X)$ by setting $\rho_X(x) = X \smallsetminus \overline{\{x\}}$ we can easily check
all the equalities corresponding to those of 3.6.2.

5.4 Sublocales and the Coframe $\mathsf{S}(L)$

Natural candidates for subobjects in a category are extremal monomorphisms. In Frm,
extremal epimorphisms are precisely the onto frame homomorphisms and hence the
extremal monomorphisms in our representation of Loc will be the adjoints to these,
namely precisely the one-to-one localic maps (recall the identities (adj3) in any adjunction; they show that onto maps correspond to one-to-one maps and vice versa).

Thus, a natural subobject of a locale L is a subposet $S \subseteq L$ that is a frame in the
induced order, such that the embedding map $j_S\colon S \subseteq L$ is a localic one. First of all,
it is closed under meets and the left adjoint of j_S is obviously given by the formula

$$v_S(x) = \bigwedge\{s \mid s \in S, \, x \leq s\}$$

(understood as a map $L \to S$ it has to be a frame homomorphism; usually, however
one considers it as a map $L \to L$ and one speaks of the *nucleus*[4] of S). By 5.2 we
have that for every $s \in S$ and every $x \in L$, $x \to s \in S$, because in this case $x \to s =
x \to j_S(s) = j_S(v_S(x) \to s) \in S$. This leads to the following definition.

A *sublocale* of a locale (frame) L is a subset $S \subseteq L$ such that

(S1) for every $M \subseteq S$, $\bigwedge M \in S$, and
(S2) for every $x \in L$ and every $s \in S$, $x \to s \in S$.

(We have already seen that if $j_S\colon S \subseteq L$ is a localic map then (S1) and (S2) hold.
On the other hand, if S satisfies (S1) and (S2) then it is closed under meets and
the Heyting operation, and hence it is a locale (with the same meets and the same

---

[4] Nuclei in L are in a one-one correspondence with onto frame homomorphisms with domain L
hence constitute an alternative representation for sublocales in L [44].

Heyting operation as in L^5). By (S1), j_S has a left adjoint v_S as above. By (S2), $x \to s = j_S(x \to s)$, and for any u, $u \leq x \to s$ iff $x \leq u \to s$ iff $v_S(x) \leq u \to s$ iff $u \leq v_S(x) \to s$ so that

$$v_S(x) \to s = x \to s \qquad \qquad \text{(nucleus)}$$

and hence $j_S(v_S(x) \to s) = x \to s = x \to j_S(s)$ and j_S is a localic map by 5.2—the condition with 1 is trivial since j_S is one-to-one.)

5.4.1 The Complete Lattice $\mathsf{S}(L)$ Obviously any intersection of sublocales is a sublocale so that we have a complete lattice

$$\mathsf{S}(L)$$

of sublocales of L. We immediately see that the join in $\mathsf{S}(L)$ is given by the formula

$$\bigvee_{i \in I} S_i = \{\bigwedge M \mid M \subseteq \bigcup_{i \in I} S_i\}$$

(every sublocale containing all S_i has to contain this set, and on the other hand this set is a sublocale by property 2.6(H2) of the Heyting operator).

By (S1) every sublocale contains the top 1. Thus, the smallest sublocale, corresponding to the classical empty subspace, is $\mathsf{O} = \{1\}$.

5.4.2 Proposition $\mathsf{S}(L)$ *is a coframe.*

Proof We need to show that $(\bigcap_{i \in I} S_i) \vee T = \bigcap_{i \in I}(S_i \vee T)$. The inclusion $\subseteq$ is obvious. Hence, consider an $x \in \bigcap_{i \in I}(S_i \vee T)$. Then for every i there are $s_i \in S_i$ and $t_i \in T$ such that $x = s_i \wedge t_i$. Set $t = \bigwedge t_i$. We have

$$x = \bigwedge(s_i \wedge t_i) = \bigwedge s_i \wedge \bigwedge t_i = (\bigwedge s_i) \wedge t \leq s_i \wedge t \leq s_i \wedge t_i = x$$

so that $x = s_i \wedge t$ for all i. Then, by rule 2.6.2(7), all the $t \to s_i$ coincide; denote by s the common value. Since $s = t \to s_i \in S_i$, $s \in \bigcap_{i \in I} S_i$ and we conclude by 2.6.2(5) that $x = t \wedge s_i = t \wedge (t \to s_i) = t \wedge s \in (\bigcap_{i \in I} S_i) \vee T$. □

---

[5] The joins $\bigsqcup$ in S are given by $\bigsqcup s_i = v_S(\bigvee s_i)$ (if $t \in S$ and $t \geq s_i$ for all i then $t \geq \bigvee s_i$ and $t = v_S(t) \geq v_S(\bigvee s_i)$).

5.5 Open and Closed Sublocales

Each element $a \in L$ is associated with a *closed* sublocale $\mathfrak{c}(a)$ and an *open* sublocale $\mathfrak{o}(a)$,[6]

$$\mathfrak{c}(a) = {\uparrow}a \quad \text{and} \quad \mathfrak{o}(a) = \{x \in L \mid a \to x = x\} = \{a \to x \mid x \in L\}$$

(the equivalence of the two expressions for $\mathfrak{o}(a)$ follows immediately from 2.6.2(6)).

5.5.1 Proposition $\mathfrak{o}(a)$ *and* $\mathfrak{c}(a)$ *are complements of each other.*

Proof If $x \in \mathfrak{o}(a) \cap \mathfrak{c}(a)$ we have $a \le x = a \to x$, hence $a = a \wedge a \le x$ and $x = a \to x = 1$ by 2.6.2(3). On the other hand, each $x \in L$ is by 2.6.2(8) equal to $(a \to x) \wedge (a \vee x) \in \mathfrak{o}(a) \vee \mathfrak{c}(a)$. $\qquad\square$

5.5.2 Proposition *We have the following formulas*

$$\mathfrak{o}(0) = \mathsf{O}, \quad \mathfrak{o}(1) = L, \quad \mathfrak{o}(a \wedge b) = \mathfrak{o}(a) \cap \mathfrak{o}(b) \quad \text{and} \quad \mathfrak{o}(\textstyle\bigvee_i a_i) = \bigvee_i \mathfrak{o}(a_i),$$

$$\mathfrak{c}(0) = L, \quad \mathfrak{c}(1) = \mathsf{O}, \quad \mathfrak{c}(a \wedge b) = \mathfrak{c}(a) \vee \mathfrak{c}(b) \quad \text{and} \quad \mathfrak{c}(\textstyle\bigvee_i a_i) = \bigcap_i \mathfrak{c}(a_i).$$

Proof We will prove the formulas for $\mathfrak{c}$, those for $\mathfrak{o}$ will then follow by De Morgan formulas. They are simple observations:

${\uparrow}0 = L$, ${\uparrow}1 = \{1\} = \mathsf{O}$, $x \ge \bigvee a_i$ iff $x \ge a_i$ for all i, and finally, $x \ge a \wedge b$ iff $x = (x \vee a) \wedge (x \vee b)$, that is, iff $x \in {\uparrow}a \vee {\uparrow}b$. $\qquad\square$

5.5.3 Proposition *A general sublocale S can be represented by open and closed sublocales as follows:*

$$S = \bigcap\{\mathfrak{c}(v_S(x)) \vee \mathfrak{o}(x) \mid x \in L\} = \bigcap\{\mathfrak{c}(y) \vee \mathfrak{o}(x) \mid v_S(x) = v_S(y)\}.$$

Proof I. If $s \in S$ then for arbitrary x, $x \to s \in S$. Hence by 2.6.2(8) and (nucleus)

$$s = (s \vee v_S(x)) \wedge (v_S(x) \to s) = (s \vee v_S(x)) \wedge (x \to s) \in \mathfrak{c}(v_S(x)) \vee \mathfrak{o}(x).$$

On the other hand, if a is in $\bigcap\{\mathfrak{c}(v_S(x)) \vee \mathfrak{o}(x) \mid x \in L\}$ then, in particular, it is in $\mathfrak{c}(v_S(a)) \vee \mathfrak{o}(a)$ and therefore $a = x \wedge (a \to y)$ with $x \ge v_S(a)$. Since $a \le a \to y$ we have $a \le y$, hence $a \to y = 1$, so that $a = x \ge v_S(a)$ and $a = v_S(a)$, that is, $a \in S$.

---

[6] The reader might have expected ${\downarrow}a$ for the definition of $\mathfrak{o}(a)$. This subset of L is not a sublocale, but the intuition is not wide from the target: ${\downarrow}a$ is isomorphic to $\mathfrak{o}(a)$ which is the image of the localic map adjoint to the map $(x \mapsto a \wedge x) \colon L \twoheadrightarrow {\downarrow}a$.

II. Since $v_S(v_S(x)) = v_S(x)$ it suffices, in view of I, to show that if $v_S(x) = v_S(y)$ then $S \subseteq \mathfrak{c}(y) \vee \mathfrak{o}(x)$. Let $s \in S$. We have $s = (s \vee y) \wedge (y \to s) = (s \vee y) \wedge (v_S(y) \to s) = (s \vee y) \wedge (v_S(x) \to s) = (s \vee y) \wedge (x \to s) \in \mathfrak{c}(y) \vee \mathfrak{o}(x)$. $\square$

5.6 Closure, Density and Isbell's Theorem. Interior

Like in spaces we have the *closure* of a sublocale, the smallest closed sublocale containing S (cf. [15]). It is determined by a particularly simple formula, namely

$$\overline{S} = \mathfrak{c}(\textstyle\bigwedge S) = {\uparrow}\textstyle\bigwedge S$$

(a closed set containing S has to contain $\bigwedge S$ and has to be an up-set). Consequently we have also an extremely simple criterion of density:

5.6.1 Observation A sublocale $S \subseteq L$ is dense in L iff it contains the bottom 0.

5.6.2 Booleanization For a frame L set

$$\mathfrak{B}(L) = \{a \in L \mid a = a^{**}\} = \{a^* \mid a \in L\}.$$

Obviously it is a sublocale: we have $a^* = a \to 0$ and $\bigwedge_i (a_i \to 0) = (\bigvee_i a_i) \to 0$ by 2.6(H4), making for (S1); (S2) follows from 2.6.2(6).

$\mathfrak{B}(L)$ is a Boolean algebra, the largest Boolean algebra in among the sublocales. It is called the *Booleanization* of L, and it is a very old construction known from algebraic logic (Glivenko [20]).

5.6.3 Theorem (Isbell's Density Theorem) *A sublocale $S \subseteq L$ is dense iff it contains $\mathfrak{B}(L)$. Thus, each locale L contains a smallest dense sublocale, namely $\mathfrak{B}(L)$.*

Proof A dense sublocale contains 0 and hence, by (S2), all the $x \to 0$, that is, all the sublocale $\mathfrak{B}(L)$ which is itself dense, since $0 = 1 \to 0$. $\square$

5.6.4 Notes (a) This fact has no counterpart in classical topology. So e.g. in the frame of reals the sublocales of rationals and irrationals have still a very rich intersection (which, then, cannot be represented as a classical subspace). In the next pages we will pay some attention to the relation of sublocales and subspaces of classical spaces.

(b) The Booleanization $\mathfrak{B}(\Omega(X))$ is an example of advantages of the point-free approach. It is in fact a very natural space, namely the space of regular open sets, that typically has no classical representation.

5.6.5 Interior Similarly one defines the *interior* of a sublocale S as the largest open sublocale contained in S. We have $\mathfrak{o}(a) \subseteq S$ iff $\mathfrak{c}(a) \supseteq S^{\#}$ iff $a \leq \bigwedge S^{\#}$ iff $\mathfrak{o}(a) \subseteq \mathfrak{o}(\bigwedge S^{\#})$ so that

$$\text{int } S = \mathfrak{o}(\bigwedge S^{\#}).$$

Note that this can be read in terms of the coHeyting difference as

$$\text{int } S = L \smallsetminus \overline{(L \smallsetminus S)}$$

in analogy with the classical relation between interior and closure (the dual formula does not hold, though; see e.g. [19] for more information).

5.7 Subspaces and Sublocales I. The Axiom T_D

This is a preparatory subsection. We will proceed in the next section after we will know more about images and preimages; now we will discuss just the correctness of point-free representation of subspaces.

5.7.1 Sublocales Induced by Subspaces Consider a space X, a subspace $Y \subseteq X$ and the embedding mapping $j_Y \colon Y \subseteq X$. Then we have the onto frame homomorphism

$$\Omega(j_Y) = (U \mapsto U \cap Y) \colon \Omega(X) \to \Omega(Y)$$

with the adjoint localic map $\kappa_Y \colon \Omega(Y) \to \Omega(X)$, an extremal monomorphism in Loc, given by

$$\kappa_Y(V) = \text{int}\,((X \smallsetminus Y) \cup V)$$

(since $U \cap Y \subseteq V$ iff $U \subseteq (X \smallsetminus Y) \cup V$ and U is open). This suggests the natural representation of Y as the sublocale

$$S_Y = \kappa_Y[\Omega(Y)].$$

Such sublocales S_Y of (locales representing) spaces are usually referred to as the *induced sublocales*, more precisely, *sublocales induced by subspaces*.

5.7.2 The Axiom T_D We have already seen that even in the most natural spaces like the Euclidean space of reals (we will learn later that in fact the contrary is rather rare) there are sublocales that are not (induced by) subspaces. This is in fact a very useful feature of point-free topology. There is, however, another hitch that has to be taken into account: the space has to have a certain very weak property to have the subspaces represented correctly.

The following property was introduced in Aull and Thron [1], and in the same year, in Thron [54], it was already used to prove one of the first results about the reconstruction of X from $\Omega(X)$. Since then it turned out to be a very important property in comparing classical and point-free theory (see e.g. [49]). A T_D-space[7] is a space X in which

$$\text{for every } x \in X \text{ there is an open } U \ni x \text{ such that } U \smallsetminus \{x\} \text{ is open.} \qquad (T_D)$$

5.7.3 **Proposition** *The representation of subspaces in $\Omega(X)$ as above is correct in the sense that distinct subspaces induce distinct sublocales iff the space X satisfies T_D.*

Proof $\Rightarrow$: Let X be a space in which T_D does not hold and let x be such that none of the $U \smallsetminus \{x\}$ with open $U \ni x$ is open. Then for $Y = X \smallsetminus \{x\}$ we have $\kappa_Y(U \cap Y) = U$ for any $U \in \Omega(X)$; indeed, if $x \in U$ then

$$\kappa_Y(U \cap Y) = \kappa_Y(U \smallsetminus \{x\}) = \text{int}(U) = U,$$

otherwise $\kappa_Y(U \cap Y) = \kappa_Y(U) = \text{int}(\{x\} \cup U) = U$. Hence $S_Y = \Omega(X) = S_X$.
$\Leftarrow$: Note that the nucleus of the S_Y is given by

$$\nu_Y(U) = \text{int}((X \smallsetminus Y) \cup (U \cap Y)) \quad (U \in \Omega(X)).$$

Let T_D hold and let Y, Z be distinct subspaces (with, say, $Y \ni x \notin Z$). Choose an open $U \ni x$ with $V = U \smallsetminus \{x\}$ open. Then $\nu_Y(U) \neq \nu_Y(V)$ while $\nu_Z(U) = \nu_Z(V)$. Hence $\nu_Y \neq \nu_Z$ and thus the corresponding sublocales S_Y and S_Z are distinct. $\qquad \square$

5.8 *Aside: Spatialization as a Sublocale*

Recall 3.7. The full subcategory of all spatial locales in Loc will be denoted by $\mathsf{Loc_{sp}}$. Let $\text{Pr}(L)$ denote the set of all primes p in a locale L and set

$$\text{Sp}(L) = \{\bigwedge A \mid A \subseteq \text{Pr}(L)\}.$$

Obviously $\text{Sp}(L) = L$ for a spatial L.

---

[7] The importance of this condition is comparable with that of sobriety. Note that in a way these two conditions are dual to each other: while sobriety requires that we cannot *add* a point to X without changing $\Omega(X)$, T_D says that we cannot *subtract* a point.

5.8.1 **Lemma** Sp(L) *is a sublocale of* L.

Proof Obviously Sp(L) is closed under meets. Now if $A \subseteq$ Sp(L) then $x \to \bigwedge A = \bigwedge_{p \in A}(x \to p) \in$ Sp(L) since $x \to p \in$ Pr(L) $\cup \{1\}$ for every $x \in L$ and $p \in$ Pr(L): if $a \wedge b \leq x \to p$ then $a \wedge b \wedge x \leq p$; if $x \leq p$ we have $x \to p = 1$, else $a \wedge b \leq p$ and then, say, $a \leq p \leq x \to p$.   □

Note further that Sp(L) $\in$ Loc$_{sp}$ since Pr(Sp(L)) = Pr(L).

5.8.2 **Lemma** *If* $f: L \to M$ *is a localic map then we have a localic map* Sp(f): Sp(L) $\to$ Sp(M) *defined by* Sp(f)(a) = $f(a)$.

Proof Since f preserves meets, we have by Observation 5.3, $f[\text{Sp}(L)] \subseteq$ Sp(M), and hence we have a map Sp(L) $\to$ Sp(M) defined as in the statement; obviously it preserves meets. Since we have for the embeddings $j_L:$ Sp(L) $\subseteq L$, $f \cdot j_L = j_M \cdot$ Sp(f), that is, $j_L^* \cdot f^* = $ Sp(f)$^* \cdot j_M^*$, we have, for any $a \in$ Sp(L) and $b \in$ Sp(M), by (nucleus),

$$\text{Sp}(f)(\text{Sp}(f)^*(b) \to a) = \text{Sp}(f)(\text{Sp}(f)^*(j_M^*(b)) \to a) = f(j_L^*(f^*(b)) \to a)$$
$$= f(f^*(b) \to a) = b \to f(a) = b \to \text{Sp}(f)(a),$$

and Sp(f) is a localic map.   □

Thus we have a functor
$$\text{Sp: Loc} \to \text{Loc}_{sp}$$

(clearly a reflection of Loc on Loc$_{sp}$). Recall the representation of the adjointness counit $\sigma_L = (a \mapsto \Sigma_a): L \to \Omega\Sigma(L)$ of the spectrum from 5.3. Restricting it to Sp(L), we get a description of the spatialization of a locale L (3.9) as a sublocale of L:

5.8.3 **Proposition** $\sigma_L:$ Sp(L) $\to \Omega\Sigma(L)$ *is a frame isomorphism.*

Proof We have already mentioned in 5.3 that $\Sigma_1 = $ Pr(L) and $\Sigma_{a \wedge b} = \Sigma_a \cap \Sigma_b$. It is also easy to check that

$$\Sigma_{\bigwedge \text{Sp}(L)} = \emptyset \quad \text{and} \quad \Sigma_{\bigsqcup_i a_i} = \Sigma_{\bigwedge \{p \in \text{Pr}(L) | \bigvee_i a_i \leq p\}} = \bigcup_i \Sigma_{a_i}$$

so that we have a frame homomorphism. σ_L is clearly one-to-one in Sp(L); it is onto since

$$\Sigma_a = \Sigma_{\bigwedge \{p \in \text{Pr}(L) | a \leq p\}} \quad \text{for every } a \in L.$$

□

6 Images and Preimages. Localic Maps as Continuous Ones. Open Maps

6.1 Proposition

Let $f: L \to M$ be a localic map. For every sublocale $S \subseteq L$ the image $f[S]$ is a sublocale of M.

Proof Trivially, $f[S]$ is closed under meets. Now take an $s \in S$ and an arbitrary $x \in M$. We have $x \to f(s) = f(f^*(x) \to s) \in f[S]$ since $f^*(x) \to s \in S$.   □

6.1.1 An (Epi, Extremal Mono) Factorization In consequence we have in Loc the factorizations

$$L \xrightarrow{\ g = (x \mapsto f(x))\ } f[L] \xrightarrow{\ j = \subseteq\ } M.$$

Indeed, g obviously preserves meets, hence it has a left adjoint, and we have $f^* = g^* j^*$ with f^* and j^* frame homomorphisms, j^* onto, and hence g^* is a frame homomorphism.

6.2 Localic Preimage

By the formula for join in $\mathsf{S}(L)$ we have for each subset $A \subseteq L$ closed under meets the sublocale
$$A_{\mathsf{sl}} = \bigvee \{S \in \mathsf{S}(L) \mid S \subseteq A\},$$
the largest sublocale contained in A.

The preimage $f^{-1}[S]$ of a sublocale is obviously closed under meets, but the condition (S2) typically fails. We set

$$f_{-1}[S] = f^{-1}[S]_{\mathsf{sl}}$$

and call this sublocale the *localic preimage* of S.

Conventions We will sometimes work with both $f^{-1}[S]$ and $f_{-1}[S]$. To avoid confusion we will speak of the former as of the *set preimage*. Further, $f^{-1}[M]$ is closed under meets for any meet-preserving f and any M that is closed under meets and hence we have a sublocale $f_{-1}[M] = f^{-1}[M]_{\mathsf{sl}}$ for any such M. We will refer to such a situation stating that $f_{-1}[M]$ *makes sense*.

6.2.1 **Proposition** *Localic preimages of closed resp. open sublocales are closed resp. open. More precisely, we have* $f_{-1}[\mathfrak{c}(a)] = f^{-1}[\mathfrak{c}(a)] = \mathfrak{c}(f^*(a))$ *and* $f_{-1}[\mathfrak{o}(a)] = \mathfrak{o}(f^*(a))$.

Proof I. $x \in f^{-1}[\uparrow a]$ iff $f(x) \geq a$ iff $x \geq f^*(a)$.
II. For a general element $f^*(a) \to x$ of $\mathfrak{o}(f^*(a))$ we have $f(f^*(a) \to x) = a \to f(x) \in \mathfrak{o}(a)$, hence $\mathfrak{o}(f^*(a)) \subseteq f^{-1}[\mathfrak{o}(a)]$.

Now let S be a sublocale contained in $f^{-1}[\mathfrak{o}(a)]$; we will show that $S \subseteq \mathfrak{o}(f^*(a))$. Set $b = f^*(a)$ and take an $s \in S$. We have $(b \to s) \to s \in S$ and hence $f((b \to s) \to s) \in \mathfrak{o}(a)$ so that, using 2.6.2(6), we compute

$$f((b \to s) \to s) = a \to f((b \to s) \to s) = f(f^*(a) \to ((b \to s) \to s))$$
$$= f((b \wedge (b \to s)) \to s) = f((b \wedge s) \to s) = f(1) = 1$$

and since for a localic map f, $f(x) = 1$ only if $x = 1$ we see that $(b \to s) \to s = 1$. But then $b \to s \leq s$, and since always $s \leq b \to s$ we conclude that $s \in \mathfrak{o}(b)$. $\qquad\square$

6.3 Proposition

For any localic map $f : L \to M$ *we have the adjunction*

$$S(L) \quad \perp \quad S(M).$$

with maps $f[-]$ *and* $f_{-1}[-]$

Hence, the image map $f[-]$ *preserves all joins and the preimage map* $f_{-1}[-]$ *preserves all meets.*

Proof We have $f[S] \subseteq T$ iff $S \subseteq f^{-1}[T]$ iff $S \subseteq f_{-1}[T]$, the first being the standard set-theoretical image-preimage adjunction, the second because S is a sublocale. $\qquad\square$

Note It can be further proved that $f_{-1}[-]$ is a coframe homomorphism that preserves complements while $f[-]$ is a colocalic map [44].

6.4 Points, Sublocales and Subspaces

Each sublocale contains, trivially, the top element, and the sublocale $\mathsf{O} = \{1\}$ plays the role of the void subspace. We have an easy

Proposition *The sublocales containing just one non-trivial element are the* $P = \{p, 1\}$ *with* p *prime.*

Proof If p is prime and if $x \to p \neq 1$ then $x \nleq p$ and since $x \wedge (x \to p) \leq p$ then $x \to p \leq p$ and hence $x \to p = p$ by 2.6.2(1).

If $x \to p \in \{p, 1\}$ and if $x \wedge y \leq p$ then either $x \to p = 1$ and $x \leq p$ or $x \to p = p$ and $y \leq x \to p = p$. □

6.4.1   One-Point Sublocales The sublocales $P = \{p, 1\}$ with p prime are called *one-point sublocales*, or simply *point sublocales*. Note that this is in agreement with the representation of points as primes. From formula (spatial) and the formula for join in $S(L)$ we immediately obtain

Observation *A frame L is spatial iff $L = \bigvee \{P \mid P$ point sublocale of $L\}$.*

In particular in a space X we have the one-point sublocales

$$P_x^X = \{X \smallsetminus \overline{\{x\}}, X\}$$

of $\Omega(X)$ and we obtain

6.4.2   Observation $\Omega(X) = \bigvee \{P_x^X \mid x \in X\}$.

(Note that here we have simply used the fact that an open $U \subseteq X$ is the intersection $\bigcap \{X \smallsetminus \overline{\{x\}} \mid x \notin U\}$. Thus, if X is not sober we actually have not needed all the prime like in the previous statement.)

6.4.3   Induced Sublocales of $\Omega(X)$ in Terms of Point Sublocales Let X be a topological space and $Y \subseteq X$ a subspace. Recall from 5.7.1 the sublocale $S_Y \subseteq \Omega(X)$ induced by Y. We have

Theorem *In $S(\Omega(X))$, $S_Y = \bigvee \{P_y^X \mid y \in Y\}$.*

Proof We have $S_Y = \kappa_Y[\Omega(Y)]$ where κ_Y is the localic map adjoint to the embedding homomorphism $\Omega(j) = (U \mapsto U \cap Y) \colon \Omega(X) \to \Omega(Y)$. From 5.7.1 we know that $\kappa_Y(V) = \text{int}((X \smallsetminus Y) \cup V)$ (the largest open $U \subseteq X$ such that $U \cap Y = V$). By 6.4.2, $\Omega(Y) = \bigvee \{P_y^Y \mid y \in Y\}$ and hence by 6.3, $S_Y = \bigvee \{\kappa_Y(P_y^Y) \mid y \in Y\}$ and it suffices to prove that $\kappa_Y(P_y^Y) = P_y^X$, that is, that for $y \in Y$,

$$\text{int}((X \smallsetminus Y) \cup (Y \smallsetminus \overline{\{y\}}^Y)) = X \smallsetminus \overline{\{y\}}$$

which, since the closure in Y is the intersection of the closure in X with Y, amounts to $\text{int}((X \smallsetminus Y) \cup (Y \smallsetminus \overline{\{y\}})) = X \smallsetminus \overline{\{y\}}$. The inclusion $\supseteq$ is trivial. Now let U be open and $U \subseteq (X \smallsetminus Y) \cup (Y \smallsetminus \overline{\{y\}})$. We have to prove that $U \subseteq X \smallsetminus \overline{\{y\}}$. Suppose the contrary. Then there is a $z \in U \cup \overline{\{y\}}$, hence $y \in U$ which is a contradiction: y is neither in $X \smallsetminus Y$ nor in $Y \smallsetminus \overline{\{y\}}$. □

6.4.4 **Note** All sublocales of spaces are induced only exceptionally. For a T_D-space (recall 5.7.2) the following statements are equivalent.

(1) *All sublocales of $\Omega(X)$ are induced by subspaces,*
(2) $S(\Omega(X))$ *is Boolean,*
(3) *X is scattered, that is, every infinite subset of X contains an isolated point.*

(See [2, 41, 46, 51].)

6.5 Geometry of Localic Maps

Localic maps were introduced in a rather formal way: in the first step the category Frm was just turned upside down to formally obtain covariance; in the second step one gained concreteness by another formal measure, namely by taking Galois adjoints. It may come as a pleasant surprise that thus formally obtained maps are characterized among non-structured maps like classical continuous maps, namely by preserving closedness and openness (the latter in some strict sense) by preimages.[8] In classical spaces it suffices to assume one, obtaining the other for free, here we will have to assume both explicitly: the complements of closed sublocales have to be formed in $S(L)$ and not set-theoretically as in classical topology, and $S(L)$ is not quite so simple as the Boolean algebra of all subsets.

6.5.1 **Lemma** *Let L, M be frames and let $f: L \to M$ be a mapping such that for every closed sublocale $B \subseteq M$ the (set-theoretical) preimage $f^{-1}[B]$ is closed. Then f preserves meets (and hence has a left adjoint).*

Proof In particular, preimages of up-sets are up-sets and hence f is continuous in the Alexandroff (quasidiscrete) topology of the posets L, M, and consequently monotone.

Next, for every $b \in M$ we have an $a \in L$ such that $f^{-1}[{\uparrow}b] = {\uparrow}a$. The a is obviously uniquely determined; let us denote it by $h(b)$. The equality ${\uparrow}h(b) = f^{-1}[{\uparrow}b]$ can be rewritten as

$$h(b) \le x \quad \text{iff} \quad b \le f(x). \tag{$*$}$$

Realizing that h is monotone (if $b \le b'$ we have ${\uparrow}b \supseteq {\uparrow}b'$ and hence ${\uparrow}h(b) \supseteq {\uparrow}h(b')$ and $h(b) \le h(b')$) we conclude that $(*)$ makes f a right Galois adjoint, hence a mapping preserving all meets. $\square$

---

[8] More precisely: a map is localic iff each closed sublocale has a closed preimage whose complement is contained in the preimage of the complement of the original sublocale, and the least sublocales are preserved.

6.5.2 Theorem *Let L, M be frames. Then a mapping $f : L \to M$ is* localic *iff*

for every closed A, $f^{-1}[A]$ is closed, $f^{-1}[O] = O$, and
for every open U, $f_{-1}[U] = f^{-1}[U^c]^c$ (and hence it is open).

(Note that because of the first condition and 6.5.1 the use of the symbol $f_{-1}[U]$ makes sense—recall the convention in 6.2.)

Proof Every localic map satisfies the conditions by 6.2.1. Thus, let $f : L \to M$ be a plain map satisfying the conditions. Since $f^{-1}[O] = O$ we have $f(a) = 1$ only if $a = 1$ and by 6.5.1 we know there is a right adjoint h, hence it remains to prove that $f(h(a) \to x) = a \to f(x)$.

Consider a $B = \uparrow a$ so that $B^c = \mathfrak{o}(a)$. Thus, $A = f^{-1}[B] = \uparrow h(a)$ and by the second assumption we have $\mathfrak{o}(h(a)) \subseteq f^{-1}[\mathfrak{o}(a)]$. Consequently

$$f(h(a) \to x) = a \to y \qquad\qquad (*)$$

for some y and we have to prove that $a \to y = a \to f(x)$, that is, by 2.6.2(7), that $a \wedge y = a \wedge f(x)$.

$\geq$: Trivially, $f(x) \leq f(h(a) \to x) = a \to y$, and hence $a \wedge f(x) \leq y$.

$\leq$: Using the adjunction inequality $\mathrm{id} \geq hf$ and $(*)$ we have

$$h(a) \to x \geq hf(h(a) \to x) = h(a \to y)$$
$$= h(\bigvee\{u \mid u \wedge a \leq y\}) = \bigvee\{h(u) \mid u \wedge a \leq y\},$$

hence $\bigvee\{h(u) \mid u \wedge a \leq y\} \leq h(a) \to x$ and so (recall rule (5) of 2.6.2)

$$\bigvee\{h(a) \wedge h(u) \mid u \wedge a \leq y\} = h(a) \wedge \bigvee\{h(u) \mid u \wedge a \leq y\}$$
$$\leq h(a) \wedge (h(a) \to x) \leq x.$$

Consequently, $h(a \wedge y) \leq h(a) \wedge h(y) \leq x$ and finally $a \wedge y \leq f(x)$.   □

6.6 Joyal-Tierney Theorem

This is a very interesting useful characterisation of *open localic maps*, that is, of localic maps $f : L \to M$ such that the image $f[\mathfrak{o}(a)]$ of every open sublocale is open.

6.6.1 A Technical Replacement It will be technically of advantage to replace embeddings of open sublocales $\mathfrak{o}(a)$ by isomorphic representations by means of the frames $\downarrow a$ as indicated in the following diagram where j_a is the embedding $\mathfrak{o}(a) \subseteq L$ and the dotted isomorphism consists of $(x \mapsto a \wedge x) \colon \mathfrak{o}(a) \to \downarrow a$ and $(x \mapsto a \to x) \colon \downarrow a \to \mathfrak{o}(a)$.

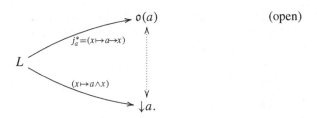

(open)

6.6.2 **Theorem (Joyal and Tierney** [35]) *A localic map* $f : L \to M$ *is open iff the adjoint frame homomorphism* $h = f^*$ *is a complete Heyting homomorphism, that is, if it preserves (also) all meets and the Heyting operation.*

Proof For each $a \in L$ we have a uniquely defined $\phi(a)$ such that $f[\mathfrak{o}(a)] = \mathfrak{o}(\phi(a))$ resulting in the decomposition

$$f \cdot j_a = j_{\phi(a)} \cdot g$$

(where $j_a : \mathfrak{o}(a) \subseteq L$ and $j_{\phi(a)} : \mathfrak{o}(\phi(a)) \subseteq M$ are the embeddings). Obviously this map $\phi : L \to M$ is monotone. In terms of the adjoining frame homomorphism we thus have $j_a^* \cdot h = g^* \cdot j_{\phi(a)}^*$. Replacing the j^*'s isomorphically as in (open) we obtain a commutative diagram

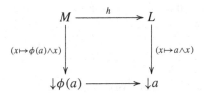

Thus the openness of f is characterized by the existence of a monotone $\phi : L \to M$ such that

$$x \wedge \phi(a) = y \wedge \phi(a) \quad \text{iff} \quad h(x) \wedge a = h(y) \wedge a$$

or, equivalently,

$$x \wedge \phi(a) \leq y \quad \text{iff} \quad h(x) \wedge a \leq h(y). \tag{$*$}$$

For $x = 1$, in particular, $\phi(a) \leq y$ iff $a \leq h(y)$ so that ϕ is a left adjoint of h and hence h preserves all meets. Further, we have by $(*)$, for arbitrary a, $a \leq h(x) \to h(y)$ iff $a \wedge h(x) \leq h(y)$ iff $x \wedge \phi(a) \leq y$ iff $\phi(a) \leq x \to y$ iff $a \leq h(x \to y)$ and hence $h(x) \to h(y) = h(x \to y)$.

On the other hand, if h preserves the Heyting operation, we have $x \wedge \phi(a) \leq y$ iff $\phi(a) \leq x \to y$ iff $a \leq h(x \to y) = h(x) \to h(y)$ iff $h(x) \wedge a \leq h(y)$, hence $(*)$. $\qquad\square$

For a thorough investigation of extensions of theorems 6.5.2 and 6.6.2 to the algebraic (not necessarily complete) setting of implicative semilattices or algebras see Erné, Picado and Pultr [18].

7 Examples

In this final section we will, first, show a few examples of point-free reasoning; in particular we present the Banaschewski-Mulvey compactification [8, 9], illustrating on the one hand that the point-free techniques can be simpler than the classical ones, and on the other hand that one can obtain better facts than in the classical setting.

At the end we will mention, without proofs, a few more examples of facts that are more satisfactory than the classical ones.

7.1 Regularity

Consider the relation between open sets of a space defined by $V \prec U$ iff $\overline{V} \subseteq U$. Thus, obviously, $x \in V \prec U$ is the same as saying that there are disjoint open sets $V \ni x$, $W \supseteq X \setminus U$, and hence the property of regularity of a space X can be expressed by stating that each open set $U \subseteq X$ is the union $\bigcup \{V \mid V \prec U\}$.

This can be extended to the point-free context. Define, in a locale L, the relation $\mathfrak{o}(b) \prec \mathfrak{o}(a)$ between open sublocales iff $\overline{\mathfrak{o}(b)} \subseteq \mathfrak{o}(a)$, and declare L as *regular* if

$$\forall a \in L, \quad \mathfrak{o}(a) = \bigvee \{\mathfrak{o}(b) \mid \mathfrak{o}(b) \prec \mathfrak{o}(a)\}.$$

Recall 5.6. Since $\overline{\mathfrak{o}(b)} = \mathfrak{c}(b^*)$, then $\mathfrak{o}(b) \prec \mathfrak{o}(a)$ iff $\mathfrak{c}(b^*) \cap \mathfrak{c}(a) = 0$, that is, $b^* \vee a = 1$. Hence L is regular iff

$$\forall a \in L, \quad a = \bigvee \{b \in L \mid b \prec a\}$$

where $b \prec a$ (*b* is *rather below* *a*) in L iff $b^* \vee a = 1$ (or, equivalently, if there is a u such that $b \wedge u = 0$ and $u \vee a = 1$).

Regularity in frames is a very expedient property, easier to work with than e.g. variants of the Hausdorff property (see below in 7.6.2) and hence it has often appeared in point-free reasoning (from its early stages) whenever one needed spaces with a "non-trivial separation". In the following we will present a few examples; technically we will typically work in frames, with localic interpretation added.

Compare the facts in 7.3.1, 7.2.3 and 7.2.4 below with the standard facts from classical topology concerning Hausdorff resp. compact Hausdorff spaces. Also in the point-free context they hold more generally, but the necessary technique is much more involved. The simplicity of the proofs presented here indicates why the regularity is so popular in point-free topology.

7.1.1   Lemma *If $a_1, a_2 \prec b$ then $a_1 \vee a_2 \prec b$, if $a \prec b_1, b_2$ then $a \prec b_1 \wedge b_2$, and if $a \prec b$ then $b^* \prec a^*$.*

Proof If $a_i^* \vee b = 0$ then $(a_1 \vee a_2)^* \vee b = (a_1^* \wedge a_2^*) \vee b = 1$. If $a^* \vee b_i = 1$ then $a^* \vee (b_1 \wedge b_2) = 1$. If $a^* \vee b = 1$ then $a^* \vee b^{**} = 1$.   □

7.1.2   Lemma *If $h \colon L \to M$ is a frame homomorphism and $x \prec y$ then $h(x) \prec h(y)$. Consequently each sublocale of a regular frame is regular.*

Proof Apply h on the equalities $x \wedge u = 0$, $u \vee y = 1$. For the second statement consider the onto homomorphism h adjoint to the embedding $S \subseteq L$. For $b = h(a) \in S$ we have $b = h(\bigvee\{x \mid x \prec a\}) = \bigvee\{h(x) \mid x \prec a\}$.   □

7.2   Dense Maps

A localic map $f \colon L \to M$ is *dense* if $\overline{f[L]} = M$, which is the same as saying that $f(0) = 0$. For the adjoint frame homomorphism we then have $h(x) = 0$ iff $x \leq f(0) = 0$ which leads to defining a *frame homomorphism* to be *dense* if $h(x) = 0$ implies that $x = 0$, and (so far for technical reasons only) as *codense* if $h(x) = 1$ implies that $x = 1$.

7.2.1   Lemma *Let a homomorphism $h \colon M \to L$ be codense and let M be regular. Then h is one-to-one.*

Proof Let $h(a) = h(b)$ and let $a \not\leq b$. Then there is an $x \prec a$, $x \not\leq b$. Since $x^* \vee a = 1$ we have $h(x^* \vee b) = h(x^* \vee a) = 1$, hence $x^* \vee b = 1$. Consequently, $x = x \wedge (x^* \vee b) = x \wedge b$ and $x \leq b$, a contradiction.   □

7.2.2   Compactness The concept of *compactness* is naturally extended to the point-free context: a *cover* of a frame L is a subset $A \subseteq L$ such that $\bigvee A = 1$, and a frame L is *compact* if every cover of L has a finite subcover.

7.2.3   Proposition *Let M be regular, L compact, and let a homomorphism $h \colon M \to L$ be dense. Then it is one-to-one. Thus, a dense localic map $f \colon L \to M$ with L compact and M regular is onto.*

Proof Let $h(a) = 1$. By regularity, $a = \bigvee\{b \mid b \prec a\}$, hence $1 = h(a) = \bigvee\{h(b) \mid b \prec a\}$ and by compactness there are b_i, $i = 1, \ldots, n$, such that $1 = \bigvee_{i=1}^{n} h(b_i)$. Set $c = \bigvee_{i=1}^{n} b_i$. Then $h(c) = 1$ and by 7.1.1 $c \prec a$, and we have $c^* \vee a = 1$. Now $h(c^*) = h(c) \wedge h(c^*) = 0$ and by density $c^* = 0$ and we obtain that $a = 1$. Hence h is also codense, and it is one-to-one by 7.2.1. $\square$

7.2.4   Corollary *Each compact sublocale of a regular frame is closed.*

(Decompose the embedding mapping $S \subseteq L$ into the embedding mappings $S \subseteq \overline{S} \subseteq L$ and use 7.1.2).

7.3   Theorem

(Banaschewski Coequalizer Theorem) *Let $h_1, h_2 \colon M \to L$ be frame homomorphisms and let M be regular. Set $c = \bigvee\{h_1(x) \wedge h_2(y) \mid x \wedge y = 0\}$. Then*

$$\gamma = (x \mapsto x \vee c) \colon L \to \mathord{\uparrow} c$$

is the coequalizer of h_1 and h_2.

In a localic formulation: the equalizer of any two localic maps $f_1, f_2 \colon L \to M$ with regular M is a closed sublocale of L, namely $\mathfrak{c}(\bigvee\{f_1^*(x) \wedge f_2^*(y) \mid x \wedge y = 0\})$.

Proof First, let us prove that $\gamma h_1 = \gamma h_2$. By symmetry, it suffices to show that $h_1(a) \le h_2(a) \vee c$. Let $x \prec a$. Then $h_1(x) \wedge h_2(x^*) \le c$ and hence $h_1(x) = h_1(x) \wedge (h_2(x^*) \vee h_2(a)) \le c \vee h_2(a)$. Since by regularity $a = \bigvee\{x \mid x \prec a\}$ we obtain $h_1(a) = \bigvee\{h_1(x) \mid x \prec a\} \le h_2(a) \vee c$.

Now let $gh_1 = gh_2$ for some homomorphism $g \colon L \to K$. Then

$$g(c) = \bigvee\{gh_1(x) \wedge gh_2(y) \mid x \wedge y = 0\} = \bigvee\{gh_1(x \wedge y) \mid x \wedge y = 0\} = 0$$

so that we can define $\widetilde{g} \colon \mathord{\uparrow} c \to K$ by setting $\widetilde{g}(x) = g(x)$ to obtain $\widetilde{g} \cdot \gamma = g$. $\square$

7.3.1   Corollary *Let M be regular and let the localic maps $f_1, f_2 \colon L \to M$ coincide on a dense sublocale of L. Then $f_1 = f_2$.*

7.4   Complete Regularity

While in the case of regularity we have just presented some parallels with the classical result, here we will be able to present an example of a considerable improvement, namely an extension of the Stone-Čech compactification that is not only technically very simple, but also choice-free (!).

In a frame L let us say that an element x is *completely below* y and write $x \lll y$ if there are x_d for diadically rationals d between 0 and 1 such that

$$x = x_0, \ y = x_1 \text{ and } x_d \prec x_e \text{ for } d < e. \tag{$*$}$$

A frame is *completely regular* if

$$\forall a \in L, \quad a = \bigvee \{b \mid b \lll a\}.$$

7.4.1 Notes (a) Again, similarly like with regularity, a space X is completely regular in the classical sense iff $\Omega(X)$ is completely regular as just defined. A continuous real function $f: X \to \mathbb{R}$ such that $f(x) = 0$ for all x in an open U and f constantly 1 on $X \smallsetminus U$ can be obtained by inserting $x \in V \lll U$ and setting $f(y) = \inf\{d \mid y \in V_d\}$ (similarly like in the construction of the function separating two closed sets in Urysohn's Lemma).

(b) Instead of the set D of diadically rational numbers we can take any countable order-dense subset D' of the unit interval. The point is just in creating an interpolative sub-relation of $\prec$ and for this D is particularly transparent.

(c) One immediately sees that $\lll$ is the largest interpolative sub-relation of $\prec$ (meaning: an R such that for aRb there is always a c with $aRcRb$—this is not necessarily true for $\prec$ itself). The construction of the largest interpolative subrelation given by the formula $(*)$ is not quite choice-free (it needs the Axiom of Countably Dependent Choice). This can be avoided by defining $\lll$ simply as *the largest interpolative sub-relation of* $\prec$ (the union of all such subrelations). All we need (in particular the properties in 7.4.2 below) can be proved for thus defined $\lll$ to obtain a fully choice-free theory (see Banaschewski and Pultr [12]); it is of course, more involved.

7.4.2 Some Properties From 7.1.1 and quite similarly like in 7.1.2 we obtain

Facts. (a) $\lll$ *is interpolative and if $a_1, a_2 \lll b$ then $a_1 \vee a_2 \lll b$, if $a \lll b_1, b_2$ then $a \lll b_1 \wedge b_2$, and if $a \lll b$ then $b^* \lll a^*$.*

(b) *Each sublocale of a completely regular frame is completely regular.*

7.5 A Point-Free Stone-Čech Compactification

(Banaschewski and Mulvey [8, 9]) For a frame L set

$$\mathfrak{J}(L) = \{J \mid J \text{ (non-empty) ideal in } L\}$$

ordered by inclusion.

7.5.1 **Lemma** $\mathfrak{J}(L)$ *is a compact frame.*

Proof Since intersections of ideals are ideals, it is a complete lattice. It is easy to check that the join in $\mathfrak{J}(L)$ is given by the formula

$$\bigvee_{i \in I} J_i = \{\bigvee M \mid M \text{ finite, } M \subseteq \bigcup_{i \in I} J_i\}.$$

Trivially, $(\bigvee J_i) \cap K \supseteq \bigvee (J_i \cap K)$ and if $x_1 \vee \cdots \vee x_n \in (\bigvee J_i) \cap K$ with $x_j \in J_{i_j}$ then $x_j \in K$ (an ideal is a down-set), hence $x_j \in J_{i_j} \cap K$, and $x_1 \vee \cdots \vee x_n \in \bigvee (J_i \cap K)$.

Finally, if $\bigvee J_i = L$ we have in particular $1 \in \bigvee J_i$, hence $1 = x_1 \vee \cdots \vee x_n$ with $x_j \in J_{i_j}$; but then $1 \in \bigvee_{j=1}^{n} J_{i_j}$ and hence $L = \bigvee_{j=1}^{n} J_{i_j}$. Thus $\mathfrak{J}(L)$ is compact. $\square$

7.5.2 **Regular Ideals** An ideal J is *regular* if for every $x \in J$ there is a $y \in J$ such that $x \prec\!\!\prec y$. In a completely regular frame L we have the regular ideals (recall the interpolation)

$$\sigma(a) = \{x \in L \mid x \prec\!\!\prec a\}$$

and we have, for every ideal J,

$$J = \bigvee \{\sigma(a) \mid a \in J\}.$$

7.5.3 **Lemma** *Let L be completely regular. Then the set*

$$\mathfrak{R}(L) = \{J \in \mathfrak{J}(L) \mid J \text{ is regular}\}$$

is a compact completely regular frame.

Proof Obviously any intersection of regular ideals is a regular ideal, and from 7.4.2 we easily infer that a join of regular ideals is regular as well. Thus, $\mathfrak{R}(L)$ is a subframe of $\mathfrak{J}(L)$, and as (obviously) a subframe of a compact frame is compact, it is compact.

It remains to be proved that it is completely regular. We have, for each $J \in \mathfrak{R}(L)$,

$$J = \bigvee \{\sigma(a) \mid a \in J\} = \bigvee \{\sigma(b) \mid \exists a \in J, \, b \prec\!\!\prec a\},$$

and since obviously $x \leq x' \prec\!\!\prec y' \leq y$ implies $x \prec\!\!\prec y$, it remains to be proved that

$$b \prec\!\!\prec a \text{ in } L \quad \Rightarrow \quad \sigma(b) \prec\!\!\prec \sigma(a) \text{ in } \mathfrak{R}(L).$$

Interpolate $b \lll x \prec y \lll a$. Then we have $y \in \sigma(a)$ and (recall 7.4.2) $x^* \in \sigma(b^*)$ and hence $1 = x^* \vee y \in \sigma(b^*) \vee \sigma(a)$ and $\sigma(b^*) \vee \sigma(a) = L$; on the other hand, trivially $\sigma(b^*) \cap \sigma(b) = \{0\}$ since $x \in \sigma(b^*) \cap \sigma(b)$ makes $x \leq b \wedge b^*$. $\qquad\square$

7.5.4 $\mathfrak{R}(L)$ **as a Compactification of** L Define $v = v_L \colon \mathfrak{R}(L) \to L$ by setting $v(J) = \bigvee J$ and consider $\sigma_L = (a \mapsto \sigma(a))$: $L \to \mathfrak{R}(L)$. We have $v(\sigma(a)) = a$ and $L \subseteq \sigma(v(L))$, hence v is a left Galois adjoint of σ, and hence it preserves all joins. Further, since J, K are down-sets,

$$v(J) \cap v(K) = \bigvee\{x \wedge y \mid x \in J, y \in K\}$$
$$\subseteq \bigvee\{z \mid z \in J \cap K\} = v(J \cap K) \subseteq v(J) \cap v(K)$$

and hence v is a frame homomorphism (and σ is a localic map). Since $\bigvee J = 0$ only if $J = \{0\}$ and v is obviously onto, we have (recall 7.2) that

each σ_L is a dense embedding of L into $\mathfrak{R}(L)$.

The construction $\mathfrak{R}$ can be extended to a functor by setting $\mathfrak{R}(h)(J) = {\downarrow}h[J]$ and it is easy to check that v is a natural transformation. Thus, to show we have here

a compactification akin to the Stone-Čech compactification of spaces

it suffices to show that for L compact the homomorphism v_L is an isomorphism, and since we already know that generally $v(\sigma(a)) = a$ and $L \subseteq \sigma(v(L))$, it suffices to prove that for a compact L, $\sigma(v(J)) \subseteq J$. Thus, let $x \in \sigma(v(J))$. In particular $x \prec \bigvee J$, hence $x^* \vee \bigvee J = 1$ and hence, by compactness, there are $y_1, \ldots, y_n$ in J such that $x^* \vee y_1 \vee \cdots \vee y_n = 1$. J is an ideal, hence $y = y_1 \vee \cdots \vee y_n \in J$, and $x^* \vee y = 1$, hence $x \leq y$ and we conclude that $x \in J$.

7.5.5 **Comments** Note that the construction is much simpler than the construction of the compactification in classical spaces. Further note that we have not used the Axiom of Choice and not even the rule of Excluded Middle. Hence the construction is fully constructive. Furthermore, it is easy to see that if a reflection is constructive then also the fact that the limits, in particular the products, are in the smaller category, is constructive. Thus, in contrast with the situation in classical spaces where the compactness of products of (in this case Hausdorff) compact spaces is compact is equivalent with a choice principle,[9] products of compact completely regular frames in Loc are compact.

Now the reader may start to doubt whether this compactification is at all closely related to the Stone-Čech one. We have been so far careful in stating that it is *akin to* that. But this was just a cagey formulation in a situation when we could not comment

---

[9] With the Boolean Ultrafilter Theorem; the compactness of products for general spaces is equivalent with the full Axiom of Choice. Even the theorem for general frames is choice-free, but this is technically much more involved [5, 27, 37].

about it properly. In fact, by the Hofmann-Lawson duality [4, 24] in particular the products of completely regular spaces exactly correspond to the products of the corresponding frames, hence the Banaschewski-Mulvey compactification does extend the Stone-Čech one. How is this possible? (Not very) roughly speaking, Tychonoff theorem does not need the choice for proving the compactness: products of compact spaces are always compact, without the Axiom of Choice, however, they can be non-spatial.

7.6   A Glimpse of Other Separation Axioms

(see [47] for more information) Normality can be immediately copied from classical topology: a frame is *normal* if

$$\forall a, b \text{ s.t. } a \vee b = 1 \; \exists u, v \text{ s.t. } a \vee u = 1, v \vee b = 1 \text{ and } u \wedge v = 0. \qquad \text{(norm)}$$

To present just a simple fact:

7.6.1   Proposition *In a normal frame the relation $\prec$ interpolates.*

Proof Let $a \prec b$. Then there is an x with $a \wedge x = 0$ and $x \vee b = 1$. By normality, there are u, v with $x \vee u = 1 = v \vee b$ and $u \wedge v = 0$ which makes $a \prec u \prec b$.   □

7.6.2   Lower Separation, in Particular Subfitness About mimicking the Hausdorff axiom let us just mention that it is a complex area, with more candidates [47]. An interesting fact is that the candidates that are conservative, that is, applied to classical spaces agree with the classical Hausdorff property, do not behave as good as the so-called strong Hausdorff property, which is not conservative but parallels very well properties of Hausdorff spaces.

Instead of T_1 we have a very expedient, weaker, subfitness. A frame is *subfit* if

$$a \not\leq b \;\; \Rightarrow \;\; \exists c, \, a \vee c = 1 \neq b \vee c. \qquad \text{(sfit)}$$

Obviously, T_1-spaces are subfit (if $x \in U \smallsetminus V$ set $W = X \smallsetminus \{x\}$ to obtain $U \cup W = X \neq V \cup W$). This property goes back to Wallman, 1938 (in a pioneering article [55] of point-free thinking, published long before point-free topology started to develop; it was later rediscovered [25], and only recently really appreciated [17, 30, 38]). To show a simple application, let us prove the following

7.6.3   Proposition *A normal subfit frame is regular and hence, by 7.6.1, completely regular.*

Proof Suppose a normal subfit L is not regular. Then there is an $a \neq b = \bigvee\{x \mid x \prec a\}$. Since $a \not\leq b$ there is a c with $a \vee c \neq b \vee c$. By normality there are u, v with $u \wedge v = 0$, $u \vee a = 1$ and $v \vee c = 1$. But then $v \prec a$, hence $v \leq b$ and we obtain a contradiction $b \vee c \geq v \vee c = 1$.   □

(In classical topology one usually speaks on normal T_1-spaces being (completely) regular, but of course subfitness suffices in the classical context as well.)

Subfitness has a lot of useful consequences. Here let us just mention a slightly surprising formula for pseudocomplement (where, in fact we will use even slightly less).

7.6.4 **Proposition** *In a subfit frame we have* $a^* = \bigwedge\{x \mid x \vee a = 1\}$.

Proof Set $b = \bigwedge\{x \mid x \vee a = 1\}$. If $x \vee a = 1$ then $a^* = a^* \wedge (x \vee a) = a^* \wedge x$, hence $a^* \leq x$ and we see that $a^* \leq b$. Thus, if $b \neq a^*$ we have $a \wedge b \neq 0$ and hence there is a $c \neq 1$ such that $c \vee (a \wedge b) = 1$. Then $c \vee a = 1$ and $c \vee b = 1$, by the former $b \leq c$, and by the latter $c = c \vee b = 1$, a contradiction.   □

Note. Thus, in a subfit frame we can compute the pseudocomplement by a formula for supplement. It is not generally a supplement, though: for that we would need the coframe distributivity. But of course we have the consequence that

> *a subfit frame that is also a coframe is a Boolean algebra.*

7.7 A Few More Examples

We will finish with a few examples of point-free facts that are more satisfactory than the classical counterparts. We will present them in an easily understandable form and provide references, but will not go into details.

7.7.1 **Choice-Free Product Compactness** In 7.5.5 we have shown, using a choice-free reflection, that compactness of the product of completely regular locales is choice-free. In fact

> *this holds for any frames whatsoever*

but the proof is more involved [5, 27, 37].

7.7.2 **Uniform Completion** The structure of frames can be naturally enriched, like that of classical spaces. Thus we have, e.g., a theory of uniform frames [44, 48], with the concept of completeness and completion quite parallel to the classical counterparts. But (after 7.7.1 not quite surprisingly)

> *completion in the point-free context is fully constructive* (see [7, 11]).

7.7.3   Paracompact Locales The reader may remember the concept of *paracompactness* that comes in many equivalent forms (the usual one requiring a locally finite refinement for every cover, another stating that the system of all covers constitutes a uniformity), has many useful applications, but behaves very badly (even a product of a paracompact space with a metric one may not be paracompact). Contrasting with this,

the subcategory of paracompact locales is reflective in Loc

(Isbell [25], cf. [10]).

There is also an elegant characterisation (not holding classically) stating that

a frame is paracompact iff it admits a complete uniformity

(Isbell [25], see also [10, 44, 50]).

7.7.4   Lindelöf Locales Also one has that

the subcategory of Lindelöf locales is reflective in Loc [39]

(the very important subcategory of *Lindelöf spaces* is not reflective in Top).

7.7.5   Localic Groups Extending the concept of a topological group one has the *localic groups* (standardly considering theory of groups over the category Loc instead of over Top) with properties similar to the classical ones (natural uniformities, etc.). But there is a fact that is fundamentally different (and somehow more satisfactory considering the classical zero group obtained as $\mathbb{R}/\mathbb{Q}$, dividing a group by an infinitely smaller one), the Closed Subgroup Theorem

every localic subgroup of a localic group L is closed in L [26, 31, 32].

7.7.6   Measures In classical measure theory, one has to restrict measure in the Euclidean space $\mathbb{R}^n$ to special measurable subsets in order to avoid Vitali and Banach-Tarski paradoxes. Instead, by enlarging the powerset $\mathscr{P}(\mathbb{R}^n)$ of subsets of $\mathbb{R}^n$ to the lattice of sublocales $S(\Omega(\mathbb{R}^n))$ (recall 5.7.3 and 6.4.4), the point-free approach produces an isometry-invariant measure on all sublocales of $\Omega(\mathbb{R}^n)$, consistent with Lebesgue measure (Simpson [52]). In particular,

every subset in $\mathbb{R}^n$ is assigned a measure

via the inclusion of $\mathscr{P}(\mathbb{R}^n)$ in $S(\Omega(\mathbb{R}^n))$. The contradictions are avoided because disjoint subsets need not be disjoint as sublocales: although the intersection of two such sublocales has no points, they nevertheless overlap in $S(\Omega(\mathbb{R}^n))$.

Acknowledgements The authors gratefully acknowledge financial support from the Centre for Mathematics of the University of Coimbra (UIDB/00324/2020, funded by the Portuguese Government through FCT/MCTES) and from the Department of Applied Mathematics (KAM) of Charles University (Prague). The first author also acknowledges the UCL project *Attractivité internationale et collaborations de recherche dans le cadre du Coimbra group 2017–2020* and an ERASMUS+ Staff Mobility Grant from the University of Coimbra that supported his visit to the Université catholique de Louvain.

References

1. Aull, C.E., Thron, W.J.: Separation axioms between T_0 and T_1. Indag. Math. **24**, 26–37 (1963)
2. Baboolal, D., Picado, J., Pillay, P., Pultr, A.: Hewitt's irresolvability and induced sublocales in spatial frames. Quaest. Math. **43**, 1601–1612 (2020)
3. Banaschewski, B.: Untersuchen uber Filterräume. Doctoral dissertation, Universität Hamburg (1953)
4. Banaschewski, B.: The duality of distributive continuous lattices. Can. J. Math. **32**, 385–394 (1980)
5. Banaschewski, B.: Another look at the localic Tychonoff theorem. Comment. Math. Univ. Carolinae **29**, 647–656 (1988)
6. Banaschewski, B.: The Real Numbers in Pointfree Topology. Textos de Matemática, vol. 12, University of Coimbra (1997)
7. Banaschewski, B., Hong, H.S., Pultr, A.: On the completion of nearness frames. Quaest. Math. **21**, 19–37 (1998)
8. Banaschewski, B., Mulvey, C.J.: Stone-Čech compactification of locales I. Houston J. Math. **6**, 301–312 (1980)
9. Banaschewski, B., Mulvey, C.J.: Stone-Čech compactification of locales II. J. Pure Appl. Algebra **33**, 107–122 (1984)
10. Banaschewski, B., Pultr, A.: Paracompactness revisited. Appl. Categ. Struct. **1**, 181–190 (1993)
11. Banaschewski, B., Pultr, A.: Cauchy points of uniform and nearness frames. Quaest. Math. **19**, 101–127 (1996)
12. Banaschewski, B., Pultr, A.: A constructive view of complete regularity. Kyungpook Math. J. **43**, 257–262 (2003)
13. Banaschewski, B., Pultr, A.: On weak lattice and frame homomorphisms. Algebra Univers. **51**, 137–151 (2004)
14. Bénabou, J.: Treillis locaux et paratopologies. Séminaire Ehresmann, 1re année, exposé 2, Paris (1958)
15. Clementino, M.M.: Separação e Compacidade em Categorias. Doctoral dissertation, University of Coimbra (1992)
16. Ehresmann, C.: Gattungen von lokalen strukturen. Jber. Deutsch. Math. Verein **60**, 59–77 (1957)
17. Erné, M.: Distributors and Wallman locales. Houston J. Math. **34**, 69–98 (2008)
18. Erné, M., Picado, J., Pultr, A.: Adjoint maps between implicative semilattices and continuity of localic maps. Algebra Universalis (to appear)
19. Ferreira, M.J., Picado, J., Pinto, S.: Remainders in pointfree topology. Topology Appl. **245**, 21–45 (2018)
20. Glivenko, V.: Sur quelque points de la logique de M. Brouwer. Bull. Acad. R. Belg. Cl. Sci. **15**, 183–188 (1929)
21. Gutiérrez García, J., Mozo Carollo, I., Picado, J.: Presenting the frame of the unit circle. J. Pure Appl. Algebra **220**, 976–1001 (2016)
22. Gutiérrez García, J., Picado, J., Pultr, A.: Notes on point-free real functions and sublocales. In: Clementino, M.M., Janelidze, G., Picado, J., Sousa, L., Tholen, W. (eds.) Categorical Methods in Algebra and Topology, vol. 46, pp. 167–200. Textos de Matemática, DMUC (2014)

23. Hausdorff, F.: Grundzüge der mengenlehre. Veit & Co., Leipzig (1914)
24. Hofmann, K.H., Lawson, J.D.: The spectral theory of distributive continuous lattices. Trans. Am. Math. Soc. **246**, 285–310 (1978)
25. Isbell, J.R.: Atomless parts of spaces. Math. Scand. **31**, 5–32 (1972)
26. Isbell, J.R., Kříž, I., Pultr, A., Rosický, J.: Remarks on localic groups. In: Borceux, F. (ed.) Categorical Algebra and Its Applications (Proceedings of the International Conference on Louvain-La-Neuve 1987). Lecture Notes in Mathematics, vol. 1348, pp. 154–172. Springer, Berlin (1988)
27. Johnstone, P.T.: Tychonoff's theorem without the axiom of choice. Fund. Math. **113**, 21–35 (1981)
28. Johnstone, P.T.: Stone Spaces. Cambridge Studies in Advanced Mathematics, vol. 3. Cambridge University Press, Cambridge (1982)
29. Johnstone, P.T.: The point of pointless topology. Bull. Am. Math. Soc. (N.S.) **8**, 41–53 (1983)
30. Johnstone, P.T.: Wallman compactification of locales. Houston J. Math. **10**, 201–206 (1984)
31. Johnstone, P.T.: A simple proof that localic groups are closed. Cahiers Topologie Géom. Différentielle Catég. **29**, 157–161 (1988)
32. Johnstone, P.T.: A constructive "closed subgroup theorem" for localic groups and groupoids. Cahiers Topologie Géom. Différentielle Catég. **30**, 3–23 (1989)
33. Johnstone, P.T.: Elements of the history of locale theory. In: Lowen, R., Aull, C.E. (eds.) Handbook of the History of General Topology, vol. 3, pp. 835–851. Kluwer Academic Publishers, Dordrecht (2001)
34. Joyal, A.: Nouveaux fondaments de l'analyse. Lectures Montréal 1973 and 1974 (unpublished)
35. Joyal, A., Tierney, M.: An Extension of the Galois Theory of Grothendieck, vol. 309. AMS: Memoirs of the American Mathematical Society, Providence (1984)
36. Klinke, O.: A presentation of the assembly of a frame by generators and relations exhibits its bitopological structure. Algebra Univers. **71**, 55–64 (2014)
37. Kříž, I.: A constructive proof of the Tychonoff's theorem for locales. Comment. Math. Univ. Carolinae **26**, 619–630 (1985)
38. Macnab, D.S.: Modal operators on Heyting algebras. Algebra Univers. **12**, 5–29 (1981)
39. Madden, J.J., Vermeer, J.: Lindelöf locales and realcompactness. In: Mathematical Proceedings of the Cambridge Philosophical Society, vol. 99, pp. 473–480 (1986)
40. Manes, E.G.: Algebraic Theories. Graduate Texts in Mathematics, vol. 26. Springer, Heidelberg (1976)
41. Niefield, S.B., Rosenthal, K.I.: Spatial sublocales and essential primes. Topology Appl. **26**, 263–269 (1987)
42. Papert, D., Papert, S.: Sur les treillis des ouverts et paratopologies. Séminaire Ehresmann (1re année, exposé 1), Paris (1958)
43. Picado, J., Pultr, A.: Locales Treated Mostly in a Covariant Way. Textos de Matemática, vol. 41. University of Coimbra (2008)
44. Picado, J., Pultr, A.: Frames and Locales: Topology Without Points. Frontiers in Mathematics, vol. 28. Springer, Basel (2012)
45. Picado, J., Pultr, A.: Notes on the product of locales. Math. Slovaca **65**, 247–264 (2015)
46. Picado, J., Pultr, A.: Axiom T_D and the Simmons sublocale theorem. Comment. Math. Univ. Carolinae **60**, 541–551 (2019)
47. Picado, J., Pultr, A.: Separation in Point-Free Topology. Birkhäuser/Springer, Cham (2021)
48. Pultr, A.: Pointless uniformities I. Complete regularity. Comment. Math. Univ. Carolinae **25**, 91–104 (1984)
49. Pultr, A., Tozzi, A.: Separation axioms and frame representation of some topological facts. Appl. Categ. Struct. **2**, 107–118 (1994)
50. Pultr, A., Úlehla, J.: Notes on characterization of paracompact frames. Comment. Math. Univ. Carolinae **30**, 377–384 (1989)
51. Simmons, H.: Spaces with Boolean assemblies. Colloq. Math. **43**, 23–29 (1980)
52. Simpson, A.: Measure, randomness and sublocales. Ann. Pure Appl. Log. **163**, 1642–1659 (2012)

53. Stone, M.H.: The theory of representations for Boolean algebras. Trans. Am. Mat. Soc. **40**, 37–111 (1936)
54. Thron, W.J.: Lattice-equivalence of topological spaces. Duke Math. J. **29**, 671–679 (1962)
55. Wallman, H.: Lattices and topological spaces. Ann. Math. **39**, 112–126 (1938)

Chapter 7
Non-associative Algebras

Tim Van der Linden

Abstract A *non-associative algebra* over a field $\mathbb{K}$ is a $\mathbb{K}$-vector space A equipped with a bilinear operation $A \times A \to A \colon (x, y) \mapsto x \cdot y = xy$. The collection of all non-associative algebras over $\mathbb{K}$, together with the product-preserving linear maps between them, forms a variety of algebras: the category $\mathsf{Alg}_{\mathbb{K}}$. The multiplication need not satisfy any additional properties, such as associativity or the existence of a unit. Familiar categories such as the varieties of associative algebras, Lie algebras, etc. may be found as subvarieties of $\mathsf{Alg}_{\mathbb{K}}$ by imposing equations, here $x(yz) = (xy)z$ (associativity) or $xy = -yx$ and $x(yz) + z(xy) + y(zx) = 0$ (anti-commutativity and the Jacobi identity), respectively.

The aim of these lectures is to explain some basic notions of categorical algebra from the point of view of non-associative algebras, and vice versa. As a rule, the presence of the vector space structure makes things easier to understand here than in other, less richly structured categories.

We explore concepts like normal subobjects and quotients, coproducts and protomodularity. On the other hand, we discuss the role of (non-associative) polynomials, homogeneous equations, and how additional equations lead to reflective subcategories.

Keywords Non-associative algebras · Categorical algebra

Math. Subj. Classification 17-01 · 18E99

Tim Van der Linden is a Research Associate of the Fonds de la Recherche Scientifique–FNRS.

T. Van der Linden (✉)
Institut de Recherche en Mathématique et Physique, Université catholique de Louvain, chemin du cyclotron 2 bte L7.01.02, 1348 Louvain-la-Neuve, Belgium
e-mail: tim.vanderlinden@uclouvain.be

© The Author(s), under exclusive license to Springer Nature Switzerland AG 2021
M. M. Clementino et al. (eds.), *New Perspectives in Algebra, Topology and Categories*, Coimbra Mathematical Texts 1,
https://doi.org/10.1007/978-3-030-84319-9_7

Introduction

An *algebra* is a vector space over a field, equipped with an additional bilinear multiplication. In practice, in the literature, and historically, this algebra multiplication is usually taken to be associative, often also commutative, or else it is asked to satisfy some other rule like the Jacobi identity. For us here, such special types of algebras will be important classes of examples. However, in these lectures about *non-associative algebras* we will *a priori* not ask that the multiplication of an algebra satisfies any rule at all, besides that it is bilinear. This makes it possible to treat many different types of algebras all at once, and interpret their common properties by means of categorical-algebraic concepts.

From the point of view of Category Theory, vector spaces are simple and extremely well behaved. (Not only is a category of vector spaces always abelian; also the fact that every vector space has a basis is important for us.) As we shall see, adding a multiplication to a vector space actually makes its behaviour worse—so for us, more interesting. The reason is, that this allows us to study categorical concepts which are often trivial for vector spaces, but whose definition makes full sense in a context which is only slightly different. In a way, it is still all Linear Algebra. The categorical properties which we shall treat here are mainly those related to the context of semi-abelian categories, which find a concrete use for instance in homology of non-abelian objects.

1 Non-associative Algebras

Throughout these lectures, we fix a field $\mathbb{K}$.

Definition 1.1 A *(non-associative) algebra* $(A, \cdot)$ *over* $\mathbb{K}$ is a $\mathbb{K}$-vector space A equipped with a bilinear operation $\cdot \colon A \times A \to A \colon (x, y) \mapsto x \cdot y$.

Recall that $\cdot$ being *bilinear* means that it is linear in both variables x and y, so that for all $\lambda \in \mathbb{K}$ and x, x', y, $y' \in A$,

$$(x + x') \cdot y = x \cdot y + x' \cdot y, \qquad x \cdot (y + y') = x \cdot y + x \cdot y',$$
$$(\lambda x) \cdot y = \lambda(x \cdot y) = x \cdot (\lambda y).$$

In other words, it induces a unique linear map $A \otimes A \to A$ sending elements of the form $x \otimes y \in A \otimes A$ to $x \cdot y \in A$.

Depending on the type of non-associative algebra which is being considered, to write the multiplication, several notations are common. We shall often drop the dot and write xy for the product $x \cdot y$. In a context related to Lie algebras, $x \cdot y$ is usually written as a bracket $[x, y]$.

Unless when this would be confusing, we write A for an algebra $(A, \cdot)$, dropping the multiplication.

2 Examples

In practice, additional conditions are imposed on the multiplication, and then only those algebras that satisfy these conditions are considered. For instance, we may ask that the multiplication of an algebra A is commutative, which means that $xy = yx$ for all elements x, y of A, and decide to study only such algebras.

The word "condition" here means any set of equations that the multiplication on an algebra should satisfy. In this section we give a number of examples of such conditions. Later we make the concept of "a condition" itself more precise and investigate it in general.

Definition 2.1 We write $\mathsf{Alg}_{\mathbb{K}}$ for the category of non-associative algebras over $\mathbb{K}$, with linear maps $f\colon A \to B$ that preserve the multiplication:

$$f(\lambda x) = \lambda f(x), \quad f(x + y) = f(x) + f(y) \quad and \quad f(x \cdot y) = f(x) \cdot f(y)$$

for $\lambda \in \mathbb{K}$ and $x, y \in A$.

Note that the identity function $1_A\colon A \to A\colon x \mapsto x$ on an algebra A is always an algebra map, and such is the composite $g \circ f\colon A \to C$ of two algebra maps $f\colon A \to B$ and $g\colon B \to C$.

Definition 2.2 Given categories $\mathscr{C}$ and $\mathscr{D}$, if the objects and arrows of $\mathscr{D}$ are also objects and arrows in $\mathscr{C}$, and if the identities and the composition in $\mathscr{D}$ agree with those in $\mathscr{C}$, then $\mathscr{D}$ is called a *subcategory* of $\mathscr{C}$.

A subcategory $\mathscr{D}$ of $\mathscr{C}$ is a *full subcategory* when the arrows in $\mathscr{D}$ are precisely the arrows in $\mathscr{C}$ between objects in $\mathscr{D}$: for each pair of objects X, Y of $\mathscr{D}$, $\mathrm{Hom}_{\mathscr{D}}(X, Y) = \mathrm{Hom}_{\mathscr{C}}(X, Y)$.

Note that a full subcategory of a given category $\mathscr{C}$ is completely determined by a choice of objects in $\mathscr{C}$.

Definition 2.3 The collection of all $\mathbb{K}$-algebras satisfying a chosen equational condition is called a *variety of non-associative algebras*. It is any class of algebras defined by a (possibly infinite) set of equations, considered as a full subcategory $\mathscr{V}$ of $\mathsf{Alg}_{\mathbb{K}}$. An object of $\mathscr{V}$ is often called a $\mathscr{V}$-*algebra*.

In other words, the morphisms in a variety of non-associative algebras are again the linear maps preserving the multiplication. In Sects. 7 and 11 we will explain what we mean by "equational condition" in Definition 2.3. Essentially it will consist of a possibly infinite system of equations $\psi_\lambda = 0$, $\lambda \in \Lambda$, where each ψ_λ is a non-associative polynomial.

Example 2.4 An algebra is said to be *commutative* if $xy = yx$ for all x and y.

Example 2.5 An algebra A is *associative* when $x(yz) = (xy)z$ for all elements x, y, z of A. The collection of all associative algebras form a variety of non-associative algebras, $\mathsf{AssocAlg}_{\mathbb{K}}$: so in these lectures, "non-associative" means "not necessarily associative".

Remark 2.6 It is easy to check that an associative $\mathbb{Z}$-algebra is the same thing as a ring (with or without unit). This doesn't fit our definition of a non-associative algebra though, because $\mathbb{Z}$ is not a field. We could have made Definition 1.1 more general, so that this example could be included. We chose not to do this, because it would make other aspects of the theory significantly more complicated.

Exercise 2.7 However, a lot can still be done. Throughout the text, try to extend the theory from vector spaces to arbitrary rings, and discover where this may make things more difficult.

Associativity may be weakened or modified, as in the next two examples.

Example 2.8 An *alternative algebra* satisfies the equations $x(xy) = (xx)y$ and $(yx)x = y(xx)$. An *anti-associative* algebra satisfies the equation $x(yz) = -(xy)z$.

We may also impose much stronger conditions, such as the next one.

Example 2.9 (Vector spaces as non-associative algebras) Any vector space V may be considered as a non-associative algebra, by imposing a trivial multiplication: $xy = 0$, the product of any x, $y \in V$ is the zero vector of V.

We write $\mathsf{Vect}_\mathbb{K}$ for the category of $\mathbb{K}$-vector spaces and linear maps between them. It is isomorphic to the variety $\mathsf{AbAlg}_\mathbb{K}$ of so-called *abelian* non-associative algebras, which are those determined by the equation $xy = 0$. This is of course not the same thing as the commutativity condition $xy = yx$ in Example 2.5.

Indeed, if the functor that equips a vector space with the trivial multiplication is written $T\colon \mathsf{Vect}_\mathbb{K} \to \mathsf{AbAlg}_\mathbb{K}$, and $U\colon \mathsf{AbAlg}_\mathbb{K} \to \mathsf{Vect}_\mathbb{K}$ is the functor which forgets the multiplication of a trivial algebra, then clearly $T \circ U = 1_{\mathsf{AbAlg}_\mathbb{K}}$ and $U \circ T = 1_{\mathsf{Vect}_\mathbb{K}}$.

Lie algebras (Example 2.11) are of a wholly different nature. First of all, they are *alternating*:

Example 2.10 An algebra A is called *alternating* when $xx = 0$ for every x in A. It is said to be *anti-commutative* when $xy = -yx$ for all x, $y \in A$.

When it exists, the smallest positive integer n such that

$$n = \underbrace{1 + \cdots + 1}_{n \text{ terms}}$$

is zero in $\mathbb{K}$ is called the *characteristic* of the field $\mathbb{K}$. If such a positive integer does not exist—which can only happen when $\mathbb{K}$ is infinite—then we say that the characteristic of $\mathbb{K}$ is 0. If the characteristic of the field $\mathbb{K}$ is different from 2, then these two conditions (being alternating, being anti-commutative) are equivalent. If $xx = 0$ for every x in A, then

$$0 = (a + b)(a + b) = aa + ab + ba + bb = ab + ba$$

for all a, $b \in A$. So *alternating implies anti-commutative*. Conversely, since we can take $x = y$, the equation $xy = -yx$ implies $xx = -xx$, hence $0 = xx + xx = (1 + 1)xx = 2xx$. So unless $0 = 2$ in the field $\mathbb{K}$, this implies that $xx = 0$.

However, they are not equivalent in general: the simplest example of a field of characteristic 2 is the field $\mathbb{Z}_2 = \{0, 1\} = \mathbb{Z}/2$ of integers modulo 2. Over $\mathbb{Z}_2$, the 2-dimensional vector space with basis $\{x, y\}$ becomes an anti-commutative algebra which is not alternating if we define its multiplication as $xx = y$ and $xy = yx = yy = 0$. Note that $xx = y = -y = -xx$.

Example 2.11 The category $\mathsf{Lie}_\mathbb{K}$ of *Lie algebras* over $\mathbb{K}$ consists of those alternating algebras satisfying the *Jacobi identity*

$$x(yz) + z(xy) + y(zx) = 0.$$

Lie algebras are notorious because of their connection with *Lie groups*, which are smooth manifolds that carry a (compatible) group structure. Actually, each Lie group induces a Lie algebra over $\mathbb{R}$, and this process gives rise to a non-trivial equivalence of suitably chosen subcategories.

Another source of Lie algebras (over any field) are those coming from associative algebras. There is a functor $G \colon \mathsf{AssocAlg}_\mathbb{K} \to \mathsf{Lie}_\mathbb{K}$ which takes an associative algebra $(A, \cdot)$ and sends it to the couple $(A, [-, -])$ where

$$[-, -] \colon A \times A \to A \colon (x, y) \mapsto [x, y] = xy - yx.$$

It is easy to check (Exercise 2.13) that this bracket does indeed define a Lie algebra structure on A. The functor G sends a morphism of associative algebras to the same linear map, now a morphism of Lie algebras, since it automatically preserves the bracket.

Note that two elements x, y of $(A, \cdot)$ commute ($xy = yx$) if and only if their bracket vanishes ($[x, y] = 0$); so the associative algebra $(A, \cdot)$ is commutative if and only if the Lie algebra $(A, [-, -])$ is abelian.

The functor G is not an equivalence of categories—the two types of structure are fundamentally different—but it has a left adjoint $\mathsf{Lie}_\mathbb{K} \to \mathsf{AssocAlg}_\mathbb{K}$ which is called the *universal enveloping algebra* functor; see Definition 6.5.

A third example of Lie $\mathbb{K}$-algebra is, for any associative $\mathbb{K}$-algebra A, the *Lie $\mathbb{K}$-algebra of derivations* $\mathsf{Der}_\mathbb{K}(A)$ of the $\mathbb{K}$-algebra A. A *derivation* of a $\mathbb{K}$-algebra A is any $\mathbb{K}$-linear mapping $D \colon A \to A$ such that $D(xy) = (D(x))y + x(D(y))$ for every x, $y \in A$. If A is any associative $\mathbb{K}$-algebra and D, D' are two derivations of A, then $DD' - D'D$ is a derivation. Thus, for any associative $\mathbb{K}$-algebra A, it is possible define the Lie $\mathbb{K}$-algebra $\mathsf{Der}_\mathbb{K}(A)$ as the subset of $\mathsf{End}_\mathbb{K}(A)$ consisting of all derivations of A with multiplication $[D, D'] := DD' - D'D$ for every D, $D' \in \mathsf{Der}_\mathbb{K}(A)$.

Exercise 2.12 Read about Lie groups and how they give rise to Lie algebras.

Exercise 2.13 Show that the bracket above defines a Lie algebra structure on A.

Exercise 2.14 Investigate the universal enveloping algebra functor.

Example 2.15 Instead of being alternating, we may ask that the multiplication of an algebra satisfying the Jacobi identity is anti-commutative ($xy = -yx$). Then this algebra is called a *quasi-Lie algebra*. The variety $\mathsf{qLie}_{\mathbb{K}}$ of quasi-Lie algebras coincides with $\mathsf{Lie}_{\mathbb{K}}$ as long as the characteristic of the field $\mathbb{K}$ is different from 2. However, when $\mathrm{char}(\mathbb{K}) = 2$, the variety $\mathsf{Lie}_{\mathbb{K}}$ is strictly smaller than $\mathsf{qLie}_{\mathbb{K}}$: the algebra over $\mathbb{Z}_2$ given in Example 2.10 is a quasi-Lie algebra which is not Lie.

Approaches to Lie algebras via *operads* usually deal with quasi-Lie algebras instead, because the repetition of the variable x which occurs in the equation $xx = 0$ cannot be expressed within that framework, so the equation $xy = -yx$ serves as a substitute.

We may further weaken or modify this definition as follows.

Example 2.16 A *Leibniz algebra* is a non-associative algebra satisfying a variation on the Jacobi identity, namely $(xy)z = x(yz) + (xz)y$. It is easy to see that an anti-commutative algebra is a Leibniz algebra if and only if it is a quasi-Lie algebra.

Example 2.17 A *Jordan algebra* is a commutative algebra ($xy = yx$) which satisfies the *Jordan identity* $(xy)(xx) = x(y(xx))$.

If a non-associative algebra is commutative and satisfies the Jacobi identity, then it is called a *Jacobi–Jordan algebra*. Over a field of characteristic 2, quasi-Lie algebras and Jacobi-Jordan algebras coincide (since commutative = anti-commutative). In particular then, they are Jordan algebras: indeed, the Jacobi identity implies that $3x(xx) = 0$, so $x(xx) = 0$; then via Example 2.16, we see that $(xy)(xx) = x(y(xx)) + (x(xx))y = x(y(xx))$.

Many more examples of varieties of non-associative algebras exist in the literature. We end with two extreme ones:

Example 2.18 The largest variety of non-associative $\mathbb{K}$-algebras is $\mathsf{Alg}_{\mathbb{K}}$ itself (no conditions) and the smallest one is the *trivial variety* 0 (consisting of the zero algebra only, satisfying all equations possible, including $x = 0$).

Example 2.19 *Unitary* (associative) algebras—those $(A, \cdot)$ which have an element 1 for which $x \cdot 1 = x = 1 \cdot x$—do not form a variety in our sense, since the *existence* of 1 cannot be expressed as an equational condition. This does not mean that an algebra cannot have a unit. On the other hand, even between algebras with units, *a priori* there is no reason why a morphism of algebras should preserve this unit.

The aim is now to explore some basic categorical concepts in the context of non-associative algebras. Most of what we are going to prove here may be seen as consequences of more general results, but to take that approach would defeat our purpose of keeping things simple.

We work in a chosen variety of non-associative algebras $\mathscr{V}$. The concepts we shall define do not depend on which variety we choose; here is a little example. Recall that a morphism $f \colon A \to B$ in a category is an *isomorphism* if and only if there exists a morphism $g \colon B \to A$ such that $f \circ g = 1_B$ and $g \circ f = 1_A$.

Lemma 2.20 *In a variety of non-associative algebras $\mathscr{V}$, a morphism is an isomorphism if and only if it is a bijection.*

Proof It follows immediately from the definition that any isomorphism of non-associative algebras is an isomorphism of its underlying sets, which makes it a bijection. Conversely, let $f : A \to B$ be a bijective morphism in $\mathscr{V}$. Then we need to show that the inverse function $g : B \to A$ is also a morphism in $\mathscr{V}$, i.e., a $\mathbb{K}$-algebra morphism. This is easy to see, using that f is an injective morphism. For instance, $g(x \cdot y) = g(x) \cdot g(y)$ for any $x, y \in B$, because $f(g(x \cdot y)) = x \cdot y = f(g(x)) \cdot f(g(y)) = f(g(x) \cdot g(y))$. □

Remark 2.21 Like several other results in these notes, this lemma is valid in the far more general context of a *variety* in the sense of universal algebra. A precise definition of this concept may be found for instance in [14]. A good exercise is to investigate which of our results generalise to that larger setting.

3 The Zero Algebra; Kernels and Cokernels

Any variety of non-associative $\mathbb{K}$-algebras contains the zero-dimensional $\mathbb{K}$-vector space 0 (whose unique element is also denoted 0) as an object. Its multiplication is the unique map $\cdot : 0 \times 0 \to 0 : (0, 0) \mapsto 0 \cdot 0 = 0$, which of course satisfies all possible equations. Categorically, this algebra is a *zero object*, and its existence makes any variety of non-associative algebras into a *pointed category*.

Definition 3.1 An object T is *terminal* in a category $\mathscr{C}$ when for each object A of $\mathscr{C}$ there is exactly one arrow $A \to T$ in $\mathscr{C}$. An object I is *initial* in $\mathscr{C}$ when for each object B of $\mathscr{C}$ there is exactly one arrow $I \to B$ in $\mathscr{C}$.

A *zero object* or *null object* in a category $\mathscr{C}$ is an object (denoted 0) which is both initial and terminal. Given any two objects A and B of $\mathscr{C}$, there is a unique *zero arrow* $0 : A \to 0 \to B$.

When a category has a terminal (or an initial) object, this object is necessarily unique up to isomorphism: for any two terminal objects T and T' there are unique arrows $T \to T'$ and $T' \to T$, which are each other's inverse, because their composites $T \to T' \to T$ and $T' \to T \to T'$ are necessarily equal to the identity morphism on T and T', respectively.

Example 3.2 In the category Set of sets and functions, the empty set is initial, and any one-element set is terminal. The same holds for topological spaces, when these sets are equipped with their unique topology.

The zero algebra is a zero object in any variety of non-associative algebras, because the unique linear map to it or from it does automatically preserve the multiplication. Between any two algebras A and B, the zero arrow is the morphism $A \to B : x \mapsto 0$.

Proposition 3.3 *Any variety of non-associative algebras is a pointed category, where the zero object is the zero algebra.*

The context of a pointed category is the right categorical environment for the definition of the concepts of the *kernel* and *cokernel* of a morphism.

Definition 3.4 In a pointed category $\mathscr{C}$, an arrow $k \colon K \to A$ is a *kernel* of an arrow $f \colon A \to B$ when $f \circ k = 0$, and every other arrow $h \colon C \to A$ such that $f \circ h = 0$ factors uniquely through k via a morphism $h' \colon C \to K$ such that $k \circ h' = h$.

$$
\begin{array}{ccc}
K & \xrightarrow{\ k\ } & \\
\wedge & & \searrow \\
\exists! h' \uparrow & & A \xrightarrow{\ f\ } B \\
\vert & \nearrow & \\
C & {}_{\forall h} &
\end{array}
$$

In other words, in the category where an object is an arrow $h \colon C \to A$ such that $f \circ h = 0$ and a morphism $g \colon h \to i$ between such arrows is a commutative triangle

$$
\begin{array}{ccc}
D & \xrightarrow{\ i\ } & \\
\wedge & & \searrow \\
g \uparrow & & A \xrightarrow{\ f\ } B, \\
\vert & \nearrow & \\
C & {}_{h} &
\end{array}
$$

a kernel of f is a terminal object. Hence kernels are unique up to isomorphism. Because of this, when a kernel exists, we sometimes speak about "the kernel" (or "the cokernel") of a morphism.

In a variety of non-associative algebras, for any arrow $f \colon A \to B$ there exists a kernel $k \colon K \to A$, namely its kernel in the vector space sense. Explicitly, the algebra K may be obtained as $\{x \in A \mid f(x) = 0\}$ with the induced operations, and $k \colon K \to A$ as the canonical inclusion. Since f is a morphism of algebras, for $x, y \in K$ and $\lambda \in \mathbb{K}$ we have $f(x + y) = f(x) + f(y) = 0$, $f(\lambda x) = \lambda f(x) = 0$ and $f(x \cdot y) = f(x) \cdot f(y) = 0$, so that K is a $\mathbb{K}$-algebra. If now $h \colon C \to A$ is a morphism such that $f \circ h = 0$, then we must define $h' \colon C \to K$ as the map which sends $x \in C$ to $h(x) \in K$—note that $f(h(x)) = 0$. This is the only function which makes the triangle commute, and it is a morphism, because h is a morphism.

Proposition 3.5 *A kernel of a morphism of non-associative algebras is computed as the kernel of the underlying linear map.*

Reversing the arrows, we find the definition of a cokernel:

Definition 3.6 In a pointed category $\mathscr{C}$, an arrow $q \colon B \to Q$ is a *cokernel* of an arrow $f \colon A \to B$ when $q \circ f = 0$, and every other arrow $h \colon B \to C$ such that $h \circ f = 0$ factors uniquely through q via a morphism $h' \colon Q \to C$ such that $h' \circ q = h$.

$$A \xrightarrow{\ f\ } B \underset{\forall h}{\overset{q}{\longrightarrow}} \begin{array}{c} Q \\ \vdots \ \exists! h' \\ \downarrow \\ C \end{array}$$

In a variety of non-associative algebras, any arrow has a cokernel, but its construction is slightly more complicated and needs some preliminary work.

4 Kernels and Ideals, Cokernels and Quotients

In varieties of non-associative algebras, arrows which are kernels can be characterised as *ideals*. An ideal of an algebra is like an ideal of a ring, or a normal subgroup of a group, both of which admit a similar categorical characterisation. This is the first place where we see that the added multiplication which distinguishes an algebra from a mere vector space makes an actual difference: *not every subalgebra can occur as a kernel.*

Definition 4.1 Given an algebra A, a *subalgebra* of A is a subspace S of A which is closed under multiplication, so $SS \subseteq S$.

An *ideal* I of A is a subalgebra such that $AI \subseteq I \supseteq IA$: for all $a \in A$ and $x \in I$, the products ax and xa in A are still elements of I.

For a given algebra in $\mathscr{V}$, all of its subalgebras are $\mathscr{V}$-algebras as well, since the multiplication on A restricts to a multiplication on S, which of course makes the same equations hold. In fact, a subset $S \subseteq A$ is a subalgebra precisely when the canonical injection $s \colon S \to A$ is a morphism in $\mathscr{V}$.

Example 4.2 Write $\mathbb{K}\langle x \rangle$ for the set of associative polynomials with zero constant term in x, which is the $\mathbb{K}$-vector space with basis $\{x, x^2, \ldots, x^n, \ldots\}$, equipped with the obvious associative multiplication determined by $x^m x^n = x^{m+n}$. Then the vector space generated by the even degrees $\{x^{2k} \mid k \geq 1\}$ of x forms a subalgebra of $\mathbb{K}\langle x \rangle$ which is not an ideal.

There is a one-to-one correspondence between the set of all ideals of A and the set of all equivalence relations $\sim$ on A compatible with the operations of A, that is, such that, for every $a, b, c, d \in A$ and every $\lambda \in \mathbb{K}$, $a \sim b$ and $c \sim d$ implies $a + c \sim b + d$, $ac \sim bd$ and $\lambda a \sim \lambda b$. The following result explains this in detail.

Proposition 4.3 *For a subalgebra K of an algebra A, let $k \colon K \to A$ denote the canonical inclusion. Let $q \colon A \to A/K \colon a \mapsto a + K$ denote the canonical linear map to the quotient vector space $A/K = \{a + K \mid a \in A\}$.*

The following conditions are equivalent:

1. *K is an ideal;*
2. *there is a unique algebra structure on A/K for which $q\colon A \to A/K$ becomes a morphism of algebras; then k is the kernel of q, and q is the cokernel of k;*
3. *k is the kernel of some morphism of algebras $f\colon A \to B$;*
4. *K is the equivalence class $[0]_\sim$ of the zero of A modulo some equivalence relation $\sim$ compatible with the operations of the algebra A.*

A map $k\colon K \to A$ satisfying these conditions is called a normal monomorphism.

Proof 3 implies 1.: if k is a kernel of f then for all $x \in K$ and $a \in A$ we have $f(xa) = f(x)f(a) = 0f(a) = 0$. Likewise, $f(ax) = 0$, so that xa and ax are in K. Hence K is an ideal of A.

4. implies 1. because if $a \sim 0$, $b \sim 0$ and $\lambda \in \mathbb{K}$ then $a + b \sim 0 + 0 = 0$, $ab \sim 00 = 0$ and $\lambda a \sim \lambda 0 = 0$.

Clearly, 2. implies 3.; let us prove that 1. implies 2. For q to become a morphism of algebras, we have no choice but to put $(a + K) \cdot (b + K) = ab + K$ for $a, b \in A$. Clearly then, k will be the kernel of q.

We need to prove that this multiplication on A/K is indeed a $\mathbb{K}$-algebra structure. First of all, it is well defined: if $a + K = a' + K$, then

$$(a + K) \cdot (b + K) = ab + K = ab + (a' - a)b + K = a'b + K = (a' + K) \cdot (b + K),$$

since $(a' - a)b \in KA$ is in the ideal K. Similarly, the multiplication does not depend on the chosen representative in the second variable. Its bilinearity follows immediately from the bilinearity of the multiplication on A. Finally, any equation which holds in A also holds in A/K, so the $\mathbb{K}$-algebra A/K is an object of $\mathcal{V}$.

Let us now check that the morphism $q\colon A \to A/K$ is indeed the cokernel of k. Suppose $h\colon A \to C$ is a morphism such that $h \circ k = 0$. Then we have no choice but to impose that $h'\colon A/K \to C$ takes a class $a + K$ and sends it to $h(a)$. The question is, whether this defines a morphism of algebras. First of all, the choice of representatives plays no role, since $a + K = a' + K$ implies $a - a' \in K$, so that $h(a) - h(a') = h(a - a') = 0$. The other properties follow easily.

Finally, 3. implies 4.: we let $a \sim b$ when $f(a) = f(b)$; then it is easily seen that $\sim$ is compatible with the operations of A. Furthermore, $a \sim 0$ if and only if a is in the kernel K of f. □

In other words, *ideals have quotients, kernels have cokernels*. What about the cokernel of an arbitrary morphism of algebras? For this we need a description of the *ideal of A generated by a subset $S \subseteq A$*, which is the smallest ideal of the algebra A that contains S.

Proposition 4.4 *Given any subset S of an algebra A, the ideal I of A generated by S exists, and may be obtained as follows:*

1. *I is the intersection of all ideals of A that contain S;*
2. *I is the union of the sequence of subspaces I_n of A, obtained inductively as follows:*

- I_0 is the subspace of A, generated by S;
- I_{n+1} is the subspace of A, generated by I_n, AI_n and I_nA.

In other words, each element of I can be obtained as a linear combination of elements of S and products of elements S with elements of A.

Proof For the proof of 1., consider the set of ideals

$$\mathfrak{J} = \{J \subseteq A \mid J \text{ is an ideal of } A \text{ containing } S\}.$$

This set is non-empty because it contains A. Any intersection of a set of subspaces of a vector space is still a subspace. If those subspaces happen to be ideals, then the result is still an ideal, since for $x \in \bigcap \mathfrak{J}$ and $a \in A$, the products ax and xa are elements of all members of the set $\mathfrak{J}$, so they are in its intersection. It follows that $\bigcap \mathfrak{J}$ is itself an element of $\mathfrak{J}$, and clearly it is the smallest element.

For 2. we have to prove that the subset $L = \bigcup_{n \in \mathbb{N}} I_n$ of A is an element of $\mathfrak{J}$. Note that the elements of I_{n+1} are linear combinations of products of elements of the form ax, xa and x, where $a \in A$ and $x \in I_n$. It is easily seen that this set L is still a subspace of A, and it is also obvious that L is an ideal. Note that *any* ideal of A that contains S necessarily also contains the other elements of L, so I and L must coincide. $\qquad\square$

Note that the ideal generated by a subset S of an algebra A may be obtained as the smallest ideal that contains the subspace of A spanned by S.

Proposition 4.5 *In any variety of non-associative algebras, given a morphism $f \colon A \to B$, a cokernel $q \colon B \to Q$ of it exists, and may be obtained as follows:*

1. *take the image $f(A) = \{f(a) \mid a \in A\}$, this is a subalgebra of B;*
2. *take the smallest ideal I of B containing $f(A)$;*
3. *let q be the quotient $B \to B/I$.*

Proof Propositions 4.3 and 4.4 already tell us that this procedure can indeed be carried out. Actually, 1. needs to be checked separately, but this is easy. We only have to show that the result q is a cokernel of f.

Since we know that q is a cokernel of the canonical inclusion $i \colon I \to B$, we only need to prove that carries out the same role for f. This amounts to showing that for any morphism $h \colon B \to C$, we have $h \circ f = 0$ if and only if $h \circ i = 0$. Note that since $f(A) \subseteq I$, there exists $f' \colon A \to I \colon a \mapsto f(a)$ such that $i \circ f' = f$. Hence $h \circ i = 0$ implies $h \circ f = h \circ i \circ f' = 0$. For the converse, we know that h vanishes on all elements of I of the form $f(a)$ for $a \in A$, and need to extend this to all of I. This is clear however, since each element of I is linear combination of elements of $f(A)$ and products of elements $f(A)$ with elements of B. $\qquad\square$

5   Short Exact Sequences and Protomodularity

One of the key notions of Homological Algebra is the concept of a *(short) exact sequence*. It can be defined in any pointed category which has kernels and cokernels, so in particular in all varieties of non-associative algebras. As it turns out, here the concept is particularly well behaved, since the *Split Short Five Lemma* holds. This implies that a variety of non-associative algebras is always a *protomodular category*—a central notion in Categorical Algebra, and part of the definition of a *semi-abelian* category.

Definition 5.1 In a pointed category, a *short exact sequence* is a couple of composable morphisms $(f: A \to B, g: B \to C)$ where f is a kernel of g and g is a cokernel of f.

This situation is usually pictured as a sequence

$$0 \longrightarrow A \xrightarrow{\ f\ } B \xrightarrow{\ g\ } C \longrightarrow 0. \qquad (\star)$$

Knowing that a couple of composable morphisms is a short exact sequence encodes certain information about the objects involved. This is precisely the type of information that is dealt with by Homological Algebra.

In a category of vector spaces, for instance, the exactness of this sequence not only says that $C \cong B/A$; it also implies that B is isomorphic to the direct sum $A \oplus C$: up to isomorphism, the outer objects completely determine the middle one. This is a consequence of the next result (Theorem 5.5), which is valid in any variety of non-associative algebras. We first need a definition.

Definition 5.2 A morphism $f: A \to B$ in a category $\mathscr{C}$ is said to be a *split epimorphism* when there exists a morphism $s: B \to A$ such that $f \circ s = 1_B$.

Dually, a morphism $s: B \to A$ in $\mathscr{C}$ is called a *split monomorphism* when there exists a morphism $f: A \to B$ such that $f \circ s = 1_B$.

As we see, split epimorphisms and split monomorphisms occur together: the splitting s of a split epimorphism f is a split monomorphism, and vice versa.

Example 5.3 In Set, any injection $s: B \to A$ where $B \neq \emptyset$ is a split monomorphism. The statement that "any surjection $f: A \to B$ is a split epimorphism" is equivalent to the Axiom of Choice, which allows us to define $s: B \to A$ by picking, for each $b \in B$, an element a in $f^{-1}(b)$, and calling this $s(b)$.

Example 5.4 Under the Axiom of Choice, every vector space has a basis. Then in $\mathsf{Vect}_{\mathbb{K}}$, every surjective linear map $g: B \to C$ is a split epimorphism. A splitting $s: C \to B$ for g may be defined as follows. Let Y be a basis of C. Then the restriction of g to a function $g^{-1}(Y) \to Y$ is a surjection. Hence it is a split epimorphism. Let $t: Y \to B$ denote a splitting, viewed as a function with codomain B. Since Y is a basis of C, the function t extends to a linear map $s: C \to B$.

Outside these two examples, split monomorphisms and split epimorphisms tend to be quite scarce. For a given morphism of non-associative algebras, to belong to one of those classes is a strong condition with serious consequences.

Theorem 5.5 (Split Short Five Lemma) *In a variety of non-associative algebras, consider the diagram*

$$
\begin{array}{ccc}
A \xrightarrow{\ f\ } B \underset{s}{\overset{g}{\rightleftarrows}} C \\
\alpha \downarrow \qquad \beta \downarrow \qquad \gamma \downarrow \\
D \xrightarrow{\ k\ } E \underset{t}{\overset{q}{\rightleftarrows}} F
\end{array}
$$

where f is a kernel of g and k is a kernel of q, where $g \circ s = 1_C$ and $q \circ t = 1_F$, and where the three squares commute: $\beta \circ f = k \circ \alpha$, $q \circ \beta = \gamma \circ g$ and $\beta \circ s = t \circ \gamma$. If α and γ are isomorphisms, then β is an isomorphism as well.

Proof By Lemma 2.20, it suffices that β is injective and surjective. Consider $e \in E$, then $e - tq(e)$ is sent to 0 by q, so there is a $d \in D$ such that $e = k(d) + tq(e)$. Take $a = \alpha^{-1}(d)$, $c = \gamma^{-1}(q(e))$ and $b = f(a) + s(c)$. Then

$$
\beta(b) = \beta(f(a) + s(c)) = k(\alpha(\alpha^{-1}(d))) + t(\gamma(\gamma^{-1}(q(e)))) = e,
$$

which proves that β is surjective.

We now use the fact that $\beta^{-1}(0) = \{0\}$ if and only if β is injective. To prove the injectivity of β, we may let $b \in B$ be such that $\beta(b) = 0$. Since then also $0 = q(\beta(b)) = \gamma(g(b))$, and since γ is an injection, $g(b) = 0$. Hence there is an $a \in A$ such that $f(a) = b$. Now $k(\alpha(a)) = \beta(f(a)) = \beta(b) = 0$, while both k and α are injections, so $a = 0$. It follows that $b = f(a) = 0$. $\qquad\square$

Note that in this proof we are only using the (additive) group structure of the algebras. This is not a coincidence, as the result is valid in contexts which are much more general than the one where we are working now.

Corollary 5.6 *In a short exact sequence of vector spaces such as $(\star)$, the object B is isomorphic to $A \oplus C$.*

Proof Being a cokernel, the morphism g is a surjection, which implies that it is a split epimorphism of vector spaces. Let $s : C \to B$ be a splitting for g, and consider the diagram

$$
\begin{array}{ccc}
A \xrightarrow{\langle 1_A, 0\rangle} A \oplus C \underset{\langle 0, 1_C\rangle}{\overset{\pi_C}{\rightleftarrows}} C \\
\| \qquad\qquad \beta \downarrow \qquad\qquad \| \\
A \xrightarrow{\ f\ } B \underset{s}{\overset{g}{\rightleftarrows}} C
\end{array}
$$

where $\beta(a, c) = f(a) + s(c)$ for $a \in A$, $c \in C$. The result now follows from Theorem 5.5. $\square$

Nothing like this is true for algebras, though. In general it is much harder to recover the middle object in a short exact sequence.

Exercise 5.7 Find an example of a short exact sequence in which the middle object does not decompose as a direct sum of the outer objects.

On the other hand, Theorem 5.5 says that *any variety of non-associative algebras is Bourn protomodular*, and as such it has important consequences, some of which we shall encounter later on.

In order for us to give the definition of a (potentially long) *exact sequence*, we need the category to have a richer structure. In this stronger context, we will also be able to extend Theorem 5.5 to general (non-split) short exact sequences.

To proceed, we need a more precise view on precisely which kind of an equation may determine a variety of non-associative algebras. It turns out that for this, the notion of a *non-associative polynomial* is crucial. We use it to determine some important adjunctions which occur in this context.

6 Polynomials and Free Non-associative Algebras

Definition 6.1 A *magma* is a set X equipped with a binary operation

$$\cdot : X \times X \to X : (x, y) \mapsto x \cdot y = xy.$$

A morphism of magmas $f : (X, \cdot) \to (Y, \cdot)$ is a function $f : X \to Y$ which preserves the multiplication: $f(x \cdot x') = f(x) \cdot f(x')$ for all x, $x' \in X$. Magmas and their morphisms form a category denoted Mag.

Like for non-associative algebras, the multiplication in magma $(X, \cdot)$ need not satisfy any additional rules such as associativity or the existence of a unit. On the other hand, any (abelian) group or (commutative) monoid has an underlying magma structure, and a non-associative algebra has two such structures (associated with $+$ and $\cdot$).

Definition 6.2 Let S be a set. A *non-associative word* in the alphabet S is a finite sequence of elements of S (called the *letters* of the word) and brackets "(" and ")" of the following kinds (and no others):

1. for every element s of S, (s) is a non-associative word;
2. if w_1 and w_2 are non-associative words, then the string $(w_1 w_2)$ is a non-associative word.

To avoid a rapidly increasing number of brackets, we will not write the outer brackets in a non-associative word.

We write $M(S)$ for the set of non-associative words in the alphabet S. The *length* of a word is the number of letters (in S, brackets don't count) that it consists of, and the *degree* of a letter the number of times it occurs in the given word. We sometimes write $\varphi(x_1, \ldots, x_n)$ for a word in which the elements $x_1, \ldots, x_n$ of S (and no others) appear.

The rule 2. allows us to define a binary operation $\cdot$ on $M(S)$ making it into a magma which we call the *free magma* on S. The canonical inclusion of S into $M(S)$ obtained from 1. is written $\eta_S \colon S \to M(S)$.

Example 6.3 When $S = \{x, y, z\}$, the strings x, xy, $(xy)z$, $x(yz)$, $x(xy)$, $(x(yz))x$ and $(xy)(zx)$ are elements of $M(S)$. The strings $x(yz)x$ and xyz are not in $M(S)$, and neither is the string $()$. (This "empty string" appears when considering free *unitary* magmas.)

Proposition 6.4 *For every magma $(X, \cdot)$ and every function $f \colon S \to X$ there exists a unique morphism of magmas $\overline{f} \colon (M(S), \cdot) \to (X, \cdot)$ such that the triangle of functions*

$$S \xrightarrow{\ \eta_S\ } M(S)$$
$$f \searrow \quad \swarrow \overline{f}$$
$$X$$

commutes.

Proof The function $\overline{f}$ must send a word (s) where $s \in S$ to $f(s)$ in order to make the triangle commute. It must preserve products, so a string $w_1 w_2$ is sent to $f(w_1) \cdot f(w_2)$. $\qquad\square$

The free magma construction conspires to a functor $M \colon \mathsf{Set} \to \mathsf{Mag}$ which sends a set S to $M(S)$, and a function $f \colon S \to T$ to the morphism of magmas $M(f) \colon M(S) \to M(T)$ induced by $\eta_T \circ f \colon S \to M(T)$. On the other hand, there is the *forgetful functor* $\mathsf{Mag} \to \mathsf{Set}$ which forgets about multiplications, taking a magma and sending it to its underlying set.

This situation fits the following general definition, of one of the key concepts in Category Theory:

Definition 6.5 Consider a pair of functors $L \colon \mathscr{C} \to \mathscr{D}$ and $R \colon \mathscr{D} \to \mathscr{C}$. Then L is said to be *left adjoint* to R, and R is said to be *right adjoint* to L, when for every object C of $\mathscr{C}$ there is a morphism $\eta_C \colon C \to R(L(C))$ in $\mathscr{C}$, such that for every object D of $\mathscr{D}$ and every morphism $f \colon C \to R(D)$ there exists a unique morphism $\overline{f} \colon L(C) \to D$ in $\mathscr{D}$ such that the triangle

$$C \xrightarrow{\ \eta_C\ } R(L(C))$$
$$f \searrow \quad \swarrow R(\overline{f})$$
$$R(D)$$

commutes in $\mathscr{C}$. This is often denoted in symbols as $L \dashv R$. When they exist, adjoints are unique (up to isomorphism), so we may say that R *has a left adjoint* or L *has a right adjoint*. The collection of morphisms $(\eta_C)_C$ is called the *unit* of the adjunction, and always forms a *natural transformation* from $1_\mathscr{C}$ to $R \circ L$, which means that for every morphism $c\colon C \to C'$ in $\mathscr{C}$, the square

$$
\begin{array}{ccc}
C & \xrightarrow{\ \eta_C\ } & R(L(C)) \\
{\scriptstyle c}\downarrow & & \downarrow{\scriptstyle R(L(c))} \\
C' & \xrightarrow[\ \eta_{C'}\ ]{} & R(L(C'))
\end{array}
$$

in $\mathscr{C}$ is commutative.

In other words, the free magma functor is left adjoint to the forgetful functor to Set, and in fact this is the reason why it carries that name: it plays the same role as the free group functor, for instance.

There are many equivalent ways to phrase adjointness, and going into the general theory of adjunctions here would lead us too far. However, we will meet several further examples, starting at once with the following one, which makes one of the relationships between non-associative algebras and magmas explicit:

Example 6.6 For any field $\mathbb{K}$, the forgetful functor $\mathsf{Alg}_\mathbb{K} \to \mathsf{Mag}$ which takes an algebra $(A, \cdot)$ and sends it to the underlying set of the vector space A, equipped with the multiplication $\cdot$, has a left adjoint denoted $\mathbb{K}[-]\colon \mathsf{Mag} \to \mathsf{Alg}_\mathbb{K}$ and called the *magma algebra functor*. (It is a variation on the *group algebra functor* which plays a similar role for groups and cocommutative Hopf algebras.)

The functor $\mathbb{K}[-]$ takes a magma $(X, \cdot)$ and sends it to the $\mathbb{K}$-vector space $\mathbb{K}[X]$ with basis X, whose elements are finite linear combinations of the elements of X, equipped with the multiplication $\cdot\colon \mathbb{K}[X] \times \mathbb{K}[X] \to \mathbb{K}[X]$ defined by

$$
\Big(\sum_{i=1}^n \lambda_i x_i, \sum_{j=1}^k \mu_j y_j\Big) \mapsto \sum_{i=1}^n \sum_{j=1}^k \lambda_i \mu_j (x_i \cdot y_j)
$$

for $x_i, y_j \in X$ and $\lambda_i, \mu_j \in \mathbb{K}$. Note that its bilinearity is obvious.

$\mathbb{K}[-]$ satisfies the universal property of a left adjoint, because for the natural inclusion $\tilde\eta_{(X,\cdot)}\colon (X, \cdot) \to (\mathbb{K}[X], \cdot)$, we have that any given morphism of magmas $f\colon X \to A$, where $(A, \cdot)$ is a non-associative algebra, extends to a unique morphism of algebras $f'\colon \mathbb{K}[X] \to A$ such that $f' \circ \tilde\eta_X = f$ in Mag. Indeed, this just follows from the fact that X is a basis of $\mathbb{K}[X]$ and the definition of the multiplication of that algebra. As in the case of magmas, this determines how the functor $\mathbb{K}[-]$ should act on morphisms.

Example 6.7 Adjunctions compose, and thus we find the construction of the *free non-associative $\mathbb{K}$-algebra* on a set.

$$\mathsf{Set} \underset{\mathrm{Forget}}{\overset{M}{\underset{\perp}{\rightleftarrows}}} \mathsf{Mag} \underset{\mathrm{Forget}}{\overset{\mathbb{K}[-]}{\underset{\perp}{\rightleftarrows}}} \mathsf{Alg}_{\mathbb{K}}$$

The functors to the left first forget the vector space structure of an algebra A, then the multiplication of the underlying magma $(A, \cdot)$, so that we obtain the underlying set of A. Looking at the diagram in Set

$$S \xrightarrow{\eta_S} M(S) \xrightarrow{\tilde{\eta}_{M(S)}} \mathbb{K}[M(S)]$$

it is easy to see that the composite functor to the right does indeed satisfy the universal property of a left adjoint.

Exercise 6.8 Use this idea to prove that *adjunctions compose* in general: given functors $L \colon \mathscr{C} \to \mathscr{D}$, $L' \colon \mathscr{D} \to \mathscr{E}$ and $R \colon \mathscr{D} \to \mathscr{C}$, $R' \colon \mathscr{E} \to \mathscr{D}$ such that $L \dashv R$ and $L' \dashv R'$, show that $L' \circ L \dashv R \circ R'$.

For a given set S, an element of $\mathbb{K}[M(S)]$ is a $\mathbb{K}$-linear combination of non-associative words in the alphabet S. In other words:

Definition 6.9 For a given set S, a *(non-associative) polynomial* with variables in S is an element of the free $\mathbb{K}$-algebra on S. For the sake of simplicity, we write $\mathbb{K}[\![S]\!]$ for the algebra $\mathbb{K}[M(S)]$.

A *monomial* in $\mathbb{K}[\![S]\!]$ is any scalar multiple of an element of $M(S)$. The *type* of a monomial $\varphi(x_1, \ldots, x_n)$ is the element $(k_1, \ldots, k_n) \in \mathbb{N}^n$ where k_i is the degree of x_i in $\varphi(x_1, \ldots, x_n)$. A polynomial is *homogeneous* if its monomials are all of the same type. Any polynomial may thus be written as a sum of homogeneous polynomials, which are called its *homogeneous components*.

Remark 6.10 Our algebras need not have units, and so our polynomials have no constant terms.

Example 6.11 When $S = \{x, y, z\}$ and $\mathbb{K}$ is any field, x, xy, xx, $(xx)x$, $-x(xx)$, $(x(yz))x$, $xy + yx$, $x(yz) - (xy)z$, $x(yz) + z(xy) + y(zx)$ and $xy - yx + (xy)z$ are polynomials over S. The first six of those are monomials, the next three are homogeneous polynomials (of respective types $(1, 1)$, $(1, 1, 1)$ and $(1, 1, 1)$ in (x, y, z)), while the last one is not homogeneous, and its homogeneous components are $xy - yx$ and $(xy)z$.

7 Varieties of Non-associative Algebras

The next step is to understand what is a variety of non-associative algebras from the categorical viewpoint. We first make Definition 2.3 fully precise. What was missing there is an explicit description of what exactly is an equational condition that determines a variety. We now see that it is given by an algebra of non-associative polynomials, an ideal of a free algebra.

We fix a countable set of variables $X = \{x_1, x_2, \ldots, x_n, \ldots\}$ and consider the free $\mathbb{K}$-algebra $\mathbb{K}[\![X]\!]$. An element ψ of this algebra is a polynomial in a finite number of variables, say $x_1, \ldots, x_n$. (In concrete examples we often prefer using letters x, y, z, etc. for the variables in a polynomial.)

Given elements $a_1, \ldots, a_n$ in an algebra A, the universal property of free algebras gives us a unique algebra morphism $z \colon \mathbb{K}[\![X]\!] \to A$ which sends x_i to a_i if $1 \le i \le n$, and to 0 otherwise. We shall write $\psi(a_1, \ldots, a_n)$ for $z(\psi)$, and say that the element $\psi(a_1, \ldots, a_n)$ of A is obtained by *substitution* of the variables $x_1, \ldots, x_n$ in the non-associative polynomial $\psi(x_1, \ldots, x_n)$ by elements $a_1, \ldots, a_n$ of A. In some sense, this process *evaluates* the polynomial in $a_1, \ldots, a_n$.

Lemma 7.1 (Algebra morphisms preserve polynomials) *Let $f \colon A \to B$ be a $\mathbb{K}$-algebra morphism and ψ a polynomial in n variables. Then for all $a_1, \ldots, a_n \in A$, we have $f(\psi(a_1, \ldots, a_n)) = \psi(f(a_1), \ldots, f(a_n))$.*

Proof This follows immediately from the definition of substitution in a polynomial. If $z \colon \mathbb{K}[\![X]\!] \to A$ determines substitution by $a_1, \ldots, a_n$ in A, then $f \circ z$ determines substitution by $f(a_1), \ldots, f(a_n)$ in B; now $f(z(\psi)) = f(\psi(a_1, \ldots, a_n))$ is $\psi(f(a_1), \ldots, f(a_n))$ by definition. $\qquad\square$

This allows us to extend Proposition 4.4 and give yet another description of the ideal of an algebra, generated by a subset.

Lemma 7.2 *Given any subset S of an algebra A, the ideal generated by S is the vector space spanned by all $\varphi(a_1, \ldots, a_n)$, where φ is a mononomial (a scalar multiple of a word) and $a_1, \ldots, a_n$ are elements of A with at least one of the a_i in S.*

Proof The vector space spanned by these elements is certainly an ideal of A, because if $\varphi(x_1, \ldots, x_n)$ is a monomial, then so are the products $x_{n+1}\varphi(x_1, \ldots, x_n)$ and $\varphi(x_1, \ldots, x_n)x_{n+1}$.

Now suppose I is an ideal of A that contains S. By Lemma 7.1, the quotient map $q \colon A \to A/I$ sends $\varphi(a_1, \ldots, a_n)$ to $\varphi(q(a_1), \ldots, q(a_n))$. This element of A/I is zero, because φ is a monomial and one of the a_i is in $S \subseteq I$. Hence $\varphi(a_1, \ldots, a_n)$ is in the kernel I of q. It follows from item 1. in Proposition 4.4 that $\varphi(a_1, \ldots, a_n)$ is a member of the ideal of A, generated by S. $\qquad\square$

Definition 7.3 A non-associative polynomial $\psi = \psi(x_1, \ldots, x_n)$ is called an *identity* of an algebra A if $\psi(a_1, \ldots, a_n) = 0$ for all $a_1, \ldots, a_n \in A$. We also say that A *satisfies the identity* ψ or that the identity ψ *is valid* in A.

Let I be a subset of $\mathbb{K}[\![X]\!]$. The class of all algebras that satisfy all identities in I is called the *variety of* $\mathbb{K}$-algebras determined by I.

As in Definition 2.3, we thus obtain a full subcategory of $\mathsf{Alg}_\mathbb{K}$. The main difference is that here we focus on polynomials instead of equations; for instance, instead of expressing commutativity as an equation $xy = yx$ that all pairs of elements x, y of an algebra A must satisfy, we write the condition as $xy - yx = 0$, and notice that the expression $xy - yx$ is a polynomial in x and y, actually an element of a free non-associative algebra.

Definition 7.4 A *T-ideal (over $\mathbb{K}$)* is an ideal of $\mathbb{K}[[X]]$ which is closed under substitution.

The set of all polynomial identities of an arbitrary variety of non-associative algebras forms a T-ideal. Conversely, given a T-ideal, we may consider the variety determined by the identities in the ideal. This gives rise to a one-to-one correspondence between the T-ideals over $\mathbb{K}$ and the varieties of non-associative algebras over $\mathbb{K}$.

Definition 7.5 Let I be a subset of $\mathbb{K}[[X]]$. For any non-associative $\mathbb{K}$-algebra A, we let $I(A)$ be the ideal of A generated by all elements of the form $\psi(a_1, \ldots, a_n)$, where $\psi \in I$ and $a_1, \ldots, a_n \in A$.

By definition, for any subset I of $\mathbb{K}[[X]]$, the ideal $I(\mathbb{K}[[X]])$ is a T-ideal. In fact, $I(\mathbb{K}[[X]])$ is the set of all identities that hold in the variety of $\mathbb{K}$-algebras determined by a given set of polynomials I.

Proposition 7.6 *Let $\mathscr{V}$ be a variety of $\mathbb{K}$-algebras determined by a set of polynomials I. Then the inclusion functor $\mathscr{V} \to \mathsf{Alg}_{\mathbb{K}}$ has a left adjoint $L: \mathsf{Alg}_{\mathbb{K}} \to \mathscr{V}$.*

Proof The functor L sends an object A to the quotient $L(A) = A/I(A)$, so that for each A we have a short exact sequence

$$0 \longrightarrow I(A) \xrightarrow{\mu_A} A \xrightarrow{\eta_A} L(A) \longrightarrow 0$$

in $\mathsf{Alg}_{\mathbb{K}}$. The algebra $A/I(A)$ is indeed an object of $\mathscr{V}$: if $\psi(x_1, \ldots, x_n) \in I$ and $a_1, \ldots, a_n \in A$, then

$$\psi(a_1 + I(A), \ldots, a_n + I(A)) = \psi(a_1, \ldots, a_n) + I(A) = I(A)$$

since $\psi(a_1, \ldots, a_n) \in I(A)$, and by the definition of the operations in $A/I(A)$. So $L(A)$ satisfies all identities of I.

Suppose B is an algebra in $\mathscr{V}$, and let $f: A \to B$ be a morphism in $\mathsf{Alg}_{\mathbb{K}}$. Then $f \circ \mu_A = 0$, because for every element $\psi(a_1, \ldots, a_n)$ of $I(A)$,

$$f(\psi(a_1, \ldots, a_n)) = \psi(f(a_1), \ldots, f(a_n)) \in B$$

by Lemma 7.1, which is zero because B satisfies all identities in I. Hence the morphism f factors uniquely through the cokernel η_A of μ_A, which proves the universal property of the left adjoint L. (Again, like in Example 6.6, the action of L on morphisms making it a functor is determined by the universal property.) □

Exercise 7.7 Prove that $A \mapsto I(A)$ determines a functor $I \colon \mathsf{Alg}_{\mathbb{K}} \to \mathsf{Alg}_{\mathbb{K}}$.

In other words, $\mathscr{V}$ is a *reflective subcategory* of $\mathsf{Alg}_{\mathbb{K}}$:

Definition 7.8 A full subcategory $\mathscr{D}$ of a category $\mathscr{C}$ is called a *reflective subcategory* when the inclusion functor $\mathscr{D} \to \mathscr{C}$ has a left adjoint.

Exercise 7.9 When its inclusion functor has a right adjoint, a subcategory is said to be *coreflective*. Look up examples of this situation.

Again using that adjunctions compose, we now see that free $\mathscr{V}$-algebras exist.

Corollary 7.10 *For any set S, the free $\mathscr{V}$-algebra on S exists, and is given by $L(\mathbb{K}[\![S]\!])$: it is the algebra of non-associative polynomials in the alphabet S, modulo the identities that determine $\mathscr{V}$.*

Proof We may proceed as in Example 6.7 and Exercise 6.8. The bottom composite functor in the diagram

$$\mathsf{Set} \underset{\mathrm{Forget}}{\overset{\mathbb{K}[\![-]\!]}{\underset{\perp}{\rightleftarrows}}} \mathsf{Alg}_{\mathbb{K}} \underset{\supseteq}{\overset{L}{\underset{\perp}{\rightleftarrows}}} \mathscr{V}$$

is the forgetful functor to Set. Its left adjoint is the above composite functor. □

Exercise 7.11 We know that the variety $\mathsf{AssocAlg}_{\mathbb{K}}$ of associative algebras is determined by the identity $x(yz) - (xy)z$. Put $I = \{x(yz) - (xy)z\}$. Prove that the free associative algebra on a singleton set $\{x\}$, which we know is the quotient of $\mathbb{K}[\![x]\!] := \mathbb{K}[\![\{x\}]\!]$ by $I(\mathbb{K}[\![x]\!])$, is isomorphic to the $\mathbb{K}$-algebra $\mathbb{K}\langle x \rangle$ of *associative* polynomials with zero constant term in x from Example 4.2.

8 Regularity, Exact Sequences

We already used the image $f(A) \subseteq B$ of a morphism of algebras $f \colon A \to B$ and noticed that it is always a subalgebra (Proposition 4.5). The existence of images can be given a general categorical treatment via the concept of a *regular category*, another key ingredient in the definition of semi-abelian categories.

Definition 8.1 In any category, a morphism $f \colon A \to B$ is said to be a *monomorphism* when for every pair of morphisms $a, b \colon X \to A$ such that $f \circ a = f \circ b$, we have $a = b$.

Exercise 8.2 A kernel is always a monomorphism.

Proposition 8.3 *In a variety $\mathscr{V}$ of non-associative algebras, a morphism is a monomorphism if and only if it is an injection.*

Proof If f is injective, then $f(a(x)) = f(b(x))$ implies that $a(x) = b(x)$. So if this happens for all $x \in X$, then $a = b$. Conversely, let $c, d \in A$ be such that $f(c) = f(d)$. Consider the free $\mathcal{V}$-algebra $X = L(\mathbb{K}[[x]])$ on a singleton set $\{x\}$, and let a and $b \colon X \to A$ be the algebra morphisms determined by $a(x) = c$ and $b(x) = d$. Then $f \circ a = f \circ b$, so $a = b$, hence $c = d$. $\square$

The "dual" concept is that of an *epimorphism*—a morphism $f \colon A \to B$ such that for every pair of morphisms $a, b \colon B \to X$ where $a \circ f = b \circ f$, we have $a = b$—but these are not well behaved in the context where we are working (see Exercise 8.8). Surjective algebra morphisms are captured by something which is slightly stronger, called a *regular epimorphism*. To define it, we first need to generalise the concept of a cokernel (Definition 3.6):

Definition 8.4 In a category $\mathscr{C}$, an arrow $q \colon B \to Q$ is a *coequaliser* of a pair of parallel arrows $f, g \colon A \to B$ when $q \circ f = q \circ g$, and every other arrow $h \colon B \to C$ such that $h \circ f = h \circ g$ factors uniquely through q via a morphism $h' \colon Q \to C$ such that $h' \circ q = h$.

$$A \underset{g}{\overset{f}{\rightrightarrows}} B \overset{q}{\underset{\forall h}{\longrightarrow}} \begin{matrix} Q \\ \mid \exists! h' \\ \downarrow \\ C \end{matrix}$$

Note that the definition of a cokernel is the special case where $g = 0$. Dually, we could define *equalisers* as a generalisation of kernels, but we shall not need those here.

Actually, in a variety of non-associative algebras, coequalisers always exist, and they can be obtained as cokernels:

Proposition 8.5 *Given a pair of parallel arrows $f, g \colon A \to B$ in a variety of non-associative algebras $\mathcal{V}$, their coequaliser $q \colon B \to Q$ may be obtained as the quotient of the ideal I of B generated by the elements of the form $f(a) - g(a)$ for $a \in A$.*

Proof Let q denote the quotient $B \to B/I$. Since q sends all elements of the form $f(a) - g(a)$ to zero, we already have that $q \circ f = q \circ g$. Now consider $h \colon B \to C$ such that $h \circ f = h \circ g$. We obtain the needed morphism h' as soon as h vanishes on all of the ideal I. By Lemma 7.2, an element of I is a linear combination of elements of the form $\psi(b_1, \dots, b_n)$, where ψ is a monomial and one of the elements $b_1, \dots, b_n \in B$ is equal to $f(a) - g(a)$ for some $a \in A$. By Lemma 7.1,

$$h(\psi(b_1, \dots, b_n)) = \psi(h(b_1), \dots, h(b_n)) = 0,$$

so h factors through the quotient q of I. $\square$

Definition 8.6 In a pointed category, a *normal epimorphism* is a cokernel of some morphism. A *regular epimorphism* is a coequaliser of some parallel pair of morphisms.

Proposition 8.7 *For any morphism* $h: B \to C$ *in a variety of non-associative algebras, the following conditions are equivalent:*

1. h *is a normal epimorphism;*
2. h *is a regular epimorphism;*
3. h *is a surjection.*

Proof 1. $\Rightarrow$ 2. is obvious from the definition, and 2. $\Rightarrow$ 1. is an immediate consequence of Proposition 8.5. Clearly, any quotient map $B \to B/I$ determined by an ideal I is a surjection, so 1. implies 3.

To show that 3. implies 1., consider a kernel $k: K \to B$ of h, and write $q: B \to Q = B/K$ for the cokernel of k. Since $h \circ k = 0$, there is the unique factorisation $h': Q \to C$ of h through q. This h' is a surjection, because h is a surjection and $h' \circ q = h$. It is also an injection: let indeed $b + K \in Q$ be such that $0 = h'(b + K) = h(b)$, then $b \in K$, so that $q(b) = 0$. By Lemma 2.20, this proves that h' is an isomorphism. $\qquad\square$

Exercise 8.8 Find an example of an epimorphism of algebras which is not a surjection. *Hint:* The canonical inclusion $\mathbb{N} \to \mathbb{Z}$ is an epimorphism of monoids.

Exercise 8.9 Prove that, in an arbitrary category, a morphism which is both a regular epimorphism and a monomorphism is an isomorphism.

In any variety of non-associative algebras, *image factorisations* exist: any morphism $f: A \to B$ may be factored into a composite $m \circ p$ of a regular epimorphism $p: A \to I$ followed by a monomorphism $m: I \to B$. This monomorphism, unique up to isomorphism, is called the *image of f*. By the above characterisations, we can simply take $I = f(A)$, with m the canonical inclusion, and p the corestriction of f to its image.

The general categorical context where image factorisations are usually defined is that of a *regular category*. For this we need one last (very important) concept:

Definition 8.10 A commutative square

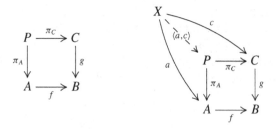

in a category $\mathscr{C}$ is called a *pullback* (of f and g) when for every pair of morphisms $a: X \to A, b: X \to C$ in $\mathscr{C}$ such that $f \circ a = g \circ c$ there exists a unique morphism $\langle a, c \rangle: X \to P$ such that $\pi_A \circ \langle a, c \rangle = a$ and $\pi_C \circ \langle a, c \rangle = c$. The object P is then usually written as a B-indexed product $A \times_B C$. The morphism π_A is called the *pullback of g along f*, and π_C is called the *pullback of f along g*.

Example 8.11 (Kernels as pullbacks) If $\mathscr{C}$ is pointed, we may consider the special case where $C = 0$. Then the square is a pullback precisely when $\pi_A : P \to A$ is a kernel of f.

Example 8.12 (Products) If $\mathscr{C}$ has a terminal object T, we may consider the special case where $B = T$. If then the square is a pullback, (P, π_A, π_B) is called a *product* of A and C.

Example 8.13 (Pullbacks of non-associative algebras) In any variety of non-associative algebras, any pair of arrows $f : A \to B$, $g : C \to B$ admits a pullback. We may let $P = A \times_B C$ be the set of couples

$$\{(a, c) \in A \times C \mid f(a) = g(c)\}$$

and π_A, π_C the canonical projections. The pointwise operations $\lambda(a, c) = (\lambda a, \lambda c)$, $(a, c) + (a', c') = (a + a', c + c')$ and $(a, c) \cdot (a', c') = (aa', cc')$ make P into an algebra, and for any two algebra morphisms a and c as above, the map which sends $x \in X$ to the couple $(a(x), c(x)) \in P$ is the needed unique morphism $\langle a, c \rangle$.

Definition 8.14 A category in which pullbacks, a terminal object, and coequalisers exist is said to be a *regular category* when any pullback of a regular epimorphism is again a regular epimorphism.

In other words, if in the square of Definition 8.10 the morphism g is a regular epimorphism, then π_A must be a regular epimorphism as well.

Proposition 8.15 *Any variety of non-associative algebras is a regular category.*

Proof By Proposition 8.7, it suffices to prove that pullbacks preserve surjections. So consider a pullback as in Definition 8.10 and assume that g is a surjection. We then use the description in Example 8.13 to prove that also π_A is a surjection. For $a \in A$, consider $c \in C$ such that $g(c) = f(a)$; then the couple (a, c) is in $A \times_B C$, and $\pi(a, c) = a$. $\qquad\square$

Exercise 8.16 Prove that Set is a regular category.

It is a theorem of Categorical Algebra that in any regular category, image factorisations exist. We shall encounter an example of how this is used in Lemma 10.2. For varieties of algebras we of course already knew this. On the other hand, a category which is pointed, regular and protomodular is called a *homological category*, and since we now know that all varieties of non-associative algebras are such, we can apply all categorical-algebraic results known to be valid for homological categories in any variety of non-associative algebras. One example is the Short Five Lemma given here below, others are famous homological diagram lemmas such as the 3×3-Lemma, a version of Noether's isomorphism theorems, etc.

Theorem 8.17 (Short Five Lemma) *In a variety of non-associative algebras, consider a commutative diagram with horizontal short exact sequences.*

$$
\begin{array}{ccccccccc}
0 & \longrightarrow & A & \xrightarrow{\ f\ } & B & \xrightarrow{\ g\ } & C & \longrightarrow & 0 \\
  &   & \downarrow{\alpha} &   & \downarrow{\beta} &   & \downarrow{\gamma} &   &   \\
0 & \longrightarrow & D & \xrightarrow[k]{} & E & \xrightarrow[q]{} & F & \longrightarrow & 0
\end{array}
$$

If α and γ are isomorphisms, then β is an isomorphism as well.

We are finally ready to extend Definition 5.1 to arbitrary exact sequences, which are the basic building blocks of Homological Algebra.

Definition 8.18 In a homological category, a pair $(f\colon A \to B, g\colon B \to C)$ of composable morphisms is called an *exact sequence* when the image of f is a kernel of g. A long sequence of composable morphisms is said to be *exact* when any pair of consecutive morphisms is exact.

Example 8.19 A sequence of morphisms such as $(\star)$ on page 218 is exact if and only if it is a short exact sequence, which explains the notation. Exactness in C means that the image of g is an isomorphism, so that g is a surjection. Exactness in A means that the kernel of f is zero, which makes f an injection. Now exactness in B says that f is the kernel of g, so that g is the cokernel of f.

9 Semi-abelian Categories

Semi-abelian categories were introduced by Janelidze–Márki–Tholen in 2002 in order to unify "old" approaches towards an axiomatisation of categories "close to the category of groups" such as the work of Higgins (1956) and Huq (1968) with "new" categorical algebra—the concepts of *Barr-exactness* and *Bourn-protomodularity*. Our aim is now to prove that all varieties of non-associative algebras are semi-abelian categories.

According to Example 8.12, a *product* of two objects A and C is a triple $(A \times C, \pi_A, \pi_C)$ that satisfies a universal property: given any pair of morphisms $a\colon X \to A$, $c\colon X \to C$, there exists a unique morphism $\langle a, c \rangle\colon X \to A \times C$ such that $\pi_A \circ \langle a, c \rangle = a$ and $\pi_C \circ \langle a, c \rangle = c$. Example 8.13 told us that in a variety of non-associative algebras, the product of two objects always exists, and may be obtained as the algebra of pairs (a, c) where $a \in A$ and $c \in C$.

The "dual" concept is that of a *coproduct*. It may be defined as a *pushout*, which is the concept dual to that of a pullback, and has cokernels for one type of examples. The direct definition goes as follows.

Definition 9.1 (Coproducts) A *coproduct* or *sum* of two objects A and C is a triple $(A + C, \iota_A, \iota_C)$ that satisfies the following universal property:

$$A \xrightarrow{\iota_A} A + C \xleftarrow{\iota_C} C$$

with a and c pointing to X via $\begin{pmatrix} a & c \end{pmatrix}$

given any pair of morphisms $a \colon A \to X, c \colon C \to X$, there exists a unique morphism $\begin{pmatrix} a & c \end{pmatrix} \colon A + C \to X$ such that $\begin{pmatrix} a & c \end{pmatrix} \circ \iota_A = a$ and $\begin{pmatrix} a & c \end{pmatrix} \circ \iota_C = c$.

Example 9.2 In the category Set, the coproduct of two sets is their disjoint union.

Example 9.3 In the variety $\mathsf{Alg}_{\mathbb{K}}$, the coproduct of two $\mathbb{K}$-algebras A and C is obtained as follows. Let $R(B)$ denote the kernel of the morphism of algebras $\varepsilon_B \colon \mathbb{K}[\![B]\!] \to B$ which sends an element b of B to itself. In the free algebra $\mathbb{K}[\![A \dot\cup C]\!]$ on the disjoint union $A \dot\cup C$ of A and C, consider the ideal J generated by the set $R(A) \dot\cup R(C)$. We claim that the quotient $\mathbb{K}[\![A \dot\cup C]\!]/J$, together with the morphisms

$$\iota_A \colon A \to \mathbb{K}[\![A \dot\cup C]\!]/J \quad \text{and} \quad \iota_C \colon C \to \mathbb{K}[\![A \dot\cup C]\!]/J$$

induced by the respective inclusions of A and C into $A \dot\cup C$, is a coproduct of A and C. Let us first show that ι_A is a morphism: for any two elements a_1 and a_2 of A, the difference $a_1 \cdot a_2 - a_1 a_2$ in $\mathbb{K}[\![A \dot\cup C]\!]$ of the product of a_1 with a_2 in $\mathbb{K}[\![A \dot\cup C]\!]$ and their product in A, viewed as an element of $\mathbb{K}[\![A \dot\cup C]\!]$, is an element of $R(A)$. The same proof works for ι_C.

By Lemma 7.2, we now only need to prove that the arrow $\mathbb{K}[\![A \dot\cup C]\!] \to X$ sending the elements of A and C to their images through a and c vanishes on the elements of $R(A)$ and $R(C)$: then it will automatically vanish on all of J, and thus factor through the quotient $\mathbb{K}[\![A \dot\cup C]\!]/J$. This, however, is immediate from the definitions of $R(A)$ and $R(C)$.

A typical element of $A + C$ is thus a polynomial with variables in A and C, for instance an expression of the form $(a_1 a_2)c$. A priori this may be interpreted in two distinct ways: either as a product $a_1 \cdot a_2$ in $\mathbb{K}[\![A \dot\cup C]\!]$ of two elements a_1 and a_2 of A, multiplied with the element c of C; or as a product of the element $a_1 a_2$ of A with the element c of C. The quotient over J ensures that these two points of view agree: $a_1 \cdot a_2 - a_1 a_2$ is an element of $R(A)$, so $(a_1 \cdot a_2 - a_1 a_2)c$ is in the ideal J.

Proposition 9.4 *In a variety of algebras $\mathscr{V}$ over a field $\mathbb{K}$, the coproduct of two algebras A and C always exists, and is obtained as the reflection into $\mathscr{V}$ of the sum $A + C$ in $\mathsf{Alg}_{\mathbb{K}}$.*

Proof This follows immediately from the definitions of sums and reflections. □

We are still missing one piece of terminology which is needed for the definition of a semi-abelian category. A regular category is said to be *Barr exact* when every internal equivalence relation is a kernel pair. A *semi-abelian category* (in the sense of Janelidze–Márki–Tholen) is then a homological category which is Barr exact and where the coproduct of any two objects exist.

We didn't introduce internal equivalence relation, though, but we can avoid those by proving that all varieties of non-associative algebras satisfy a condition which is equivalent to Barr exactness in any homological category: *the direct image of a kernel along a regular epimorphism is always a kernel*. This means that whenever we have a regular epimorphism $f : A \to B$ and a normal monomorphism $k : K \to A$,

$$
\begin{array}{ccc}
K & \xrightarrow{\ p\ } & I \\
\downarrow{\scriptstyle k} & & \downarrow{\scriptstyle i} \\
A & \xrightarrow{\ f\ } & B
\end{array}
$$

the image $i : I \to B$ of the composite $f \circ k$ is again a normal monomorphism.

Theorem 9.5 *Any variety of non-associative algebras is a semi-abelian category.*

Proof We only need to prove that the direct image of an ideal along a surjective algebra morphism is not just any subalgebra, but again an ideal. Let us consider the commutative square above, where K is an ideal of A and $I = f(K)$. Since f is surjective, for any $b \in B$ there is an $a \in A$ such that $f(a) = b$. Now $bI = f(a)f(K) = f(aK) \subseteq f(K) = I$. $\qquad\square$

Exercise 9.6 The requirement that f is a regular epimorphism is essential here. Give an example where k is an ideal, but i is not. *Hint:* Use Example 4.2.

As a consequence, typical constructions and results, valid in semi-abelian categories, hold in any variety of non-associative algebras. Examples are, for instance, the snake lemma, or the fact that homology of simplicial objects is captured by a Quillen model structure. Others are results in radical theory, commutator theory, or cohomology.

10 Birkhoff Subcategories

We find a simple version of a famous theorem by Birkhoff.

Definition 10.1 A *Birkhoff subcategory* $\mathscr{D}$ of a semi-abelian category $\mathscr{C}$ is a full reflective subcategory, closed under subobjects and quotients.

Closure under subobjects means that whenever we have a monomorphism $m : M \to D$ in $\mathscr{C}$ where D is an object of $\mathscr{D}$, the object M is also in $\mathscr{D}$. *Closure under quotients* means that whenever we have a regular epimorphism $q : D \to Q$ in $\mathscr{C}$ where D is an object of $\mathscr{D}$, the object Q is also in $\mathscr{D}$.

Lemma 10.2 *Let $\mathscr{D}$ be a full reflective subcategory of a regular category $\mathscr{C}$. Let $L : \mathscr{C} \to \mathscr{D}$ be the left adjoint of the inclusion functor. Then $\mathscr{D}$ is closed under subobjects in $\mathscr{C}$ if and only if each component $\eta_C : C \to L(C)$ of the unit η of the adjunction is a regular epimorphism.*

Proof ⇒ Factor $\eta_C\colon C \to L(C)$ as a regular epimorphism $p\colon C \to M$ followed by a monomorphism $m\colon M \to L(C)$ in $\mathscr{C}$. Then closure under subobjects tells us that M is an object of $\mathscr{D}$, which by the universal property of L gives us a unique morphism $p'\colon L(C) \to M$ such that $p = p' \circ \eta_C$. We see that p' is an inverse to m, hence m is an isomorphism.

⇐ Let D be an object of $\mathscr{D}$ and $m\colon M \to D$ a monomorphism in $\mathscr{C}$. Applying the functor L, we find the commutative square

$$
\begin{array}{ccc}
M & \xrightarrow{\ m\ } & D \\
{\scriptstyle \eta_M}\downarrow & & \downarrow{\scriptstyle \eta_D} \\
L(M) & \xrightarrow[L(m)]{} & L(D)
\end{array}
$$

in $\mathscr{C}$. It is easy to check that η_D is an isomorphism; hence η_M is both a monomorphism and a regular epimorphism, so that it is an isomorphism. It follows that M is an object of $\mathscr{D}$. $\qquad\square$

Exercise 10.3 Prove that if $\mathscr{D}$ is a Birkhoff subcategory of $\mathscr{C}$ and $\mathscr{E}$ is a subcategory of $\mathscr{D}$, then $\mathscr{E}$ is a Birkhoff subcategory of $\mathscr{D}$ iff it is a Birkhoff subcategory of $\mathscr{C}$.

Theorem 10.4 (Birkhoff) *A variety of non-associative* $\mathbb{K}$*-algebras* $\mathscr{V}$ *is the same thing as a Birkhoff subcategory of* $\mathsf{Alg}_{\mathbb{K}}$.

Proof Given a variety on non-associative $\mathbb{K}$-algebras $\mathscr{V}$, it is reflective by Proposition 7.6 and closed under quotients by Proposition 4.5. Closure under subobjects is a consequence of Lemma 10.2 and the first step in the proof of Proposition 7.6.

For a proof of the converse, let $L\colon \mathsf{Alg}_{\mathbb{K}} \to \mathscr{V}$ denote the left adjoint to the inclusion functor $\mathscr{V} \to \mathsf{Alg}_{\mathbb{K}}$. Fix a countable set of variables $X = \{x_1, x_2, \ldots, x_n, \ldots\}$ and take the free $\mathbb{K}$-algebra $\mathbb{K}[\![X]\!]$. Then the induced morphism

$$\eta_{\mathbb{K}[\![X]\!]}\colon \mathbb{K}[\![X]\!] \to L(\mathbb{K}[\![X]\!])$$

is a regular epimorphism by Lemma 10.2. Taking the kernel of $\eta_{\mathbb{K}[\![X]\!]}$, we find a short exact sequence

$$0 \longrightarrow I \longrightarrow \mathbb{K}[\![X]\!] \xrightarrow{\ \eta_{\mathbb{K}[\![X]\!]}\ } L(\mathbb{K}[\![X]\!]) \longrightarrow 0.$$

We shall prove that the set of polynomials $I \subseteq \mathbb{K}[\![X]\!]$ determines $\mathscr{V}$. That is to say, a non-associative $\mathbb{K}$-algebra A is in $\mathscr{V}$ if and only if it satisfies the identities in I.

First suppose that A is in $\mathscr{V}$. Let $\psi(x_1, \ldots, x_n)$ be a polynomial in I, and consider elements $a_1, \ldots, a_n$ of A. The morphism of $\mathbb{K}$-algebras $z\colon \mathbb{K}[\![X]\!] \to A$ which sends x_i to a_i if $0 \le i \le n$ and to 0 otherwise necessarily factors through $L(\mathbb{K}[\![X]\!])$ as $z = z' \circ \eta_{\mathbb{K}[\![X]\!]}$, since A is in $\mathscr{V}$. Hence $\psi(a_1, \ldots, a_n) = z(\psi) = z'(\eta_{\mathbb{K}[\![X]\!]}(\psi)) = 0$, because ψ is in the kernel of $\eta_{\mathbb{K}[\![X]\!]}$.

Conversely, suppose that A is a $\mathbb{K}$-algebra which satisfies the identities in I. We consider the surjective morphism $\varepsilon_A \colon \mathbb{K}[\![A]\!] \to A$ which sends $a \in A$ to itself, as well as the short exact sequence

$$0 \longrightarrow J(A) \xrightarrow{\ \kappa_A\ } \mathbb{K}[\![A]\!] \xrightarrow{\ \eta_{\mathbb{K}[\![A]\!]}\ } L(\mathbb{K}[\![A]\!]) \longrightarrow 0.$$

We prove that $\varepsilon_A \circ \kappa_A = 0$; then ε_A factors over $\eta_{\mathbb{K}[\![A]\!]}$ as a regular epimorphism $\varepsilon_A' \colon L(\mathbb{K}[\![A]\!]) \to A$ such that $\varepsilon_A' \circ \eta_{\mathbb{K}[\![A]\!]} = \varepsilon_A$. It follows that A is a quotient of $L(\mathbb{K}[\![A]\!])$, hence an object of $\mathscr{V}$.

Consider an element of $J(A)$: it is a polynomial $\psi(a_1, \ldots, a_n)$ in n variables a_1, $\ldots$, $a_n \in A$. Let $Y \subseteq X$ be the subset $\{x_1, \ldots, x_n\}$ of X and consider the morphism of $\mathbb{K}$-algebras $y \colon \mathbb{K}[\![Y]\!] \to \mathbb{K}[\![A]\!]$ which sends x_i to a_i for $0 \leq i \leq n$. In particular, it sends $\psi = \psi(x_1, \ldots, x_n)$ to $\psi(a_1, \ldots, a_n)$. Note that it is a split monomorphism, because any functor preserves split monomorphisms; a splitting may be defined which sends a_i to x_i and all other $a \in A$ to 0. On the other hand, the inclusion of Y into X induces a split monomorphism $\mathbb{K}[\![Y]\!] \to \mathbb{K}[\![X]\!]$. Since L is a functor, we find the following two vertical split monomorphisms of short exact sequences in $\mathsf{Alg}_{\mathbb{K}}$.

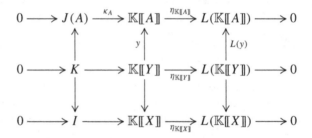

We have that $\psi(a_1, \ldots, a_n) = y(\psi)$, so

$$0 = \eta_{\mathbb{K}[\![A]\!]}(\psi(a_1, \ldots, a_n)) = L(y)(\eta_{\mathbb{K}[\![Y]\!]}(\psi)).$$

Since $L(y)$ is an injection, ψ is an element of K. Hence it is also in I. Taking $z \colon \mathbb{K}[\![X]\!] \to A$ as above and $z' \colon \mathbb{K}[\![X]\!] \to \mathbb{K}[\![A]\!]$ the morphism defined by the same rules, we have

$$\varepsilon_A(\kappa_A(\psi(a_1, \ldots, a_n))) = \varepsilon_A(z'(\psi)) = z(\psi) = \psi(a_1, \ldots, a_n) = 0,$$

because ψ is in I and $I(A) = 0$. (Note that the $\psi(a_1, \ldots, a_n)$ on the left is a polynomial in the variables $a_1, \ldots, a_n$, while the $\psi(a_1, \ldots, a_n)$ on the right is an element of A: the evaluation of this polynomial.) It follows that A is in $\mathscr{V}$. $\square$

Example 10.5 The trivial variety is determined by $I = \mathbb{K}[\![X]\!]$, while $\mathsf{Alg}_{\mathbb{K}}$ is determined by $I = 0$.

Any variety of non-associative $\mathbb{K}$-algebras contains the variety $\mathsf{AbAlg}_{\mathbb{K}}$ of abelian $\mathbb{K}$-algebras (Example 2.9), which is determined for instance by the set $\{x_1 x_2\}$. Note that the set $\{x_3 x_4\}$ determines the same variety of algebras, even though the ideals I and J generated by these two sets are different. On the other hand, the kernel of the unit $\eta_{\mathbb{K}[\![X]\!]}\colon \mathbb{K}[\![X]\!] \to L(\mathbb{K}[\![X]\!])$, where $L\colon \mathsf{Alg}_{\mathbb{K}} \to \mathsf{AbAlg}_{\mathbb{K}}$ is left adjoint to the inclusion, is equal to both $I(\mathbb{K}[\![X]\!])$ and $J(\mathbb{K}[\![X]\!])$. In other words, they generate the same T-ideal (Definition 7.4).

11 Homogeneous Identities

Recall from Definition 6.9 that the *type* of a monomial $\varphi(x_1, \dots, x_n)$ is the element $(k_1, \dots, k_n) \in \mathbb{N}^n$ where k_i is the degree of x_i in $\varphi(x_1, \dots, x_n)$. So for each of the variables $x_1, \dots, x_n$, it keeps track of the number of times this variable occurs in the monomial $\varphi(x_1, \dots, x_n)$. Then a polynomial is said to be *homogeneous* if its monomials are all of the same type, and any polynomial may thus be written as a sum of homogeneous polynomials, which are called its *homogeneous components*.

We shall now prove Theorem 11.1 which says that, over an infinite field $\mathbb{K}$ (in particular, over any field of characteristic zero), when a polynomial is an identity of a variety of $\mathbb{K}$-algebras $\mathscr{V}$, then its homogeneous components are also identities of $\mathscr{V}$. So for instance, the singleton set $\{x(yz) - (xy)z + xy - yx\}$ already determines the variety of associative commutative algebras. As we shall see in the next section, this result has some strong categorical-algebraic consequences.

Let $\psi(x_1, \dots, x_n)$ be an identity of a variety of $\mathbb{K}$-algebras $\mathscr{V}$. Write $\psi = \phi_0 + \phi_1 + \cdots + \phi_k$ where ϕ_i is the sum of all monomials in ψ which are of degree i in x_1. Now consider $k + 1$ distinct elements $\alpha_1, \dots, \alpha_{k+1}$ of the (infinite) field $\mathbb{K}$. Then the Vandermonde determinant

$$d = \begin{vmatrix} 1 & \alpha_1 & \alpha_1^2 & \cdots & \alpha_1^k \\ 1 & \alpha_2 & \alpha_2^2 & \cdots & \alpha_2^k \\ \vdots & \vdots & \vdots & \ddots & \vdots \\ 1 & \alpha_k & \alpha_k^2 & \cdots & \alpha_k^k \\ 1 & \alpha_{k+1} & \alpha_{k+1}^2 & \cdots & \alpha_{k+1}^k \end{vmatrix} = \prod_{1 \le i < j \le k+1} (\alpha_j - \alpha_i)$$

is non-zero. Let $a_1, \dots, a_n$ be elements of an algebra A of $\mathscr{V}$. Write $\phi_i(a) = \phi_i(a_1, \dots, a_n)$. Then for all $j \in \{1, \dots, k+1\}$ we have

$$\phi_0(a) + \alpha_j \phi_1(a) + \cdots + \alpha_j^k \phi_k(a)$$
$$= \phi_0(a_1, \dots, a_n) + \alpha_j \phi_1(a_1, \dots, a_n) + \cdots + \alpha_j^k \phi_k(a_1, \dots, a_n)$$
$$= \phi_0(\alpha_j a_1, \dots, a_n) + \phi_1(\alpha_j a_1, \dots, a_n) + \cdots + \phi_k(\alpha_j a_1, \dots, a_n)$$
$$= \psi(\alpha_j a_1, a_2, \dots, a_n) = 0$$

where the first equality holds by definition of the $\phi_i(a)$, the second because ϕ_i is of degree i in x_1, and the third since $\psi = \phi_0 + \phi_1 + \cdots + \phi_k$. So, in other words,

Fig. 1 Some categorical-algebraic conditions, with currently known implications between them: a semi-abelian category

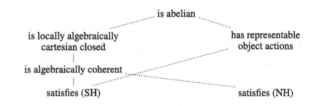

$$
\begin{pmatrix} 0 \\ 0 \\ \vdots \\ 0 \end{pmatrix} = \begin{pmatrix} 1 & \alpha_1 & \alpha_1^2 & \cdots & \alpha_1^k \\ 1 & \alpha_2 & \alpha_2^2 & \cdots & \alpha_2^k \\ \vdots & \vdots & \vdots & \ddots & \vdots \\ 1 & \alpha_k & \alpha_k^2 & \cdots & \alpha_k^k \\ 1 & \alpha_{k+1} & \alpha_{k+1}^2 & \cdots & \alpha_{k+1}^k \end{pmatrix} \cdot \begin{pmatrix} \phi_0(a) \\ \phi_1(a) \\ \vdots \\ \phi_k(a) \end{pmatrix}
$$

where the matrix in the middle is invertible because $d \neq 0$. It follows that

$$
\phi_0(a) = \phi_1(a) = \cdots = \phi_k(a) = 0.
$$

Since A was an arbitrary $\mathscr{V}$-algebra and $a_1, \ldots, a_n$ were arbitrary elements of A, each ϕ_i is an identity of $\mathscr{V}$. Note that the monomials in ϕ_i are all of the same degree in x_1. So repeating this process for the variables $x_2, \ldots, x_n$, in the end we find that the homogeneous components of ψ are identities of $\mathscr{V}$. Thus we proved:

Theorem 11.1 (Zhevlakov–Slin'ko–Shestakov–Shirshov) *In a variety of algebras $\mathscr{V}$ over an infinite field $\mathbb{K}$, if $\psi(x_1, \ldots, x_n)$ is an identity of $\mathscr{V}$, then each of its homogeneous components $\phi(x_{i_1}, \ldots, x_{i_m})$ is again an identity of $\mathscr{V}$.*

12 Some Recent Results

For certain applications (in Homological Algebra, for example) the axioms of semi-abelian categories are too weak. With the aim of including such applications in the theory, over the last 15 years or so, a whole tree of interdependent additional conditions has been investigated. As a rule, such a condition strengthens the context so that it becomes closer to the abelian setting, while at the same time excluding certain examples.

(An instance of this process, unfortunately not within the scope of this text, is the description of the derived functors of the abelianisation functor, which in its simplest form is only valid when an additional condition holds that excludes the semi-abelian category of loops.)

See Fig. 1 for an overview of some of the conditions in this tree. Two such additional conditions turn out to be particularly relevant for us in the present context: *algebraic coherence* [6] and *local algebraic cartesian closedness* [11]. There is no space here to explain what these conditions are useful for, how they were discovered, or what their consequences (SH) and (NH) mean. On the other hand, we can briefly sketch the unexpected interpretation they gain in the current setting of non-associative algebras: the former amounts to a weak associativity rule, while the latter gives *a categorical characterisation of the concept of a (quasi-)Lie algebra.*

The following is not the original definition, but it is suitable for us:

Definition 12.1 Given objects B and X in a semi-abelian category $\mathscr{C}$, take their coproduct and then the kernel of the induced split epimorphism $\begin{pmatrix} 1_B & 0 \end{pmatrix} : B + X \to B$ in order to obtain the short exact sequence

$$0 \longrightarrow B\flat X \longrightarrow B + X \overset{\begin{pmatrix} 1_B & 0 \end{pmatrix}}{\longrightarrow} B \longrightarrow 0.$$

Fixing B, this process determines a functor $B\flat(-) \colon \mathscr{C} \to \mathscr{C}$. (We shall not explore this aspect here, but the functor $B\flat(-)$ occurs in the definition of an *internal B-action in* $\mathscr{C}$: it is part of the monad whose algebras are the internal actions.) For any two objects X and Y, we have a canonical comparison morphism

$$\begin{pmatrix} B\flat\iota_X & B\flat\iota_Y \end{pmatrix} : B\flat X + B\flat Y \to B\flat(X + Y).$$

The category $\mathscr{C}$ is called *algebraically coherent* when for all B, X, Y in $\mathscr{C}$, the morphism $\begin{pmatrix} B\flat\iota_X & B\flat\iota_Y \end{pmatrix}$ is a regular epimorphism; $\mathscr{C}$ is said to be *locally algebraically cartesian closed (LACC)* when each $\begin{pmatrix} B\flat\iota_X & B\flat\iota_Y \end{pmatrix}$ is an isomorphism.

We have the following two results, of which we shall sketch part of the proofs:

Theorem 12.2 [9] *Let $\mathbb{K}$ be an infinite field. If $\mathscr{V}$ is a variety of non-associative $\mathbb{K}$-algebras, then $\mathscr{V}$ is algebraically coherent iff there exist $\lambda_1, \ldots, \lambda_{16} \in \mathbb{K}$ such that the equations*

$$z(xy) = \lambda_1 y(zx) + \lambda_2 x(yz) + \lambda_3 y(xz) + \lambda_4 x(zy)$$
$$+ \lambda_5 (zx)y + \lambda_6 (yz)x + \lambda_7 (xz)y + \lambda_8 (zy)x$$

and

$$(xy)z = \lambda_9 y(zx) + \lambda_{10} x(yz) + \lambda_{11} y(xz) + \lambda_{12} x(zy)$$
$$+ \lambda_{13} (zx)y + \lambda_{14} (yz)x + \lambda_{15} (xz)y + \lambda_{16} (zy)x$$

hold in $\mathscr{V}$.

Exercise 12.3 It follows easily that this is equivalent to $\mathscr{V}$ being a *2-variety* in the sense of [18]: for any ideal I of an algebra A, the subalgebra I^2 of A is again an ideal. As an immediate consequence, it may now be seen that in the context of varieties of non-associative algebras over an infinite field, algebraic coherence is equivalent to *normality of Higgins commutators* in the sense of [7]—the condition (NH) in Fig. 1.

Exercise 12.4 These conditions are also equivalent to $\mathscr{V}$ being an *Orzech category of interest* [15].

The implication $\Rightarrow$ of Theorem 12.2 becomes a straightforward consequence of Theorem 11.1 once we have a sufficiently explicit interpretation of the objects $B\flat X$. Let B, X and Y be free $\mathbb{K}$-algebras with a single generator b, x and y, respectively. The ideal $B\flat X$ of $B + X$ is generated by monomials of the form $\psi(b, x)$ in which x occurs at least once. If now B, X and Y are free $\mathscr{V}$-algebras, and $B\flat X$ is computed in $\mathscr{V}$, then the only difference is that we need to take classes of such polynomials, modulo the identities of $\mathscr{V}$.

Algebraic coherence now means that the element $b(xy)$ of $B\flat(X + Y)$ may be obtained as the image of some polynomial $\psi(b_1, x, b_2, y)$ in $B\flat X + B\flat Y$ through the function $\begin{pmatrix} a & c \end{pmatrix}$. Note that this polynomial cannot contain any monomials obtained as a product of a b_i with xy or yx. This allows us to write, in the sum $B + X + Y$, the element $b(xy)$ as

$$\lambda_1 y(bx) + \lambda_2 x(yb) + \lambda_3 y(xb) + \lambda_4 x(by)$$
$$+ \lambda_5 (bx)y + \lambda_6 (yb)x + \lambda_7 (xb)y + \lambda_8 (by)x$$
$$+ \nu\phi(b, x, y)$$

for some $\lambda_1, \ldots, \lambda_8, \nu \in \mathbb{K}$, where $\phi(b, x, y)$ is the part of the polynomial in b, x and y which is not in the homogeneous component of $b(xy)$. Since $B + X + Y$ is the free $\mathscr{V}$-algebra on three generators b, x and y, from Theorem 11.1 we deduce that the first equation in Theorem 12.2 is again an identity in $\mathscr{V}$. Analogously, for $(xy)b$ we deduce the second equation.

Theorem 12.5 [10] *Let $\mathbb{K}$ be an infinite field. Let $\mathscr{V}$ be a non-abelian (LACC) variety of $\mathbb{K}$-algebras. Then*

1. $\mathscr{V} = \mathsf{Lie}_{\mathbb{K}} = \mathsf{qLie}_{\mathbb{K}}$ *when* $\mathrm{char}(\mathbb{K}) \neq 2$;
2. $\mathscr{V} = \mathsf{Lie}_{\mathbb{K}}$ *or* $\mathscr{V} = \mathsf{qLie}_{\mathbb{K}}$ *when* $\mathrm{char}(\mathbb{K}) = 2$.

Let us first consider the special case where $xy = -yx$ in $\mathscr{V}$. Then we can reduce the first equation of Theorem 12.2 to $z(xy) = \lambda y(zx) + \mu x(yz)$, for some λ, $\mu \in \mathbb{K}$. Considering now $y = z$, and then $x = z$, we deduce that either $\lambda = \mu = -1$, or $z(zx) = 0$ is an identity of $\mathscr{V}$. The first case is exactly the Jacobi identity. In the second case, we see that $0 = (x + y)((x + y)b) = x(yb) + y(xb)$. Therefore, the comparison map sends $x(yb_2) - y(xb_1) \in B\flat X + B\flat Y$ to zero in $B\flat(X + Y)$, which via (LACC) and Theorem 11.1 implies that $x(yz)$ is an identity of $\mathscr{V}$. Variations on these ideas allow us to prove that when $\mathscr{V}$ is (LACC) and $x(yz) = 0$ in $\mathscr{V}$, the variety is necessarily abelian; hence if $\mathscr{V}$ is non-abelian, then the Jacobi identity must hold.

In general, not assuming anti-commutativity, this result is much harder to prove. Our current strategy involves a proof by computer, which shows that a certain system of polynomial equations is inconsistent.

In this context, many basic questions currently remain unanswered. In view of the above result, we may ask questions such as, for instance: *What is associativity?* Can it be captured in terms of a categorical-algebraic condition?

13 Bibliography

The aim of these notes is to provide a gentle introduction to two subjects at the same time, semi-abelian categories and non-associative algebras, focusing on the connections between them. "Gentle" here is supposed to mean that little prior knowledge of either subject is required to understand most of the results. There is nothing new here: the content is essentially a rearrangement of known results, often not in their most general form, with easier proofs preferred over more efficient ones. The notes are as self-contained as possible, which necessarily means that they are incomplete in many different ways. Here follows an overview of some works which go beyond what is presented in this text.

- Non-associative algebras: [1, 17]
- Basic category theory: [2, 14]
- Protomodular, regular, exact, semi-abelian categories: [3–5, 13]
- (Co)homological applications: [8, 16]
- Birkhoff subcategories: [12]
- Recent results: [6, 9–11].

Acknowledgements Many thanks to Xabi García-Martínez, who introduced me to non-associative algebras and made me want to understand them better. Thanks to the participants of the *Summer School in Algebra and Topology*, and to François Renaud, Corentin Vienne and the anonymous referees for their feedback on the text.

References

1. Bahturin, Y.A.: Identical Relations in Lie Algebras. VNU Science Press, Utrecht (1987)
2. Borceux, F.: Handbook of Categorical Algebra 1: Basic Category Theory, Encyclopedia of Mathematics and its Applications, vol. 50. Cambridge University Press, Cambridge (1994)
3. Borceux, F.: A survey of semi-abelian categories. In: Janelidze, G., Pareigis, B., Tholen, W. (eds.) Galois Theory, Hopf Algebras, and Semiabelian Categories, Fields Institute Communications, vol. 43, pp. 27–60. American Mathematical Society (2004)
4. Borceux, F., Bourn, D.: Mal'cev, Protomodular, Homological and Semi-Abelian Categories, Mathematics and Its Applications, vol. 566. Kluwer Academic Publishers, Dordrecht (2004)
5. Bourn, D., Gran, M.: Regular, protomodular, and abelian categories. In: Pedicchio, M.C., Tholen, W. (eds.) Categorical Foundations: Special Topics in Order, Topology, Algebra and Sheaf Theory, Encyclopedia of Math. Appl., vol. 97, pp. 165–211. Cambridge Univ. Press (2004)

6. Cigoli, A.S., Gray, J.R.A., Van der Linden, T.: Algebraically coherent categories. Theory Appl. Categ. **30**(54), 1864–1905 (2015)
7. Cigoli, A.S., Gray, J.R.A., Van der Linden, T.: On the normality of Higgins commutators. J. Pure Appl. Algebra **219**, 897–912 (2015)
8. Everaert, T., Gran, M., Van der Linden, T.: Higher Hopf formulae for homology via Galois Theory. Adv. Math. **217**(5), 2231–2267 (2008)
9. García-Martínez, X., Van der Linden, T.: A characterisation of Lie algebras amongst anti-commutative algebras. J. Pure Appl. Algebra **223**, 4857–4870 (2019)
10. García-Martínez, X., Van der Linden, T.: A characterisation of Lie algebras via algebraic exponentiation. Adv. Math. **341**, 92–117 (2019)
11. Gray, J.R.A.: Algebraic exponentiation in general categories. Appl. Categ. Structures **20**, 543–567 (2012)
12. Janelidze, G., Kelly, G.M.: Galois theory and a general notion of central extension. J. Pure Appl. Algebra **97**(2), 135–161 (1994)
13. Janelidze, G., Márki, L., Tholen, W.: Semi-abelian categories. J. Pure Appl. Algebra **168**(2–3), 367–386 (2002)
14. Mac Lane, S.: Categories for the Working Mathematician, Graduate Texts in Mathematics, vol. 5, 2nd edn. Springer, New York (1998)
15. Orzech, G.: Obstruction theory in algebraic categories I and II. J. Pure Appl. Algebra **2**, 287–314 and 315–340 (1972)
16. Rodelo, D., Van der Linden, T.: Higher central extensions and cohomology. Adv. Math. **287**, 31–108 (2016)
17. Zhevlakov, K.A., Slin'ko, A.M., Shestakov, I.P., Shirshov, A.I.: Rings that are Nearly Associative. Academic Press, New York (1982)
18. Zwier, P.J.: Prime ideals in a large class of nonassociative rings. Trans. Amer. Math. Soc. **158**(2), 257–271 (1971)

Printed in the United States
by Baker & Taylor Publisher Services

Printed in the United States
by Baker & Taylor Publisher Services